NUMBERS AND FUNCTIONS
STEPS INTO ANALYSIS

R. P. BURN

NUMBERS AND FUNCTIONS

STEPS INTO ANALYSIS

CAMBRIDGE
UNIVERSITY PRESS

Published by the Press Syndicate of the University of Cambridge
The Pitt Building, Trumpington Street, Cambridge CB2 1RP
40 West 20th Street, New York, NY 10011-4211, USA
10 Stamford Road, Oakleigh, Melbourne 3166, Australia

First published 1992
First paperback edition (with corrections) 1993

Printed and bound in Great Britain at
the University Press, Cambridge

A catalogue record of this book is available from the British Library

Library of Congress cataloguing in publication data

Burn, R. P.
Numbers and functions: steps into analysis/R. P. Burn.
    p. cm.
Includes bibliographical references and index.
1. Mathematical analysis.   I. Title.
QA300.B87 1992
515′–dc20   91–22839   CIP

ISBN 0 521 41086 X hardback
ISBN 0 521 45773 4 paperback

KT

Contents

Preface

This text is written for those who have studied calculus in the sixth form at school, and are now ready to review that mathematics rigorously and to seek precision in its formulation. The question sequence given here tackles the key concepts and ideas one by one, and invites a self imposed precision in each area. At the successful conclusion of the course, a student will have a view of the calculus which is in accord with modern standards of rigour, and a sound springboard from which to study metric spaces and point set topology, or multi-dimensional calculus.

Generations of students have found the study of the foundations of the calculus an uncomfortable business. The reasons for this discomfort are manifold.

(1) The student coming from the sixth form to university is already familiar with Newtonian calculus and has developed confidence in the subject by using it, and experiencing its power. Its validity has been established for him/her by reasonable argument and confirmed by its effectiveness. It is not a source of student uncertainty and this means that an axiomatic and rigorous presentation seems to make heavy weather of something which is believed to be sound, and criticisms of Newtonian calculus seem to be an irritating piece of intellectual nit-picking.

(2) At an age when a student's critical capacity is at its height, an axiomatic presentation can have a take-it-or-leave-it quality which *feels* humiliating: axioms for the real numbers have none of the 'let's-play-a-game' character by which some simpler systems appeal to the widespread interest in puzzles. The bald statement of axioms for the real numbers covers up a significant process of decision-making in their choice, and the Axiom of Completeness,

which lies at the heart of most of the main results in analysis, seems superfluous at first sight, in whatever form it is expressed.

(3) Even when the axiom system has been accepted, proofs by contradiction can be a stumbling block, either because the results are unbelievable, as in the case of irrationality or uncountability, or because they make heavy weather of such seemingly obvious results as the Intermediate Value Theorem or Rolle's Theorem.

(4) Definitions, particularly those of limits and continuity, appear strangely contrived and counter-intuitive.

There are also discomforts of lesser moment which none the less make the subject indigestible:

(5) the abstract definition of a function (when most students have only used the word function to mean a formula),

(6) the persistent use of inequalities in argument to tame infinity and infinitesimals,

(7) and proofs by induction (which play an incidental rôle in most school courses).

(8) The student who has overcome these hurdles will find that some of the best textbooks will present him/her with exercises at the end of each chapter which are so substantial that it could be a term's work to complete even those associated with three hours of lectures.

(9) The student who seeks help in the bibliography of his/her current text may find that the recommended literature is mostly for 'further reading'.

Most of these difficulties are well-attested in the literature on mathematical education. (See for example articles published in *Educational Studies in Mathematics* throughout the 1980s or the review article by David Tall (1992).)

With these difficulties in mind we may wonder how any students have survived such a course! The questions which they ask analysis lecturers may reveal their methods. Although I believed, as an undergraduate, that I was doing my best, I remember habitually asking questions about the details of the lecturer's exposition and never asking about the main ideas and results. I realise now that I was exercising only secretarial skills in the lecture room, and was not involved in the overall argument. A more participatory style of learning would have helped. Many of those who have just completed a degree in mathematics will affirm that 0.9 recurring is not the limit of a sequence, but an ordinary number less than 1! This is not evidence of any lack of intelligence: through the nineteenth century the best mathematicians stumbled because of the difficulty of imagining dense but incomplete sets of points and the seeming unreality of continuous but non-differentiable functions. There was real discomfort too in banning the language of infinitesimals from the discussion of limits. In their difficulties today, students have much in common with the best mathematicians of the nineteenth century.

It has been said that the most serious deficiency in undergraduate mathematics is the lack of an existence theorem for undergraduates! Put in other words, by an eminent educationalist, 'If I had to reduce all of educational psychology to just one principle, I would say this: the most important single factor influencing learning is what the learner already knows. Ascertain this, and teach him accordingly.' (Ausubel, 1968). There is a degree of recognition of this principle in virtually every elementary text on analysis, when, in the exercises at the end of a chapter, the strictly logical order of presentation is put aside and future results anticipated in order that the student should better understand the points at issue. The irrationality of π may be presumed before the number itself has been defined. The trigonometric, exponential and logarithmic functions almost invariably appear in exercises before they have been formally or analytically defined. In this book, my first concern, given the subject matter, has been to let students *use what they already know* to generate new concepts, and to explore situations which invite new definitions. In working through each chapter of the book the student will come to formulate, in a manner which respects modern standards of rigour, part of what is now the classical presentation of analysis. The formal achievements of each chapter are listed at the end of that chapter, but on the way there is no reluctance to use notions which will be familiar to students from their work in the sixth form. This, after all, is the way the subject developed historically. Sixth-form calculus operates with the standards of rigour which were current in the middle of the eighteenth century and it was from such a standpoint that the modern rigorous analysis of Bolzano, Cauchy, Riemann, Weierstrass, Dedekind and Cantor grew.

So the first principle upon which this text has been constructed is that of involving the student in the generation of new concepts by using ideas and techniques which are already familiar. The second principle is that generalisation is one of the least difficult of the new notions which a student meets in university mathematics. The judiciously chosen special case which may be calculated or computed, provides the basis for a student's own formulation of a general theorem and this sequence of development (from special case to general theorem) keeps the student's understanding active when the formulation of a general theorem on its own would be opaque.

The third principle, on which the second is partly based, is that every student will have a pocket calculator with 'scientific' keys, and access to graph drawing facilities on a computer. A programmable calculator with graphic display will possess all the required facilities.

There is a fourth principle, which could perhaps be better called an ongoing tension for the teacher, of weighing the powerful definition and consequent easy theorem on the one hand, against the weak definition (which seems more meaningful) followed by the difficult theorem on the other. Which is the better teaching strategy? There is no absolute rule here. However I have chosen 'every infinite decimal is convergent' as the axiom of completeness, and then established the convergence of monotonic bounded sequences adapting an argument given by the American mathematician W. F. Osgood in a German textbook of 1907. Certainly to assume that monotonic bounded sequences are convergent and to deduce the convergence of the sequence for an infinite decimal takes less paper than the converse, but I believe that, more often than has usually been allowed, the combination of weaker definition and harder theorem keep the student's feet on the ground and his/her comprehension active.

Every mathematician knows theorems in which propositions are proved to be equivalent but in which one implication is established more easily than the other. A case in point is the neighbourhood definition of continuity compared with the convergent sequence definition. After considerable experience with both definitions it is clear to me that the convergent sequence definition provides a more effective teaching strategy, though it is arguable that this is only gained by covert use of the Axiom of Choice. (A detailed comparison of different limit definitions from the point of view of the learner is given in ch. 14 of Hauchart and Rouche.) As I said earlier, the issue is not one of principle, but simply an acknowledgement that the neat piece of logic which shortens a proof *may* make that proof and the result *less* comprehensible to a beginner. More research on optimal teaching strategies is needed.

It sometimes seems that those with pedagogical concerns are soft on mathematics. I hope that this book will contradict this impression. If anything, there is more insistence here than is usual that a student be aware of which parts of the axiomatic basis of the subject are needed at which juncture in the treatment.

The first two chapters are intended to enable the student who needs them to improve his/her technique and his/her confidence in two aspects of mathematics which need to become second nature for anyone studying university mathematics. The two areas are those of mathematical induction and of inequalities. Ironically, perhaps, in view of what I have written above, the majority of questions in each of these chapters develop rudimentary properties of the number system from stated axioms. These questions happen to be the most effective learning sequences I know for generating student skills in these areas. My debt to Landau's *Foundations of Analysis* and to Thurston's *The Number-system* will be evident. There is another skill which these exercises will foster, namely that of distinguishing between what is familiar and what has been proved. This is perhaps the key distinction to be drawn in the transition from school to university mathematics, where it is expected that everything which is to be assumed without proof is to be overtly stated. Commonly, a first course in analysis contains the postulational basis for most of a degree course in mathematics. This justifies lecturers being particularly fussy about the reasoning used in proofs in analysis. In the second chapter we also establish various classical inequalities to use in later work on convergence.

In the third chapter we step into infinite processes and define the convergence of sequences. It is the definition of limit which is conventionally thought to be the greatest hurdle in starting analysis, and we define limits first in the context of null sequences, the preferred context in the treatments of Knopp (1928) and Burkill (1960). In order to establish basic theorems on convergence we assume Archimedean order. When the least upper bound postulate is used as a completeness axiom, it is common to deduce Archimedean order from this postulate. Unlike most properties established from a completeness axiom, Archimedean order holds for the rational numbers, and indeed for any subfield of the real numbers. So this proof can mislead. The distinctive function of Archimedean order in banishing infinite numbers and infinitesimals, whether the field is complete or not, is often missed. The Archimedean axiom expressly forbids ∞ being a member of the number field. Now that non-standard analysis is a live option, clarity is needed at this point. In any case, the notion of completeness is such a hurdle to students that there is good reason for proving as much as possible without it. We carry this idea through the book by studying sequences *without* completeness in chapter 3 and *with* completeness in chapters 4 and 5; continuous functions and limits *without* completeness in chapter 6 and *with* completeness in chapter 7; differentiation *without* completeness in chapter 8 and *with* completeness in chapter 9.

The fourth chapter is about the completeness of the real numbers. We identify irrational numbers and contrast the countability of the rationals with the uncountability of the set of infinite decimals. We adopt as a completeness axiom the property that every infinite decimal is convergent. We deduce that bounded monotonic sequences are convergent, and thereafter standard results follow one by one. Of the possible axioms for completeness, this is the only one which relates directly to the previous experience of the students. With completeness under our belt, we are ready to tackle the convergence of series in chapter 5.

The remaining chapters of the book are about real functions and start with a section which shows why the consideration of limiting processes requires the modern definition of a function and why formulae do not provide a sufficiently rich diet of possibilities. By adopting Cantor's sequential definition of continuity, a broad spectrum of results on continuous functions follows as a straightforward consequence of theorems about limits of sequences. The second half of chapter 6 is devoted to reconciling the sequential definitions of continuity and limit with Weierstrass' neighbourhood definitions, and deals with both one- and two-sided limits. These have been placed as far on in the course as possible. There is a covert appeal to the Axiom of Choice in the harder proofs. Chapter 6 builds on chapter 3, but does not depend on completeness in any way. In chapter 7 we establish the difficult theorems about continuity on intervals, all of which depend on completeness. Taking advantage of the sequential definition of continuity, we use completeness in the proofs by claiming that a bounded sequence contains a convergent subsequence. There is again a covert appeal to the Axiom of Choice.

Chapters 8 and 9, on differentiation, are conventional in content, except in stressing the distinction between those properties which do not depend on completeness, in chapter 8 (the definition of derivative and the product, quotient and chain rules), and those which do, in chapter 9 (Rolle's Theorem to Taylor's Theorem). The differentiation of inverse functions appears out of place, in chapter 8. Chapter 10, on integration, starts with the computation of areas in ways which were, or could have been, used before Newton, and proceeds from these examples of the effective use of step functions to the theory of the Riemann integral. Completeness is used in the definition of upper and lower integrals.

The convention of defining logarithmic, exponential and circular functions either by neatly chosen integrals or by power series is almost universal. This seems to me to be an excellent procedure in a *second* course. But the origins of exponentials and logarithms lie in the use of indices and that is the starting point of our development in chapter 11. Likewise our development of circular functions starts by investigating the length of arc of a circle. The treatment is necessarily more lengthy than is usual, but offers some powerful applications of the theorems of chapters 1 to 10.

A chapter on uniform convergence completes the book. Some courses, with good reason, postpone such material to the second year. This chapter rounds off the problem sequence in two senses: firstly, by the discussion of term-by-term integration and differentiation, it completes a university-style treatment of sixth-form calculus; and secondly by discussing the convergence of functions it is possible to see the kind of questions which provoked the rigorous analysis of the late nineteenth century.

The interdependence of chapters is illustrated below.

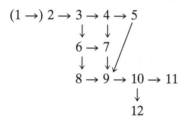

One review of my *Pathway into Number Theory* (*Times Higher Educational Supplement*, 3.12.82) suggested that a pathway to analysis would be of more value than a pathway to number theory. While not disagreeing with the reviewer, I could not, and still today cannot, see the two tasks as comparable. The subtlety of the concepts and definitions of undergraduate analysis is of a different order. But the reviewer's challenge has remained with me, and the success I have seen students achieve with my pathways to number theory and geometrical groups has spurred me on. None the less, I offer these steps (notice the cautious claim by comparison with the earlier books) aware that they contain more reversals of what I regard as an optimal teaching sequence than the earlier pathways.

It may be helpful to clarify the differences between the present text and other books on analysis which have the word 'Problem' in their titles. I refer firstly to the books of the Schaum series. Although their titles read *Theory and problems of* . . . the books consist for the most part of solutions. Secondly, a book with the title *Introductory Problem Course in Analysis and Topology* written by E. E. Moise consists of a list of theorems cited in logical sequence. Thirdly, the book *Problems and Propositions in Analysis* by G. Klambauer provides an enriching supplement to any analysis course, and, in my opinion, no lecturer in the subject should be without a copy. And finally there is the doyen of all problem books, that by Pólya and Szegö (1976), which expects greater maturity than the present text but, again, is a book no analysis lecturer should be without.

Sometimes authors of mathematics books claim that their publications are 'self-contained'. This is a coded claim which may be helpful to an experienced lecturer, but can be misleading to an undergraduate. It is never true that a book of university mathematics can be understood without experience of other mathematics. And when the concepts to be studied are counter-intuitive, or the proofs tricky, it is not just that one presentation is better than another, but that all presentations are problematic, and that whichever presentation a student meets *second* is more likely to be understood than the one met *first*. For these reasons I persistently encourage the consultation of other treatments. While it is highly desirable to recommend a priority course book (lest the student be entirely at the mercy of the lecturer) the lecturer needs to use the ideas of others to stimulate and improve his/her own teaching; and, because students are different from one another, no one lecturer or book is likely to supply quite what the student needs.

There is a distinction of nomenclature of books on this subject, with North American books tending to include the word 'calculus' in their titles and British books the word 'analysis'. In the nineteenth century the outstanding series of books by A. L. Cauchy clarified the distinction. His first volume was about convergence and continuity and entitled 'Course of Analysis' and his later volumes were on the differential and integral calculus. The general study of convergence thus preceded its application in the context of differentiation (where limits are not reached) and in integration (where the limiting process occurs by the refinement of subdivisions of an interval). British university students have met a tool-kit approach to calculus at school and the change in name probably assists the change in attitude required in transferring from school to university. But the change of name is unhelpful to the extent that it is part of the purpose of every first course of analysis in British universities to clarify the concepts of school calculus and to put familiar results involving differentiation and integration on a more rigorous foundation.

I must express gratitude and indebtedness to many colleagues: to my own teacher Dr J. C. Burkill for his pursuit of simplicity and clarity of exposition; to Hilary Shuard my colleague at Homerton College whose capacity to put her finger on a difficulty and keep it there would put a terrier to shame (an unpublished text on analysis which she wrote holds an honoured place in my filing cabinet; time and again, when I have found all the standard texts unhelpful, Hilary has identified the dark place, and shown how to sweep away the cobwebs); to Alan Beardon for discussions about the subject over many years; to Dr D. J. H. Garling for redeeming my mistakes, and for invaluable advice on substantial points; to Dr T. W. Körner for clarifying the relationship between differentiability and invertibility of functions for me by constructing a differentiable bijection $\mathbb{Q} \to \mathbb{Q}$ whose inverse is not continuous; and to Dr F. Smithies for help with many historical points. At a fairly late stage in the book's production I received a mass of detailed and pertinent advice from Dr Tony Gardiner of Birmingham University which has led to many improvements. This book would not yet be finished but for the sustained encouragement and support of Cambridge University Press. The first book which made me believe that a humane approach to analysis might be possible was *The Calculus, a Genetic Approach* by O. Toeplitz. The historical key to the subject which he picked up can open more doors than we have yet seen. I still look forward to the publication of an undergraduate analysis book, structured by the historical development of the subject during the nineteenth century. Historians, to date, have told us much about Newton and Leibniz, and about the definition of limit, but remarkably little about the notion of completeness. (Since writing this paragraph, I have come across the work of P. Dugac, who has taken steps to repair this deficiency, and I have been told that there is valuable material, in this regard, written in Russian by P. Medvedev.)

I would like to be told of any mistakes which students or lecturers find in the book.

School of Education, R. P. Burn
Exeter University, EX1 2LU

Glossary

qn 27 refers to question 27 of the same chapter.
qn 6.27 refers to question 27 of chapter 6.
6.27^+ refers to what immediately follows qn 6.27.
6.27^- refers to what immediately precedes qn 6.27.

$x \in A$	x is an element of the set A; for example, $x \in \{x, \ldots\}$
$x \notin A$	x is not an element of the set A
$\{x \mid x \in A\}$ or $\{x : x \in A\}$	the set A
$A \subseteq B$	A is a subset of B every member of A is a member of B $x \in A \Rightarrow x \in B$
$A \cup B$	the union of A and B $\{x \mid x \in A \ or \ x \in B\}$
$A \cap B$	the intersection of A and B $\{x \mid x \in A \ and \ x \in B\}$
$A \setminus B$	$\{x \mid x \in A \ and \ x \notin B\}$
$A \times B$	the cartesian product $\{(a, b) \mid a \in A, b \in B\}$
$f : A \to B$	the function f, a subset of $A \times B$, $\{(a, f(a)) \mid a \in A, f(a) \in B\}$ A is the domain and B is the co-domain of the function f
$f(A)$	$\{f(x) \mid x \in A\} \subseteq B$
$x \mapsto f(x)$	the function f, x is an element of the domain of f
$\mathbb{N}$	the set of counting numbers or natural numbers $\{1, 2, 3, \ldots\}$

$\mathbb{Z}$	the set of integers $\{0, \pm 1, \pm 2, \ldots\}$
$\mathbb{Z}^+$	the set of positive integers $\{1, 2, 3, \ldots\}$
$\mathbb{Q}$	the set of rational numbers $\{p/q \mid p \in \mathbb{Z}, q \in \mathbb{Z} \setminus \{0\}\}$
$\mathbb{Q}^+$	the set of positive rationals $\{x \mid x \in \mathbb{Q}, 0 < x\}$
$\mathbb{R}$	the set of real numbers, or infinite decimals
$\mathbb{R}^+$	the set of positive real numbers $\{x \mid x \in \mathbb{R}, 0 < x\}$
$[a, b]$	closed interval $\{x \mid a \leq x \leq b, x, a, b \in \mathbb{R}\}$
(a, b)	open interval $\{x \mid a < x < b, x, a, b \in \mathbb{R}\}$, the symbol may also denote the ordered pair, (a, b), or the two coordinates of a point in $\mathbb{R}^2$
$[a, b)$	half-open interval $\{x \mid a \leq x < b, x, a, b \in \mathbb{R}\}$
$[a, \infty)$	closed half-ray $\{x \mid a \leq x, a, x \in \mathbb{R}\}$
(a, ∞)	open half-ray $\{x \mid a < x, a, x \in \mathbb{R}\}$
(a_n)	the sequence $\{a_n \mid n \in \mathbb{N}, a_n \in \mathbb{R}\}$ that is, a function: $\mathbb{N} \to \mathbb{R}$
$(a_n) \to a$ as $n \to \infty$	For any $\varepsilon > 0$, $\lvert a_n - a \rvert < \varepsilon$ for all $n > N$
$\dbinom{n}{r}$	$\dfrac{n!}{(n-r)!r!}$ when $n \in \mathbb{Z}^+$
$\dbinom{a}{r}$	$\dfrac{a(a-1)(a-2)\ldots(a-r+1)}{r!}$ the binomial coefficient when $a \in \mathbb{R}$.
$\displaystyle\sum_{n=1}^{n=N} a_n$	$a_1 + a_2 + a_3 + \ldots + a_N$

1

The counting numbers and mathematical induction

Preliminary reading: Rosenbaum, Pólya ch. 7.
Concurrent reading: Sominskii.
Further reading: Stewart and Tall ch. 8, Thurston, Levi.

The set of counting numbers $\{1, 2, 3, \ldots, n, \ldots\}$ is usually denoted by $\mathbb{N}$ and is also called the set of natural numbers.

Mathematical induction

1 $\quad 1^2 + 2^2 + 3^2 + \ldots + n^2 = \dfrac{n(n + 1)(2n + 1)}{6}.$

Is this proposition true or false?
Test it when $n = 1$, $n = 2$ and $n = 3$.
How many values of n should you test if you want to be sure it is true for all counting numbers n?
If we write

$$f(n) = \frac{n(n + 1)(2n + 1)}{6},$$

show that $f(n) + (n + 1)^2 = f(n + 1)$.
Now suppose that the proposition with which we started holds for some particular value of n: add $(n + 1)^2$ to both sides of the equation and deduce that the proposition holds for the next value of n.
Since we have already established that the proposition holds for $n = 1, 2$ and 3, the argument we have just formulated shows that it must hold for $n = 4$, and then by the same argument for $n = 5$, and so on.

This is a classic example of proof by *mathematical induction*, establishing a result for all the counting numbers.

2 When n is a counting number,

$$6^n - 5n + 4$$

is divisible by 5.
Check this proposition for $n = 1$, 2 and 3.
By examining the difference between this number and

$$6^{n+1} - 5(n + 1) + 4,$$

show that if the proposition holds for one value of n, it holds for the succeeding value of n.
When you have done this, you have established the two components of the proof of the proposition by mathematical induction.

3 By induction or otherwise, prove the following propositions:

(i) $1 + 2 + 3 + \ldots + n = \frac{1}{2}n(n + 1)$,

(ii) $a + (a + d) + (a + 2d) + \ldots + (a + (n - 1)d)$
$= \frac{1}{2}n(2a + (n - 1)d)$,

(iii) $1^3 + 2^3 + 3^3 + \ldots + n^3 = [\frac{1}{2}n(n + 1)]^2$,

(iv) $1 \cdot 2 + 2 \cdot 3 + 3 \cdot 4 + \ldots + n(n + 1)$
$= \frac{1}{3}n(n + 1)(n + 2)$,

(v) $1 + 2 + 4 + \ldots + 2^{n-1} = 2^n - 1$,

(vi) $1 + x + x^2 + \ldots + x^{n-1} = \dfrac{x^n - 1}{x - 1}$, provided $x \neq 1$,

(vii) $1 + 2x + 3x^2 + \ldots + nx^{n-1} = \dfrac{nx^n}{x - 1} - \dfrac{x^n - 1}{(x - 1)^2}$,
provided $x \neq 1$.

4 Pascal's triangle, shown here, is defined inductively, each entry being the sum of the two (or one) entries in the preceding row nearest to the new entry.
The $(r + 1)$th entry in the nth row is denoted by $\binom{n}{r}$

```
                1   1
              1   2   1
            1   3   3   1
          1   4   6   4   1
        1   5  10  10   5   1
      1  ...  ...  ...  ...  ...   1
```

and called 'n choose r' because it happens to count the number of ways of choosing r objects from a set of n objects. From the portion of Pascal's triangle which has been shown, for example,

$$\binom{3}{0} = 1 \quad \text{and} \quad \binom{3}{1} = 3.$$

By definition we have $\binom{n}{r-1} + \binom{n}{r} = \binom{n+1}{r}$.

Prove by induction that $\binom{n}{r} = \dfrac{n!}{r!(n-r)!}$, taking $0! = 1$.

Verify that this formula gives $\binom{n}{r} = \binom{n}{n-r}$.

What aspect of Pascal's triangle does this reflect?

5   *The Binomial Theorem for positive integral index*
Prove by induction that

$$(1 + x)^n = \binom{n}{0} + \binom{n}{1}x + \binom{n}{2}x^2 + \ldots + \binom{n}{r}x^r + \ldots + \binom{n}{n}x^n.$$

6   Examine each of the propositions:

   (i) $2^{n-1} \leqslant n^2$,

   (ii) $n^2 + n + 41$ is a prime number.

Find values of n for which these propositions hold and also the least value of n for which each is false.
If you were to attempt a proof of either of these propositions by induction, where would the proof break down?

7   Examine the numbers $6^n - 5n + 1$ and $6^{n+1} - 5(n+1) + 1$.
Find their difference.
Deduce that if the first of these numbers were divisible by 5 then the second would also be divisible by 5. Deduce also that if the second were divisible by 5 then the first would also be divisible by 5.
Is the first number divisible by 5 when $n = 1$?
Are there any values of n for which these numbers are divisible by 5?
If you were to attempt a proof by induction that the first number was divisible by 5, where would the proof break down?

Questions 6 and 7 establish the independence of the two parts of an inductive proof.

Axioms for the counting numbers

8   A first attempt to define the counting numbers from the property that the numbers start with 1, and each number has a successor which is new, might consist of the following rules.
Rule 1. The counting numbers include the number 1.
Rule 2. Each counting number n, has a unique successor n'.
(Think of this as meaning $n + 1$, but we don't actually need the idea of addition to construct a counting sequence.)
Rule 3. $m' = n'$, only if $m = n$.
Rule 4. 1 is not the successor of any number.
If the arrows mean 'has as its successor',
which rule is broken by $1 \to 2,$

$$\uparrow \quad \downarrow$$

$$4 \leftarrow 3$$

which rule is broken by $1 \to 2 \to 3,$

$$\uparrow \quad \downarrow$$

$$5 \leftarrow 4$$

and which rule is broken by $1 \to 2 \to 3 \to 4?$
Is there a rule which is broken by the pair of non-overlapping endless chains

$$\begin{cases} 1 \to 2 \to 3 \to 4 \to 5 \to \dots \\ a \to b \to c \to d \to e \to \dots \end{cases}$$

or by the chain together with the cycle

$$\begin{cases} 1 \to 2 \to 3 \to 4 \to 5 \to \dots \\ a \to b \end{cases}$$

$$\uparrow \quad \downarrow$$

$$d \leftarrow c?$$

9   Find some sets of numbers which satisfy the four rules of qn 8, but are not just the counting numbers. Keep taking $n' = n + 1$ in order to make comparisons between your sets and thereby to see what special properties we conventionally expect the counting numbers to have which distinguishes them from other sets which

satisfy the four rules. Consider the set $\{x \mid 1 \leqslant x, x \in \mathbb{Q}\}$ and also the set $\mathbb{N} \cup \{x + \frac{1}{2} \mid x \in \mathbb{Z}\}$.

10 Use the ideas of an inductive proof to specify two properties of a collection, M, of counting numbers, which would be sufficient to guarantee that this collection is only the set of counting numbers, that is, $M = \mathbb{N}$.

We will call these two properties rule 5, and now rules 1, 2, 3, 4, 5 form the *Peano postulates*, the conventional axiomatic definition of the set of counting numbers. The Peano postulates are formally listed in the summary of this chapter. Is the induction axiom, rule 5, broken by the pair of non-overlapping chains at the end of qn 8, or by the chain and cycle at the end of that question?

The set $\mathbb{N}$ is commonly called the set of *natural numbers*, but there is ambiguity in the literature on the question of whether $\mathbb{N}$ contains 0 or not. Those who let $0 \in \mathbb{N}$ are those who want to use $\mathbb{N}$ to count the elements of any finite set, including the empty set. We have used the term *counting numbers* to avoid any ambiguity and to echo the psychological origins of arithmetic. Having made the point, we will use the two descriptions interchangeably with zero *not* regarded as a *natural* or *counting number*.

Where question numbers appear in parentheses, this is an indication that the question is not part of the core sequence of the book, and may be omitted without prejudice to the study of later chapters.

Questions 11–34 have been included to demonstrate the crucial rôle of mathematical induction in the development of the number system, and to give practice in constructing inductive arguments.

Full development of the natural numbers

Addition

⋮	⋮	⋮	⋮		⋮	⋮
$1 + y'$	$2 + y'$	$3 + y'$	$4 + y'$	… $x + y'$	$x' + y'$	…
$1 + y$	$2 + y$	$3 + y$	$4 + y$	… $x + y$	$x' + y$	…
⋮	⋮	⋮	⋮		⋮	⋮
$1 + 4$	$2 + 4$	$3 + 4$	$4 + 4$	… $x + 4$	$x' + 4$	…
$1 + 3$	$2 + 3$	$3 + 3$	$4 + 3$	… $x + 3$	$x' + 3$	…
$1 + 2$	$2 + 2$	$3 + 2$	$4 + 2$	… $x + 2$	$x' + 2$	…
$1 + 1$	$2 + 1$	$3 + 1$	$4 + 1$	… $x + 1$	$x' + 1$	…

We show that addition is uniquely definable as the only way of putting together two counting numbers x and y so that

(a) $x + 1 = x'$, and

(b) $x + y' = (x + y)'$.

(11) If such an operation can be defined for some value a of x, the operation can be shown to be unique by supposing another operation, $\oplus$, exists such that

   (a) $a \oplus 1 = a'$, and

   (b) $a \oplus y' = (a \oplus y)'$,

   and defining

   $M = \{y \mid a + y = a \oplus y\}$.

   Now you can show that $M = \mathbb{N}$, and this ensures that there is at most one such operation.

(12) Now that we know there is at most *one* way of defining $x + y$ for a given x to satisfy (a) and (b) above, we check whether

   $1 + y = y'$,

   provides such a definition for $x = 1$.
   (i) Does this definition satisfy (a) and (b)?
   (ii) Does it provide a definition for all y?

(13) Now we suppose that $x + y$ has been successfully defined to satisfy (a) and (b), for the value $x = a$.
   Does the definition

   $a' + y = (a + y)'$

   satisfy (a) and (b) for $x = a'$?
   If so, we will have established the existence and uniqueness of $x + y$, satisfying (a) and (b) for all x and $y \in \mathbb{N}$.

(14) Define $1' = 2, 2' = 3, 3' = 4$, and prove the theorem

   $2 + 2 = 4$.

(15) Prove that $x + (y + z) = (x + y) + z$, by considering the set

   $M = \{z \mid x + (y + z) = (x + y) + z, \text{ for all } x, y \in \mathbb{N}\}$.

   This is called the *associative law* for addition.

(16) Prove that $x + y = y + x$, by considering the set

   $M = \{y \mid x + y = y + x, \text{ for all } x \in \mathbb{N}\}$.

   This is called the *commutative law* for addition.

(17) Prove that $y + x = z + x$ implies $y = z$, by considering the set

$M = \{x \mid y + x = z + x \text{ implies } y = z\}$.

This is called the *cancellation law* for addition.

Order

(18) Prove that every element of $\mathbb{N}$, other than 1, is a successor by considering the set

$M = \{1\} \cup \{x \mid x = u' \text{ for some } u \in \mathbb{N}\}$.

(19) Prove that there are no values of x or y such that $x + y = 1$, by first considering the possibility that $y = 1$, and secondly considering the possibility that $y \neq 1$, and in each case obtaining a contradiction.

(20) If two numbers x and $y \in \mathbb{N}$ are given, prove that at least one equation of the following three kinds holds:

$y = x$,

$y = x + u$, for some $u \in \mathbb{N}$,

$x = y + v$, for some $v \in \mathbb{N}$.

First show that this is true if $x = 1$, by considering the set

$M = \{y \mid y = 1 \text{ or } y = 1 + u \text{ or } 1 = y + v\}$.

Then show that this is true if $x = a \neq 1$, by using qn 18 and considering the set

$M = \{y \mid y = a \text{ or } y = a + u \text{ or } a = y + v\}$.

(21) Why is $1' \neq 1$?
Show that $x' \neq x$, for any $x \in \mathbb{N}$, using qn 17 to contradict $x' = x$.
Show that $x + u \neq x$, for any $x \in \mathbb{N}$.

We *define* $x < y$, and say; x is less than y, when $y = x + u$ for some $u \in \mathbb{N}$.

(22) If two numbers x and $y \in \mathbb{N}$ are given, prove that the three possibilities of qn 20 are mutually exclusive.
Interpret this result in terms of the 'less than' relation, to give the *trichotomy law* (*either* $x < y$, *or* $x = y$ *or* $y < x$).

(23) (i) Prove that if $x < y$, then $x + s < y + s$.

(ii) Prove that if $x + s < y + s$, then $x < y$.

(iii) Prove that if $x < y$ and $y < z$, then $x < z$.

(iv) Deduce from (i) and (iii) that if $x < y$ and $s < t$, then $x + s < y + t$.

Multiplication

The next step is to show that multiplication is uniquely definable as the only way of putting together two natural numbers x and y so that $x \cdot y \in \mathbb{N}$, with

(a) $x \cdot 1 = x$, and

(b) $x \cdot y' = x \cdot y + x$.

Notice the connection with the idea of repeated addition.

(24) If such an operation can be defined for some value a of x, the operation can be shown to be unique by supposing that another operation, $\times$, exists, such that

(a) $a \times 1 = a$

(b) $a \times y' = a \times y + a$

and considering the set $M = \{y \mid a \cdot y = a \times y\}$.

(25) Now that we know there is at most *one* way of defining $x \cdot y$ for a given x, to satisfy (a) and (b) above, we check whether

$1 \cdot y = y$

provides such a definition for $x = 1$.
Does this definition satisfy (a) and (b)?

(26) Now we suppose that $x \cdot y$ has been successfully defined to satisfy (a) and (b), for the value $x = a$. Does the definition

$a' \cdot y = a \cdot y + y$

satisfy (a) and (b) for $x = a'$?
If so, we shall have established the existence and uniqueness of $x \cdot y$ satisfying (a) and (b) for all x and $y \in \mathbb{N}$.

(27) Use the definitions of qn 14 to prove that $2 \cdot 2 = 4$.

(28) Prove that $x \cdot y = 1$, only when $x = y = 1$.
First suppose $y = 1$, and then suppose $y \neq 1$ and use qn 18.

(29) Prove that $x \cdot (y + z) = x \cdot y + x \cdot z$, by considering the set

$M = \{x \mid x \cdot (y + z) = x \cdot y + x \cdot z$ for all y and $z \in \mathbb{N}\}$.

This is one of the *distributive laws*.

(30) Prove that $(x \cdot y) \cdot z = x \cdot (y \cdot z)$, by considering the set

$M = \{z \mid (x \cdot y) \cdot z = x \cdot (y \cdot z)$ for all x and $y \in \mathbb{N}\}$.

This is the *associative law* for multiplication.

(31) Prove that $x \cdot y = y \cdot x$, by considering the set

$M = \{y \mid x \cdot y = y \cdot x$ for all $x \in \mathbb{N}\}$.

This is the *commutative law* for multiplication.

(32) Show that $(x + y) \cdot z = x \cdot z + y \cdot z$.
This is the other *distributive law*. See qn 29.

(33) Show that $x \cdot y = x \cdot z$ implies that $y = z$, by considering the set

$M = \{x \mid x \cdot y = x \cdot z$ implies $y = z\}$.

This is the *cancellation law* for multiplication

(34) If $y < z$, prove that $x \cdot y < x \cdot z$.

From this point, mathematical induction ceases to play an overt rôle in the development. The rest of this chapter is here to satisfy the curiosity of those who wish to see how the set of integers, $\mathbb{Z}$, and the set of rational numbers, $\mathbb{Q}$, may be constructed from the counting numbers, $\mathbb{N}$, and how their properties may be established.
    That these constructions can be accomplished using the set of counting numbers together with the two set-theoretic notions of cartesian product (to make the set of ordered pairs $\mathbb{N} \times \mathbb{N}$ and later $\mathbb{Z} \times \mathbb{Z} \backslash \{0\}$) and equivalence relations shows that the integers and the rational numbers are as 'real' and as logically consistent as the counting numbers. In the early nineteenth century the legitimacy of negative numbers was in question because they did not correspond to measurements. The methods we use for the constructions are not memorable, but the fact that the job can be done is reassuring. The final step of constructing the real numbers from the rationals is much harder and will not be attempted in this book. It is done thoroughly in the texts of Landau (1960), Thurston (1967), and Cohen and Ehrlich (1963). Those who are not familiar with equivalence relations and with the reflexive, symmetric and transitive laws will find a

free-standing treatment of them in the sixth chapter of my book
Groups.

Difference

When x and y are counting numbers and $x < y$, we know that
$x + v = y$, for some unique $v \in \mathbb{N}$, and in these circumstances we may
define $v = y - x$. However, when $x < y$, $x - y$ has no meaning
among the counting numbers, and to give a meaning to the operation
of subtraction between *any* two numbers, we will have to extend our
number system. The extension of the counting numbers to the
positive and negative integers is based on a property of subtraction
which holds when subtraction *can* be defined on the counting
numbers. When $x < y$, not only can $y - x$ be defined but, from
qn 23, $x + u < y + u$, for every $u \in \mathbb{N}$ and, in fact,

$$(y + u) - (x + u) = y - x.$$

So there is an infinite family of pairs of counting numbers with equal
difference, and we base our extension of the counting numbers on
this property.

Integers

We start building our extension of the counting numbers by
looking at ordered pairs of counting numbers. We will think of the
pair (x, y) as the difference $x - y$ and claim that two pairs are
equivalent when their differences are equal. For example, the pair
$(x + u, y + u)$ is equivalent to the pair (x, y) in the sense that the
two pairs have the same difference when that difference can be
defined.

We now define any two pairs (a, b) and (c, d) to be equivalent
when $a + d = b + c$, and we write this

$(a, b) \cong (c, d)$ precisely when $a + d = b + c$.

Although this definition is designed to correspond to equality of
difference, it is important to notice that the definition is meaningful
whether or not the difference exists within the counting numbers.

Now we can say that for any a, b, and $u \in \mathbb{N}$,
$(a + u, b + u) \cong (a, b)$.
To get a clearer idea of what pairs are equivalent, we note that, if
$a = b + v$, and $(a, b) \cong (c, d)$, then $c = d + v$, and also that
$(a, b) \cong (x + v, x)$ for every $x \in \mathbb{N}$.

If, on the other hand, $b = a + u$, and $(a, b) \cong (c, d)$, then $d = c + u$, and $(a, b) \cong (x, x + u)$ for every $x \in \mathbb{N}$.

Because we can establish the *reflexive law*, $(a, b) \cong (a, b)$; the *symmetric law*, that if $(a, b) \cong (c, d)$, then $(c, d) \cong (a, b)$; and the *transitive law*, that if $(a, b) \cong (c, d)$, and $(c, d) \cong (e, f)$, then $(a, b) \cong (e, f)$; we have the three conditions which make '$\cong$' an *equivalence relation* on the set of ordered pairs $\mathbb{N} \times \mathbb{N}$, and this means that the ordered pairs are partitioned into disjoint equivalence classes by this relation.

We illustrate the equivalence classes below.

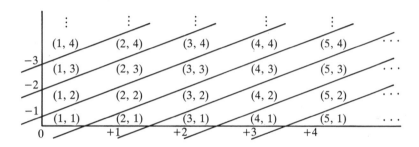

If the equivalence class containing (a, b) is denoted by $[a, b]$, then the equivalence class $[x + v, x]$, which is independent of x, will be called the integer $+v$. The equivalence class $[x, x + v]$ will be called the integer $-v$, and the equivalence class $[x, x]$ will be called the integer 0. We have now given a meaning to the symbols 0, $+1$, -1, $+2$, -2, $+3$, -3, ... The set of all integers will be denoted by $\mathbb{Z}$, the first letter of the German word *Zahlen* for numbers.

Defining operations on the integers

In order to define addition, multiplication and order on the integers, we start by defining these operations, and this relation, on the ordered pairs of counting numbers $\mathbb{N} \times \mathbb{N}$, building on our constructions within $\mathbb{N}$ itself.

We choose the definitions

$(a, b) + (c, d) = (a + c, b + d)$

$(a, b) < (c, d)$ only when $a + d < b + c$

$(a, b) \cdot (c, d) = (ac + bd, ad + bc)$

to be compatible with our understanding of (a, b) as the difference $a - b$ when this exists.

To show that these definitions may be applied to the integers we need to establish that if

$(a, b) \cong (p, q)$ and $(c, d) \cong (x, y)$, then

 (i) $(a, b) + (c, d) \cong (p, q) + (x, y)$,

 (ii) $(a, b) < (c, d)$ implies $(p, q) < (x, y)$, and

(iii) $(a, b) \cdot (c, d) \cong (p, q) \cdot (x, y)$.

Of these three results, (i) and (ii) are easy consequences of the definitions. The third is more troublesome.

Since $(a, b) \cong (p, q)$ means $a + q = b + p$, we have

$$c(a + q) + y(p + q) + bd = c(b + p) + y(p + q) + bd$$

so, using the associative law for addition, the distributive law and the commutative laws for addition and multiplication,

$$ac + bd + q(c + y) + yp = bc + p(c + y) + yq + bd.$$

So,

$$ac + bd + q(d + x) + yp = bc + p(d + x) + yq + bd,$$

since $(c, d) \cong (x, y)$ means $c + y = d + x$.

So, $ac + bd + py + qx + qd = bc + px + qy + (p + b)d$

$$= bc + px + qy + (a + q)d$$

$$= ad + bc + px + qy + qd.$$

Now, using the cancellation law for addition,

$$ac + bd + py + qx = ad + bc + px + qy,$$ and (iii) is proved.

Results (i), (ii) and (iii) show that definitions of addition, multiplication and order on the integers, given by

$[a, b] + [c, d] = [a + c, b + d]$,

$[a, b] < [c, d]$ only when $a + d < b + c$, and

$[a, b] \cdot [c, d] = [ac + bd, ad + bc]$,

are consistent.

When we start using single letters for integers, we shall need to say that, if the integer $x = [a, b]$, we denote the integer $[b, a]$ by $-x$, and it follows at once that $-(-x) = x$, and also, exceptionally, $-0 = 0$.

The familiar theorems for integers can now be established.

Addition
i. If x and y are integers, then $x + y$ is an integer. *Closure*.
ii. $x + 0 = 0 + x = x$, for all integers x. That is, 0 is *a zero* for addition.
iii. $x + (-x) = (-x) + x = 0$, for all integers x. That is, $-x$ is the *additive inverse* or *negative* of x.
iv. $x + (y + z) = (x + y) + z$. The *associative law* for addition.
v. If $x + y = z$, then $x = z + (-y)$ and $y = (-x) + z$.
vi. If $x + y = x$ for some integers, x, y then $y = 0$. So 0 is the *unique zero* for addition.
vii. $x + y = y + x$. The *commutative law* for addition.

Order
viii. Either $x < 0$, or $x = 0$, or $0 < x$. The *trichotomy law*.
ix. If $x < y$, then $x + z < y + z$.
x. If $x < y$ and $y < z$, then $x < z$. The *transitive law*.
xi. If $x < y$ and $s < t$, then $x + s < y + t$.
xii. If $0 < x$ and $0 < y$, then $0 < x + y$. *Closure*.
xiii. If $x < y$, then $-y < -x$.

Multiplication
xiv. If x and y are integers, then $x \cdot y$ is an integer. *Closure*.
xv. $x \cdot (+1) = (+1) \cdot x = x$, for all integers x. That is, $+1$ is *an identity* or *unit* for multiplication.
xvi. $x \cdot (y \cdot z) = (x \cdot y) \cdot z$. The *associative law* for multiplication.
xvii. $x \cdot y = y \cdot x$. The *commutative law* for multiplication.
xviii. $x \cdot 0 = 0 \cdot x = 0$, for all integers x.
xix. $x \cdot (y + z) = x \cdot y + x \cdot z$. The *distributive law*.
xx. $x \cdot (-y) = -(x \cdot y)$. [The proof comes from considering $x \cdot (y + (-y))$ and uses xviii and v.]
xxi. $(-x) \cdot (-y) = x \cdot y$.
xxii. If $0 < x$ and $0 < y$, then $0 < x \cdot y$. [Consider the product $(a + u, a) \cdot (b + v, b)$.]
xxiii. If $x \neq 0$, then $0 < x \cdot x$. [Consider the two products $x \cdot x$ and $(-x) \cdot (-x)$ and use viii, xxi and xxii.]
xxiv. If $0 < x \cdot y$ and $0 < x$, then $0 < y$. [Consider $y = 0$ and $0 < -y$.]
xxv. If $x \cdot y = 0$, then either x or $y = 0$. [Take x, $y \neq 0$ and consider the products $x \cdot y$, $x \cdot (-y)$, $(-x) \cdot y$, $(-x) \cdot (-y)$.]

Positive integers

The integers contain a subset which is like the set of counting numbers in every respect.

$$1 \leftrightarrow [a + 1, a] = +1,$$
$$2 \leftrightarrow [a + 2, a] = +2,$$
$$\vdots$$
$$n \leftrightarrow [a + n, a] = +n,$$
$$m \leftrightarrow [a + m, a] = +m,$$
$$\vdots$$

$$m + n \leftrightarrow [a + m + n, a] = +m + +n,$$
$$m \cdot n \leftrightarrow [a + m \cdot n, a] = (+m) \cdot (+n).$$

The subset of the integers which is thus matched with the counting numbers is precisely the set $\{x \mid x \in \mathbb{Z}, 0 < x\}$ and for this reason is called the set of *positive integers*. This set is denoted by $\mathbb{Z}^+$. Notice xii and xxii in this context. Because the set of positive integers has the same properties and structure as the set of counting numbers, the principle of mathematical induction may be applied to the positive integers.

Subtraction
We may define

$$x - y = x + (-y)$$

for any two integers x and y.

xxvi. $x < y$ if and only if $0 < y - x$. [Use ix.]

xxvii. If $x \cdot y = x \cdot z$ and $x \neq 0$, then $y = z$. The *cancellation law*.
[Consider $x \cdot (y - z)$ with xxv.]

Rational numbers

We have constructed the set of integers, $\mathbb{Z}$, and seen that the three binary operations $+$, $-$ and $\cdot$, and the relation $<$, can be defined on it in such a way as to give the basic properties that we expect. Division, however, is sometimes possible and sometimes not, and a further extension of the number system from the integers to the rational numbers is needed if division, other than by 0, is to be possible. To make this further extension, we consider the set of ordered pairs of integers, where the ordered pair (a, b) will be intended to correspond to the fraction a/b. The second element in the ordered pair must not be 0, and so the cartesian product we consider is $\mathbb{Z} \times \mathbb{Z} \backslash \{0\}$. On this set we define $(a, b) \simeq (c, d)$ when $ad = bc$. By proving the reflexive law, the symmetric law and the transitive law for this relation, we can establish that it is an equivalence relation, and the equivalence classes will be the rational numbers. The set of rational numbers is denoted by $\mathbb{Q}$ for *quotient*.

It is necessary to show that
addition defined by $(a, b) + (c, d) = (ad + bc, bd)$,
multiplication defined by $(a, b).(c, d) = (ac, bd)$, and
order defined by $(0, d) < (a, b)$ when $0 < ab$
may be used to give consistent definitions of addition, multiplication and order on the rational numbers. And then the properties of the rational numbers can be established.

These properties make $\mathbb{Q}$ an *ordered field*, and it is with the axioms for an ordered field that chapter 2 begins.

Summary

The principle of mathematical induction
A proposition relating to a positive integer n is valid for all positive integers n if
(a) the proposition is valid for $n = 1$,
(b) the proposition for n implies the proposition for $n + 1$.

The counting numbers, or natural numbers, $\mathbb{N}$, are defined by the Peano postulates:

1. $1 \in \mathbb{N}$,

2. for each $n \in \mathbb{N}$, there exists a unique $n' \in \mathbb{N}$,

3. $m' = n' \Rightarrow m = n$,

4. $1 \neq n'$ for any $n \in \mathbb{N}$,

5. If $M \subseteq \mathbb{N}$, $1 \in M$, and $n \in M \Rightarrow n' \in M$, then $M = \mathbb{N}$.
Addition, multiplication and order can be defined on $\mathbb{N}$ with their familiar properties in exactly one way.

The integers, $\mathbb{Z}$, are defined as equivalence classes on the set of ordered pairs $\mathbb{N} \times \mathbb{N}$ with (a, b) equivalent to (c, d) when $a + d = b + c$.

Addition, subtraction, multiplication and order can be defined on $\mathbb{Z}$ with their familiar properties.

The integers contain a subset (the set of positive integers) with the same structure and properties as $\mathbb{N}$.

The rational numbers, $\mathbb{Q}$, are defined as equivalence classes on the set of ordered pairs $\mathbb{Z} \times \mathbb{Z} \setminus \{0\}$ with (a, b) equivalent to (c, d) when $ad = bc$.

Addition, subtraction, multiplication, division (except by zero), and order can be defined on $\mathbb{Q}$ with their familiar properties.

The rational numbers contain a subset with the same structure and properties as $\mathbb{Z}$.

Historical note

The general form of a triangular number is to be found in Euclid (IX.36). The sum of the first n squares was known to Archimedes (250 B.C.) and the sum of the first n cubes to the Arabs (A.D. 1010). The justification of these forms uses induction implicitly.

The earliest use of proof by mathematical induction in the literature is by Maurolycus in his study of polygonal numbers published in Venice in 1575. Pascal knew the method of Maurolycus and used it for work on the binomial coefficients (c. 1657). In 1713, Jacques Bernoulli used an inductive proof to make rigorous the claim of Wallis that

$$\sum_{r=0}^{r=n} r^k / n^{k+1} \rightarrow 1/(k + 1) \quad \text{as } n \rightarrow \infty.$$

By the latter part of the eighteenth century, induction was being used by several authors. The name 'mathematical induction' as distinct from scientific induction is due to De Morgan (1838). The use of inductive definitions long predates this method of proof.

The well-ordering principle, that every non-empty set of natural numbers has a least member, is logically equivalent to mathematical induction (see Ernest 1982). Well-ordering was used by Euclid (VII.31) to show that every integer has a prime factor and by Fermat (1637) in his proof of the non-existence of integral solutions to $x^4 + y^4 = z^2$. However, the recognition that well-ordering and induction are equivalent is recent and, from an historical point of view, awareness of the two principles developed independently.

In his *Paradoxes of the Infinite*, published posthumously in 1851, B. Bolzano recognised that the proof of the theorem $5 + 7 = 12$ depended on the existence of a '+1' successor function and the

associative law for addition. In 1861, H. Grassmann published a
formal development of the integers, similar to that which G. Peano
was to give, but he omitted from his postulates the properties we
have called rules 3 and 4 in our treatment. Frege gave a development
of the number system in 1884, but his work was ignored until
B. Russell drew attention to it in 1901. Both R. Dedekind and Peano
developed different formal constructions of the natural numbers at
about the same time. Dedekind's (1887) and Peano's (1889) work are
readily available in English. [See Heijenoort (1967) for Peano.]
E. Landau gave the first presentation of Peano's work for
undergraduates in 1929. The earliest use of ordered pairs to construct
an extension of the number system was by W. R. Hamilton (1833) in
his construction of the complex numbers. In 1872, E. Heine used
ordered pairs to construct the rational numbers from the integers.

Answers

2 $5|5$, 30, 205. Difference $= 5 \cdot 6^n - 5$.

3   (i) $\frac{1}{2}n(n + 1) + (n + 1) = \frac{1}{2}(n + 1)(n + 2)$.

   (ii) $\frac{1}{2}n[2a + (n - 1)d] + (a + nd) = \frac{1}{2}(n + 1)(2a + nd)$.

   (iii) $[\frac{1}{2}n(n + 1)]^2 + (n + 1)^3 = [\frac{1}{2}(n + 1)(n + 2)]^2$.

   (iv) $\frac{1}{3}n(n + 1)(n + 2) + (n + 1)(n + 2) = \frac{1}{3}(n + 1)(n + 2)(n + 3)$.

   (v) $(2^n - 1) + 2^n = 2^{n+1} - 1$.

   (vi) $\dfrac{x^n - 1}{x - 1} + x^n = \dfrac{x^{n+1} - 1}{x - 1}$.

   (vii) $\dfrac{nx^n}{x - 1} - \dfrac{x^n - 1}{(x - 1)^2} + (n + 1)x^n = \dfrac{(n + 1)x^{n+1}}{x - 1} - \dfrac{x^{n+1} - 1}{(x - 1)^2}$.

4   Use the definition to prove that $\dbinom{n + 1}{r} = \dfrac{(n + 1)!}{r!(n + 1 - r)!}$.

   Symmetry about a vertical axis.

5   The coefficient of x^r in $(1 + x)^{n+1} = \dbinom{n}{r - 1} + \dbinom{n}{r}$.

6   (i) True for $n \leqslant 6$, false for $n \geqslant 7$.
   (ii) True for $n \leqslant 39$, false for $n = 40$ or a multiple of 41.
   The validity of the proposition for n does not imply the validity of the proposition for $n + 1$.

7 $6^n - 5n + 1$ divisible by $5 \Leftrightarrow 6^{n+1} - 5(n + 1) + 1$ divisible by 5, but both statements are false because the first statement does not hold when $n = 1$.

8   Rule 4, rule 3, rule 2. None of the rules is broken by either of the last two illustrations.

9   For example $\{x \mid 1 \leqslant x, x \in \mathbb{R}\}$, and the two sets mentioned at the end of the question.

10 Rule 5. Given $M \subseteq \mathbb{N}$. Provided

   (i) $1 \in M$, and

   (ii) $n \in M \Rightarrow n' \in M$,

   then $M = \mathbb{N}$.
   The induction axiom puts the natural numbers into a single chain.

11 Use the induction axiom.

12 (i) Agrees with (a) for $y = 1$.

Agrees with (b) because $1 + y' = (y')'$ from the definition

$$= (1 + y)' \text{ using the definition again.}$$

(ii) Show that $M = \{y \mid 1 + y \text{ defined by } 1 + y = y'\} = \mathbb{N}$.

13 Satisfies (a) because $a' + 1 = (a + 1)'$ from the definition

$$= (a')' \text{ since } x = a \text{ satisfies (a).}$$

Satisfies (b) because $a' + y' = (a + y')'$ from the definition

$$= \big((a + y)'\big)' \text{ since } x = a \text{ satisfies (b),}$$

$$= (a' + y)' \text{ from the definition.}$$

14 $\quad 2 + 2 = 2 + 1' \qquad$ by definition

$\qquad = (2 + 1)' \qquad$ by (b)

$\qquad = (2')' \qquad$ by (a)

$\qquad = 3' \qquad$ by definition

$\qquad = 4 \qquad$ by definition.

15 $1 \in M$ using (a) to give (b).

$z \in M \Rightarrow z' \in M$ by repeated use of (b).

16 $1 \in M$ from (a) and the definition in qn 12.

$y \in M \Rightarrow x + y = y + x \Rightarrow (x + y)' = (y + x)' \Rightarrow$

$x + y' = y' + x$ from (b) and definition $12 \Rightarrow y' \in M$.

17 $1 \in M$ by (a) and rule 3.

Suppose $x \in M$, $y + x' = z + x' \Rightarrow (y + x)' = (z + x)'$ from (b)

$\Rightarrow y + x = z + x$ from rule $3 \Rightarrow y = z$ by hypothesis. So $x' \in M$.

19 If $y = 1$, $x + 1 = x' \neq 1$ by rule 4. If $y \neq 1$, $y = v'$ (say, by qn 18), and $x + y = x + v' = (x + v)' \neq 1$, by rule 4.

20 True for $x = 1$, for either $y = 1$ or $y = u'$.

$y = a \Rightarrow y + 1 = a'$, $y = a + 1 \Rightarrow y = a'$,

$y = a + c' \Rightarrow y = a' + c$,

$y + v = a \Rightarrow y + v' = a'$.

21 Rule 4. $x' = x \Rightarrow 1 + x' = 1 + x \Rightarrow 1' + x = 1 + x \Rightarrow 1' = 1$.

$x + u = x \Rightarrow (x + u)' = x' \Rightarrow x + u' = x + 1 \Rightarrow u' = 1$

against rule 4.

22 If $y = x$ then $y \neq x + u$ by 21. If $x = y$, then $x \neq y + v$ by qn 21. If $y = x + u$ then $y + v = (x + u) + v = x + (u + v) \neq x$ by qn 21.

23 (i) If $y = x + u$, then $y + s = (x + u) + s = x + (u + s) = x + (s + u) = (x + s) + u$, so $x + s < y + s$.
(ii) Cancellation.
(iii) Associativity.

24, 25, 26 Similar argument and method to qns 11, 12, 13.

27 $2 \cdot 2 = 2 \cdot 1' = 2 \cdot 1 + 2 = 2 + 2$. Now see qn 14.

28 If $y = 1$, then $x \cdot y = x$ from (a), so $x = 1$.
If $y \neq 1$, $y = v'$. $x \cdot y = x \cdot v' = x \cdot v + x \neq 1$ by qn 19.

29 $1 \in M$ by the definition in qn 25. $a \in M \Rightarrow a' \in M$ by repeated use of qn 26.

30 $1 \in M$ by (a). If $a \in M$, $(xy)a' = (xy)\,a + xy = x(ya) + xy = x(ya + y) = x(ya') \Rightarrow a' \in M$.

31 $1 \in M$ from (a) and qn 25. If $a \in M$,

$a'x = ax + x = xa + x = xa'$.

33 $1 \in M$ from qn 25. If $a \in M$, $a'y = a'z \Rightarrow ay + y = az + z$ by qn 26, $\Rightarrow y = z$ by cancellation, so $a' \in M$.

34 If $z = y + u$, then $x \cdot z = x(y + u) = x \cdot y + x \cdot u$.

2

Order
Arithmetic with inequalities

Preliminary reading: Beckenbach and Bellman.
Concurrent reading: Korovkin, Kazarinoff.
Further reading: Thurston, chs F and G, Hardy, Littlewood and
Pólya ch. 2.

The rational numbers,

$$\mathbb{Q} = \left\{ \frac{p}{q} \,\middle|\, p, q \in \mathbb{Z}, q \neq 0 \right\}.$$

We begin this chapter by listing some properties of the rational
numbers. We do this partly because the exercise of constructing the
rational numbers from the integers and establishing their properties
was suggested as an *option* in the last chapter and the experience of
making this construction is not integral to this book. More
importantly, these properties hold for number systems other than the
rational numbers, notably for the real numbers, and so they carry a
significance beyond that which was established in chapter 1.

Properties of addition, $+$

i. *Closure:* if $a, b \in \mathbb{Q}$, then $a + b \in \mathbb{Q}$.

ii. *The associative law:* if $a, b, c \in \mathbb{Q}$, then
$a + (b + c) = (a + b) + c$.

iii. *Zero:* there is an element $0 \in \mathbb{Q}$ such that, for any $a \in \mathbb{Q}$,
$a + 0 = a$.

iv. *Negatives:* for each $a \in \mathbb{Q}$, there exists $-a \in \mathbb{Q}$ such that
$a + (-a) = 0$.

v. *The commutative law:* for any $a, b \in \mathbb{Q}$, $a + b = b + a$.

These five properties are conventionally summed up by saying that $(\mathbb{Q}, +)$ is an *abelian group*. The five properties are the axioms for an abelian group. There are four important consequences of these properties which will be offered as the first four (optional) exercises in this chapter. Please read these questions carefully whether you attempt them as exercises or not.

(1) Using the associative law, show that any two ways of inserting brackets in the sum $a + b + c + d$ give equal numbers.
This freedom to insert brackets arbitrarily, without changing the value of the expression, extends to any finite sum, but not always to infinite sums.

(2) Use the definition of zero and the commutative law to show that there is *only one zero* in an abelian group.

(3) Show that the negative of each element is unique. That is, if $a + (-a) = 0$ and $a + (\sim a) = 0$, then $-a = \sim a$.

(4) Show that $-(-a) = a$; that is, that the negative of $-a$ is a.

5 Does the set of natural numbers, $\mathbb{N}$, form an abelian group in the sense that we could replace $\mathbb{N}$ for $\mathbb{Q}$ in the five listed properties of addition? Does the set of integers, $\mathbb{Z}$, form an abelian group in this sense? Find a proper subset of the integers which forms an abelian group under addition. Find a set of rational numbers which contains all the integers and more, but does not contain all the rational numbers, and which forms an abelian group under addition.

(6) If $\mathbb{Q}[s]$ denotes the set of numbers $\{x + sy \mid x, y \in \mathbb{Q}\}$ where s is not a rational number, show how to add two elements of $\mathbb{Q}[s]$ together in such a way that that addition makes $\mathbb{Q}[s]$ an abelian group.

There is a universal convention of writing $a + (-b)$ as $a - b$.

Properties of multiplication, ·

vi. *Closure:* if $a, b \in \mathbb{Q}$, then $a \cdot b \in \mathbb{Q}$.

vii. *The associative law:* if $a, b, c \in \mathbb{Q}$, then
$a \cdot (b \cdot c) = (a \cdot b) \cdot c$.

viii. *Identity or unit:* there is an element $1 \in \mathbb{Q}\backslash\{0\}$ such that
$a \cdot 1 = a$ for all $a \in \mathbb{Q}$.

ix. *Inverses:* for each $a \in \mathbb{Q}\backslash\{0\}$, there exists
$1/a \in \mathbb{Q}\backslash\{0\}$ such that $a \cdot (1/a) = 1$.

x. *The commutative law:* for any $a, b \in \mathbb{Q}$, $a \cdot b = b \cdot a$.

7 What property of an abelian group cannot be claimed for $\mathbb{Q}\backslash\{0\}$ under multiplication, using only the information on multiplication given here in vi–x?

We now claim the crucial link between addition and multiplication which originates in the notion of multiplication of the counting numbers as repeated addition.

The distributive law
xi. If $a, b, c \in \mathbb{Q}$, then $a \cdot (b + c) = a \cdot b + a \cdot c$.

Properties i–xi have been stated for the rational numbers $\mathbb{Q}$. These properties may hold for other sets of numbers, for a finite set, such as the integers modulo a prime number, and for other infinite sets such as the real numbers or complex numbers.

Any set of numbers satisfying *these eleven conditions* is called a *field*, the eleven conditions forming the **axioms for a field**. A result deduced from some combination of these axioms will hold in any field.

(8) The integers modulo 2 (with just two elements 0 and 1) can be exhibited thus:

+	0	1		·	0	1
0	0	1		0	0	0
1	1	0		1	0	1

Check the eleven axioms i–xi for this field.

9 By considering the equation $a \cdot a = a \cdot (a + 0)$, prove that in any field $a \cdot 0 = 0$. Deduce that $0 \cdot a = 0$.

10 By considering the equation $a \cdot 0 = a \cdot \big(b + (-b)\big)$, prove that in any field $a \cdot (-b) = -(a \cdot b)$. Deduce that $(-a) \cdot b = -(a \cdot b)$. Use qn 4 to prove that $(-a) \cdot (-b) = a \cdot b$.

11 From $(-1) \cdot (-1) = 1$ deduce that $1/(-1) = -1$, and from $(-1) \cdot (-a) = a$, deduce that $1/(-a) = -(1/a)$ provided $a \neq 0$.

12 If a and b are elements of a field and $a \cdot b = 0$, suppose that $a \neq 0$ and use ix and qn 9 to prove that $b = 0$.
Deduce that $a \cdot b = 0$ only when either $a = 0$ or $b = 0$.

13 Must the non-zero elements of a field form an abelian group under multiplication?

There is a universal convention of writing $a \cdot (1/b)$ as a/b.

The final group of fundamental properties of the rational numbers contains those concerned with inequalities, that is, with the 'greater than' and 'less than' relations. We will state the basic properties and draw the immediate conclusions with just one of these relations, for the sake of clarity, but that is not meant to discourage the use of both in later work. There is no logical reason for treating one of these relations as more fundamental than the other. We will start working with 'less than' for the reason that the order of display in this case coincides with the conventional order on a displayed number line, or the 'x-axis'.

The properties of the 'less than' relation are in the first instance properties of the rational numbers, but we will state them for elements of an arbitrary field, F. A field on which a 'less than' relation is defined, satisfying these properties, is called an *ordered field*.

Properties of order

xii. *The trichotomy law:* for any $a \in F$, either $0 < a$, or $0 = a$, or $0 < -a$; and only one of these three holds.

xiii. *Positive closure:* if $0 < a$ and $0 < b$, then $0 < a + b$, and $0 < a \cdot b$.

xiv. $a < b$ means precisely $0 < b - a$.

Properties i–xiv are the **axioms for an ordered field**.

Algebra of inequalities

14 In an ordered field, prove that $a < 0$ when $0 < -a$.
Deduce that the trichotomy law may be expressed in the form: either $0 < a$, or $0 = a$, or $a < 0$.

15 Use qn 4 to deduce that $-a < 0$ when $0 < a$.

16 Prove that $-b < -a$ if and only if $a < b$.

17 *Extended trichotomy law.* Prove that, for any a, b in an ordered field, either $a < b$, or $a = b$, or $b < a$, and only one of these three holds.

18 Prove that, if $a < b$, then $a + c < b + c$.

19 *Transitive law.* Prove that, if $a < b$ and $b < c$, then $a < c$.

20 Prove that, if $a < b$ and $c < d$, then $a + c < b + d$.

21 Prove that, if $a < b$ and $0 < c$, then $a \cdot c < b \cdot c$.

22 Prove that, if $0 < a < b$, then $a^2 < b^2$.

23 Prove that, if $0 < a$ and $b < 0$, then $a \cdot b < 0$.

24 The definition of 1 (in property viii) excludes the possibility that $0 = 1$, so either $0 < 1$ or $0 < -1$ from qn 14.
Now use xiii and qn 10 to prove that $0 < 1$ from the equation
$1 \cdot 1 = (-1) \cdot (-1)$.

25 For any element a in an ordered field show that either $0 = a$ or $0 < a^2$.

26 Prove that, if $0 < a$, then $0 < 1/a$. Use trichotomy, and qns 23 and 24.

27 Prove that, if $0 < a < b$, then $0 < 1/b < 1/a$.

28 Prove that, if $a < 0$, then $1/a < 0$.

29 Prove that, if $a < b < 0$, then $1/b < 1/a < 0$.

30 Prove that, if $0 < a$, $0 < b$ and $a^2 < b^2$, then $a < b$.
Use trichotomy and qn 22.

31 Show that 0, 1, $1 + 1$, $1 + 1 + 1$, ... must all be different, or a contradiction to qn 24 would arise. Deduce that an ordered field must be infinite.
It follows that any ordered field must contain a copy of $\mathbb{N}$, and thus from iv a copy of $\mathbb{Z}$, and so from ix a copy of $\mathbb{Q}$.

32 Show that no field containing a square root of -1 can be ordered, since the existence of such an element will lead to a contradiction between qns 24 and 25.

By universal convention, $b > a$ means the same as $a < b$. It is also conventional to write $a \leq b$ when either $a < b$ or $a = b$ and to write $a \geq b$ when either $a > b$ or $a = b$. In an ordered field, F, the subset $\{a \mid 0 < a\}$ is known as the set of *positive* elements, and denoted by F^+. The positive integers may likewise be denoted by $\mathbb{Z}^+$.

33 (*Bernoulli's inequality*, 1689) Prove by induction, using qn 21, that if $-1 < x$, then $1 + nx \leqslant (1 + x)^n$, for all positive integers n. For positive values of x, this could also have been deduced from the binomial theorem.

34 With reference to qns 22 and 30, show that, for positive a and b, $a < b$ if and only if $a^3 < b^3$.

35 Prove by induction for $0 < a$ and $0 < b$, that $a < b$ if and only if $a^n < b^n$ for all positive integers n.
   If the four strict inequalities '$<$' in this proposition were replaced by '$\leqslant$', would the analogous proposition still hold?

   The result in qn 35 is important in its own right and we will use it in our study of real functions. But this result has another implication in our present context. In an arbitrary ordered field there is no guarantee that a positive element has an nth root, or even a square root, in the field. For example the number 2 has no square root in the field of rational numbers as we will see in 4.17. There are famous inequalities which include nth roots and which thereby appear to be meaningless in, say, the field of rational numbers. But qn 35 establishes that inequalities involving nth roots have meaningful equivalents in any ordered field: for example, for positive numbers, $a < \sqrt[n]{b}$ is equivalent to $a^n < b$. In the rest of this chapter we will use nth roots without further apology, knowing that an equivalent and generally meaningful inequality could be constructed in each case. We will demonstrate the existence of nth roots of positive real numbers in qn 4.39. In order to make the best use of qn 35 we will always presume that $\sqrt[n]{a}$ is positive when a is positive (so $\sqrt{4} < \sqrt{9}$, because $\sqrt{9}$ *never* means -3).

Arithmetic mean and geometric mean

36 If 2, A, 8 are three consecutive terms of an arithmetic progression, what is the number A? A is called the *arithmetic mean* of 2 and 8.
   If 2, G, 8 are three consecutive terms of a geometric progression, what is the number G? G is called the *geometric mean* of 2 and 8.
   Which is the greater, A or G?

37 Find the arithmetic mean and also the geometric mean of the numbers 2 and 3. Show that the inequality $0 < (\sqrt{3} - \sqrt{2})^2$ establishes which is the greater.

38 (*Euclid*) If a and b are positive, find the arithmetic mean and geometric mean of the numbers a^2 and b^2. Which of the means is greater? Under what circumstances are the two means equal?

39 If a and b are positive, prove that

$$\sqrt{(ab)} \leq \tfrac{1}{2}(a + b).$$

Under what circumstances can there be equality?

40 For any positive number a, prove that $2 \leq a + (1/a)$.

41 For a given perimeter, what shape of rectangle gives the greatest area?

42 If x and a are both positive numbers, what are the arithmetic and the geometric means of the numbers x and a^2/x?
If $x_1 > 0$ and x_n is defined by $x_{n+1} = \tfrac{1}{2}(x_n + a^2/x_n)$, show that $a \leq x_{n+1} \leq x_n$ for $n \geq 2$.

43 Prove that, if a_1 and b_1 are given positive numbers with $a_1 < b_1$, and two sequences are defined by

$$a_{n+1} = \sqrt{(a_n b_n)} \quad \text{and} \quad b_{n+1} = \tfrac{1}{2}(a_n + b_n),$$

then

(i) $0 < a_n < a_{n+1} < b_{n+1} < b_n$,

(ii) $0 < b_{n+1} - a_{n+1} < b_{n+1} - a_n = \tfrac{1}{2}(b_n - a_n)$,

(iii) $0 < b_{n+1} - a_{n+1} < (\tfrac{1}{2})^n (b_1 - a_1)$.

So the two means, a_n and b_n, get closer together as n increases.

44 If a, b, c, d are all positive numbers, justify the claim that

$$\sqrt[4]{(abcd)} = \sqrt{((\sqrt{ab}) \cdot (\sqrt{cd}))} \leq \tfrac{1}{2}(\sqrt{(ab)} + \sqrt{(cd)})$$
$$\leq \tfrac{1}{2}(\tfrac{1}{2}(a + b) + \tfrac{1}{2}(c + d)) = \tfrac{1}{4}(a + b + c + d).$$

Under what circumstances do the strict equalities hold?

For any four positive numbers a_1, a_2, a_3, a_4, the number $\tfrac{1}{4}(a_1 + a_2 + a_3 + a_4)$ is called the arithmetic mean, A, and the number $\sqrt[4]{(a_1 a_2 a_3 a_4)}$ is called the geometric mean, G. In qn 44 we have proved that $G < A$ unless all four numbers are equal, in which case $G = A$.

For any n positive numbers $a_1, a_2, \ldots, a_n$, the number

$$\frac{(a_1 + a_2 + \ldots + a_n)}{n}$$

is called the arithmetic mean, and the number

$$\sqrt[n]{(a_1 a_2 \ldots a_n)}$$

is called the geometric mean.

In qns 39 and 44 we have shown that, for $n = 2$ and $n = 4$, the geometric mean is less than or equal to the arithmetic mean with equality only when all n numbers are equal.

The method by which qn 39 was used to give the result of qn 44 can be generalised.

45 Suppose that for any n positive numbers, not all equal, we know that the geometric mean is less than the arithmetic mean. Suppose further that the arithmetic mean of the n positive numbers $a_1, a_2, \ldots, a_n$, is A and the geometric mean is G, and that the arithmetic mean of the n positive numbers b_1, $b_2, \ldots, b_n$, is B and the geometric mean is H. Use

$$\sqrt[2n]{(a_1 a_2 \ldots a_n b_1 b_2 \ldots b_n)} = \sqrt{(GH)}$$

$$\leq \tfrac{1}{2}(G + H) \leq \tfrac{1}{2}(A + B)$$

$$= (a_1 + a_2 + \ldots + a_n + b_1 + b_2 + \ldots + b_n)/(2n),$$

to prove that the geometric mean of $2n$ positive numbers, not all equal, is less than the arithmetic mean by showing that at least one of the inequalities must be strict.

We can now claim for $n = 2$, 2^2, 2^3 and, by induction, *any* power of 2, that for n positive numbers, not all equal, the geometric mean is less than the arithmetic mean.

What about the values of n which are *not* powers of 2?

46 Let a_1, a_2 and a_3 be three positive numbers which are not all equal, and let $A = \tfrac{1}{3}(a_1 + a_2 + a_3)$. Apply the result of qn 44 to the four positive numbers a_1, a_2, a_3 and A, which are not all equal, to prove that

$$\sqrt[3]{(a_1 a_2 a_3)} < A.$$

47 Generalise the result of qn 46 to show that if for any n positive numbers, not all equal, the geometric mean is less than the arithmetic mean, then the same is true for any $n - 1$ positive numbers, not all equal.

48 If $1 < a$, prove that

(i) $1 < a < a^2 < \ldots < a^n$,

(ii) $\dfrac{1 + a + a^2 + \ldots + a^{n-1}}{n}$

$$< \dfrac{1 + a + a^2 + \ldots + a^{n-1} + a^n}{n + 1},$$

(iii) $\dfrac{a^n - 1}{n} < \dfrac{a^{n+1} - 1}{n + 1}$ (from qn 1.3(vi)),

(iv) if $b = a^{1/n(n+1)}$, $(n + 1)(\sqrt[n+1]{b} - 1) < n(\sqrt[n]{b} - 1)$.

The sequence $(1 + 1/n)^n$

49 Apply the geometric mean–arithmetic mean inequality to the $n + 1$ positive numbers,

$$1, 1 + \frac{1}{n}, 1 + \frac{1}{n}, \ldots, 1 + \frac{1}{n},$$

which are not all equal, to prove that

$$\left(1 + \frac{1}{n}\right)^n < \left(1 + \frac{1}{(n + 1)}\right)^{n+1}$$

for any positive integer n.

50 Apply the geometric mean–arithmetic mean inequality to the $n + 1$ positive numbers,

$$1, 1 - \frac{1}{n}, 1 - \frac{1}{n}, \ldots, 1 - \frac{1}{n},$$

which are not all equal, to prove that

$$\left(1 - \frac{1}{n}\right)^n < \left(1 - \frac{1}{(n + 1)}\right)^{n+1},$$

for any positive integer n.

51 Prove that $(1 + 1/n)^{n+1} = 1/(1 - 1/(n + 1))^{n+1}$ and, with the help of qns 49 and 50, prove that

$$\left(1 + \frac{1}{n}\right)^n < \left(1 + \frac{1}{(n + 1)}\right)^{n+1}$$

$$< \left(1 + \frac{1}{(n + 1)}\right)^{n+2} < \left(1 + \frac{1}{n}\right)^{n+1}.$$

Check that, for $n = 5$, $(1 + 1/n)^{n+1} < 3$ and deduce that
$2 \le (1 + 1/n)^n < 3$ for all positive integers n.

Question 51 establishes that in any ordered field

$$2 < \left(1 + \frac{1}{2}\right)^2 < \left(1 + \frac{1}{3}\right)^3 < \left(1 + \frac{1}{4}\right)^4 < \ldots < \left(1 + \frac{1}{n}\right)^n < 3.$$

We return to this example in qn 4.35.

52 Deduce from qn 51 that, if $n \ge 3$, then

$$(1 + 1/n)^n < n \text{ and so } \sqrt[n+1]{(n + 1)} < \sqrt[n]{n}.$$

The Cauchy–Schwarz inequality

53 Use the identity $a^2x^2 + abx + ac = (ax + \frac{1}{2}b)^2 + ac - \frac{1}{4}b^2$ to
show that, if $0 < a$, then

$$c - b^2/4a \le ax^2 + bx + c, \quad \text{for all values of } x.$$

Is equality possible? Illustrate on a graph.

54 If $0 < a$, prove that $0 \le ax^2 + bx + c$, for all values of x, if and
only if $b^2 \le 4ac$. If $b^2 < 4ac$, is it possible to have
$ax^2 + bx + c = 0$?
What is the analogous claim if $a < 0$?

55 Express $(a_1x + b_1)^2 + (a_2x + b_2)^2$ as a quadratic polynomial in
x, and then apply qn 54 to prove that

$$(a_1b_1 + a_2b_2)^2 \le (a_1^2 + a_2^2)(b_1^2 + b_2^2),$$

for any numbers a_1, a_2, b_1, b_2.
Under what circumstances does equality hold?

56 If, in the plane $\mathbb{R} \times \mathbb{R}$, A is the point with coordinates (a_1, a_2), B
is the point with coordinates (b_1, b_2) and O is the origin, show
that the inequality of qn 55 is equivalent to the triangle inequality
$AB \le OA + OB$.

57 Construct an appropriate quadratic polynomial in x which will
enable you to generalise the work of qn 55 and show that

$$(a_1b_2 + a_2b_2 + \ldots + a_nb_n)^2$$
$$\le (a_1^2 + a_2^2 + \ldots + a_n^2)(b_1^2 + b_2^2 + \ldots + b_n^2)$$

for any numbers $a_1, a_2, \ldots, a_n, b_1, b_2, \ldots, b_n$.
Under what circumstances does equality hold?

Absolute values

We define

$$|a| = \begin{cases} a & \text{when } 0 \leqslant a, \text{ and} \\ -a & \text{when } a < 0. \end{cases}$$

The results that follow on absolute values may be scanned rather than proved. The results are essential for the rest of the book, but the proofs are not notably instructive. We will repeatedly appeal to qns 65 and 67.

58 Prove that $0 \leqslant |a|$ for all a.

59 Prove that $a \leqslant |a|$ and $-|a| \leqslant a$ for all a.

60 Prove that $|a| = |-a|$ for all a.

61 Prove that $|ab| = |a| \cdot |b|$ for all a, b.

62 Prove that $|1/a| = 1/|a|$, provided $a \neq 0$.

63 Prove that $|a/b| = |a|/|b|$, provided $b \neq 0$.

64 Prove that, if $|a| \leqslant b$, then $-b \leqslant a \leqslant b$;
and that, if $-b \leqslant a \leqslant b$, then $|a| \leqslant b$.
To see what this means, explore the possible positions of a and b on a number line.

65 Use qn 59 to show that

$$-|a| - |b| \leqslant a + b \leqslant |a| + |b|,$$

and qn 64 to deduce that $|a + b| \leqslant |a| + |b|$, for all a and b. This is called the *triangle inequality* because $|a - b|$ is the distance between the points a and b on the number line, and the inequality is equivalent to $|b - c| \leqslant |b - a| + |a - c|$. See qn 56.

66 Deduce from qn 65 that $|a - b| \leqslant |a| + |b|$, for all a and b.

67 By putting $b = c - a$ in qn 65, the triangle inequality, prove that $|c| - |a| \leqslant |c - a|$, and deduce that

$$|a| - |c| \leqslant |a - c| = |c - a|.$$

Deduce further that $\big||c| - |a|\big| \leqslant |c - a|$.

The contrast between qns 65 and 67 is striking: adding absolute values may only increase a sum; subtracting absolute values may only diminish a difference.

Summary

Axioms for an ordered field $(F, +, \cdot, <)$

Properties of addition, $+$

i. *Closure:* if $a, b \in F$, then $a + b \in F$.

ii. *The associative law:* if $a, b, c \in F$, then
$a + (b + c) = (a + b) + c$.

iii. *Zero:* there is an element $0 \in F$, such that, for any $a \in F$,
$a + 0 = a$.

iv. *Negatives:* for each $a \in F$, there exists $-a \in F$ such that
$a + (-a) = 0$.

v. *The commutative law:* for any $a, b \in F$, $a + b = b + a$.

Properties of multiplication, $\cdot$

vi. *Closure:* if $a, b \in F$, then $a \cdot b \in F$.

vii. *The associative law:* if $a, b, c \in F$, then
$a \cdot (b \cdot c) = (a \cdot b) \cdot c$.

viii. *Identity or unit:* there is an element $1 \in F \backslash \{0\}$ such that
$a \cdot 1 = a$ for all $a < F$.

ix. *Inverses:* for each $a \in F \backslash \{0\}$, there exists
$1/a \in F \backslash \{0\}$ such that $a \cdot (1/a) = 1$.

x. *The commutative law:* for any $a, b \in F$, $a \cdot b = b \cdot a$.

The distributive law

xi. If $a, b, c \in F$, then $a \cdot (b + c) = a \cdot b + a \cdot c$.

Properties of order

xii. *The trichotomy law:* for any $a \in F$, either $0 < a$, or $0 = a$, or
$0 < -a$; and only one of these three holds.

xiii. *Positive closure:* if $0 < a$ and $0 < b$, then
$0 < a + b$, and $0 < a \cdot b$.

xiv. $a < b$ if and only if $0 < b - a = b + (-a)$.

Theorem   In an ordered field, every squared element is
qn 25   positive or zero.

Theorem qn 31	An ordered field contains a subset with the same properties and structure as the set of rational numbers, $\mathbb{Q}$.
Theorem qn 35	For positive numbers a and b, $a < b$ if and only if $a^n < b^n$ for all positive integers n.
Theorem qn 47	The geometric mean of n positive numbers is less than the arithmetic mean of those numbers, unless all n numbers are equal. The statement of this result presumes the existence of nth roots of positive numbers.

Theorem
qn 48

If $1 < a$, then $(n + 1)(\sqrt[n+1]{a} - 1) < n(\sqrt[n]{a} - 1)$.

Theorem
qn 51

$(1 + 1/n)^n$ increases with n, and $2 \leqslant (1 + 1/n)^n < 3$.

Cauchy–Schwarz inequality
qn 57

$$(a_1 b_2 + a_2 b_2 + \ldots + a_n b_n)^2$$

$$\leqslant (a_1^2 + a_2^2 + \ldots + a_n^2)(b_1^2 + b_2^2 + \ldots + b_n^2)$$

for any numbers $a_1, a_2, \ldots, a_n, b_1, b_2, \ldots, b_n$.

Definition

$|a| = a$, when $0 \leqslant a$, $|a| = -a$, when $a < 0$.

Triangle inequality
qn 65

$|a + b| \leqslant |a| + |b|$, for all $a, b \in F$.

Theorem
qn 67

$\big||a| - |b|\big| \leqslant |a - b|$, for all $a, b \in F$.

Historical note

Early steps towards the arithmetical axiomatisation of the number system were taken by G. Peacock in 1830 and 1845, with his principle of permanence of forms. Precise arguments using the total ordering of the positive real numbers pervade Euclid's *Elements*. Hilbert's construction of a non-archimedean ordered field in his *Foundations of Geometry* (1899, trans. Open Court 1902) paved the way for the study of axiomatically defined ordered fields. The notation $<$ and $>$ was introduced by Harriot and first appeared in print in 1631. The symbols $\leqslant$ and $\geqslant$ were first used in France about 100 years later. The arithmetic mean–geometric mean inequality for $n = 2$ appears in Euclid's *Elements* (II.5, V.25 and lemma between X.59 and X.60). The general form of this inequality appears in A. L. Cauchy's *Cours d'Analyse* (1821) which also contains the so-called Cauchy–Schwarz inequality. Schwarz (1885) used an analogous inequality for integrals (see qn 10.38) which was first given by Buniakowsky in 1859. The sequence $(1 + 1/n)^n$ was first investigated by L. Euler (1736) who named its limit with his own initial. The notation for absolute value $|x|$ was due to Weierstrass who used it in his lectures in Berlin from 1859. It did not appear in print until 1877. Bolzano (1817) wrote $x < \pm 1$, where we would write $|x| < 1$. Harnack (1881) wrote abs m and Jordan (1882) wrote mod m for $|m|$.

Answers

1 $\big((a + b) + c\big) + d = (a + b) + (c + d) = a + \big(b + (c + d)\big)$, etc.

2 If 0 and Ø were both zeros, $Ø = Ø + 0 = 0 + Ø = 0$.

3 $a + (-a) = a + (\sim a) \Rightarrow \quad (-a) + a + (-a) = (-a) + a + (\sim a)$

$\Rightarrow \qquad\qquad 0 + (-a) = 0 + (\sim a)$

$\Rightarrow \qquad\qquad\qquad -a = \sim a.$

4 The inverse of $-a$ is unique by qn 3.

5 $(\mathbb{Z}, +)$ is an abelian group, but $(\mathbb{N}, +)$ is not because there are no inverses. The even integers form an abelian group under addition; so does the set $\{m/2^n \mid m, n \in \mathbb{Z}\}$.

6 If $a, b, x, y \in \mathbb{Q}$, let $(a + sb) + (x + sy) = (a + x) + s(b + y)$.

7 Closure of $\mathbb{Q}\backslash\{0\}$ is not given. Nor does it follow from vi–x. For example, if $A = \mathbb{Q}\backslash\{0\}$, then $(A, \cdot)$ is an abelian group and so satisfies vi, vii and x. If, for viii and ix, we consider $A\backslash\{-1)$ we find that $A\backslash\{-1\}$ is not closed under multiplication.

9 Add $-(a \cdot a)$ to both sides of the equation. Use the commutative law.

10 Use qn 9 and add $-(a \cdot b)$ to both sides of the equation. Notice in particular that $(-1) \cdot b = -(1 \cdot b) = -b$. Use the commutative law.

$(-a) \cdot (-b) = -\big(a \cdot (-b)\big) = -\big(-(a \cdot b)\big) = a \cdot b.$

11 Multiply both sides of the equation by $1/(-1)$.
Multiply both sides of $(-1) \cdot (-a) = a$ by $\big((1/(-a)\big) \cdot (1/a)$.

12 $a \cdot b = 0 \Rightarrow (1/a) \cdot a \cdot b = (1/a) \cdot 0 \Rightarrow 1 \cdot b = 0.$

13 From qn 12, $a \neq 0$, $b \neq 0 \Rightarrow a \cdot b \neq 0$, so $\mathbb{Q}\backslash\{0\}$ is closed. The other necessary properties are in vi–x.

14 $a < 0 \Leftrightarrow 0 < 0 - a \Leftrightarrow 0 < -a$ by xiv. In the development of the ordered field axioms from the definition of the rational numbers at the end of chapter 1, order axiom xiv is needed as a definition.

15 $0 < a \Leftrightarrow 0 < 0 - (-a) \Leftrightarrow -a < 0.$

16 $a < b \Leftrightarrow 0 < b - a \Leftrightarrow 0 < -a - (-b) \Leftrightarrow -b < -a.$

17 For given a and b, either $0 < b - a$, or $0 = b - a$, or $b - a < 0$; and only one of these three holds.

18 $a < b \Leftrightarrow 0 < b - a = (b + c) - (a + c) \Leftrightarrow a + c < b + c.$

19 $a < b \Leftrightarrow 0 < b - a$ and $b < c \Leftrightarrow 0 < c - b$.
$a < c \Leftrightarrow 0 < c - a = (b - a) + (c - b)$ which follows from xiii.

20 $a < b \Leftrightarrow a + c < b + c$. $c < d \Leftrightarrow c + b < d + b$.
So transitivity gives $a + c < b + d$.

21 $a < b \Leftrightarrow 0 < b - a$, with xiii, $0 < (b - a) \cdot c \Leftrightarrow a \cdot c < b \cdot c$.

22 $0 < a < b$ gives $0 < a^2 < a \cdot b < b^2$ from qn 21. Result from transitivity.

23 $b < 0 \Leftrightarrow 0 < -b$, with xiii, $0 < a \cdot (-b) = -(a \cdot b)$, from qn 10, so
$a \cdot b < 0$.

24 Suppose $0 < -1$ then, from xiii, $0 < (-1) \cdot (-1) = 1$, from qn 10. With
xii, this contradicts the hypothesis.

25 If $a \neq 0$, either $0 < a$ or $0 < -a$. So either $0 < a \cdot a$ or $0 < (-a) \cdot (-a)$.
But $a \cdot a = (-a) \cdot (-a) = a^2$.

26 $1/a < 0 \Rightarrow a \cdot (1/a) < 0$ from qn 23, so $1 < 0$, which contradicts qn 24.

27 $0 < a < b \Rightarrow 0 < a \cdot (1/a) \cdot (1/b) < b \cdot (1/a) \cdot (1/b)$.

28 Suppose $0 < 1/a$, then $0 < (1/a) \cdot (-a) = -((1/a) \cdot a) = -1$, which
contradicts qn 24.

29 $a < b < 0 \Leftrightarrow 0 < -b < -a \Rightarrow 0 < 1/(-a) < 1/(-b)$ from qn 27.

$\Rightarrow 0 < -(1/a) < -(1/b)$ from qn 11.

$\Rightarrow 1/b < 1/a < 0$.

30 If $a = b$ then $a^2 = b^2$. If $b < a$, then $b^2 < a^2$, from qn 22.

31 If a sum of m 1s = a sum of n 1s, for some m and n, and n is the
smaller, then a sum of $m - n$ 1s = 0 and from xiii this contradicts qn 24.
Thus an ordered field contains a sequence of sums of 1s without
repetition, and this forms a copy of $\mathbb{N}$.

32 Suppose an ordered field contained an element i such that $i^2 = -1$. From
qn 25, $0 < i^2$ but, from qn 24, $-1 < 0$. Contradiction.

33 Proposition valid for $n = 1$.
$-1 < x \Leftrightarrow 0 < 1 + x$, so $1 + nx \leqslant (1 + x)^n$
$\Rightarrow (1 + nx)(1 + x) \leqslant (1 + x)^{n+1}$.
Now $1 + nx + x + nx^2 \leqslant (1 + x)^{n+1} \Rightarrow 1 + (n + 1)x \leqslant (1 + x)^{n+1}$.

34 $0 < a < b \Rightarrow a^3 < b^3$ since $0 < a^3 < a^2b < ab^2 < b^3$, from qn 21. Use
trichotomy for the converse.

35 $0 < a < b$ and $a^n < b^n$ gives $a^{n+1} < a^n b < b^{n+1}$. Use trichotomy for the converse. Proposition valid with $\leqslant$.

36 $A = 5$. $G = 4$.

37 $\sqrt{6} < 5/2$.

38 $ab \leqslant \frac{1}{2}(a^2 + b^2)$. Equality only when $a = b$.

39 $0 \leqslant (\sqrt{a} - \sqrt{b})^2$ from qn 25.

$\Rightarrow 0 \leqslant a - 2\sqrt{a} \cdot \sqrt{b} + b$

$\Rightarrow \sqrt{a} \cdot \sqrt{b} \leqslant \frac{1}{2}(a + b)$.

Equality in the first line only when $\sqrt{a} = \sqrt{b}$.

40 Put $b = 1/a$ in qn 39.

41 If constant perimeter $= p$ and one side $= a$, then other side $= \frac{1}{2}p - a$, so area $= a(\frac{1}{2}p - a)$. For the numbers a and $\frac{1}{2}p - a$, geometric mean (GM) $= \sqrt{(\text{area})}$ and arithmetic mean (AM) $= \frac{1}{4}p$ which is constant. Max GM $=$ AM, with equality when $a = \frac{1}{2}p - a$ and the rectangle is a square.

42 GM $= a$, AM $= \frac{1}{2}(x + a^2/x)$. GM $\leqslant$ AM $\Rightarrow a \leqslant x_{n+1} \leqslant x_n$, except possibly for $n = 1$. This is the basis of Heron's method for finding square roots.

43 (i) needs GM $<$ AM and induction.
(iii) needs (ii) and induction.
Such pairs of sequences were investigated by Gauss.

44 The first inequality is strict unless $\sqrt{(ab)} = \sqrt{(cd)}$, or $ab = cd$.
The second inequality is strict unless both $a = b$ and $c = d$.
So one or other inequality is strict unless all four numbers are equal.

45 The first inequality is strict unless $G = H$, and the second is strict unless $G = A$ and $H = B$. Equality holds for the second case only when the as are all equal and the bs are all equal. Equality holds in both cases only when the as and bs are all equal. There is, as yet, no basis for the induction if n is not a power of 2.

46 If a_1, a_2, and a_3 are not all equal, then the four numbers a_1, a_2, a_3 and A are not all equal, so

$$\sqrt[4]{(a_1 a_2 a_3 A)} < \frac{1}{4}(a_1 + a_2 + a_3 + A) = \frac{1}{4}(3A + A).$$

So $\sqrt[4]{(a_1 a_2 a_3 A)} < A$, and $a_1 a_2 a_3 A < A^4$.

47 If a_1, a_2, ..., a_{n-1} are $n - 1$ positive numbers which are not all equal, let $A = (a_1 + a_2 + \ldots + a_{n-1})/(n - 1)$, and argue as in qn 46 with the n positive numbers a_1, a_2, ..., a_{n-1} and A which are not all equal.

48 (i) follows from qn 21.

51 $(1 + 1/5)^6 = (6/5)^6 = 46\,656/15\,625 < 46\,875/15\,625 = 3.$

52 $\big((n + 1)/n\big)^n < n \Leftrightarrow (n + 1)^n < n^{n+1}.$

53 Equality is only possible when $ax + \frac{1}{2}b = 0$.

54 If $0 < a$, then $0 \leqslant ax^2 + bx + c \Leftrightarrow 0 \leqslant a^2x^2 + abx + ac$.
This holds for all x, when $0 \leqslant ac - \frac{1}{4}b^2$ or $b^2 \leqslant 4ac$.
If $b^2 < 4ac$, then $0 < ac - \frac{1}{4}b^2$, so $0 < a^2x^2 + abx + ac$ and so
$0 < ax^2 + bx + c$.
If $a < 0$, then $ax^2 + bx + c \leqslant 0$ provided $b^2 \leqslant 4ac$.

55 $(a_1x + b_1)^2 + (a_2x + b_2)^2$

$$= (a_1^2 + a_2^2)x^2 + 2(a_1b_1 + a_2b_2)x + (b_1^2 + b_2^2)$$

$$0 \leqslant (a_1x + b_1)^2 + (a_2x + b_2)^2 \text{ for all } x.$$

Equality only when both $a_1x + b_1 = 0$ and $a_2x + b_2 = 0$. That is, when the bs are a constant multiple of the as.

56 $AB = \surd\big((a_1 - b_1)^2 + (a_2 - b_2)^2\big), \ OA = \surd(a_1^2 + a_2^2),$

$$OB = \surd(b_1^2 + b_2^2)$$

$$AB^2 = a_1^2 + a_2^2 + b_1^2 + b_2^2 - 2(a_1b_1 + a_2b_2)$$

$$(OA + OB)^2 = a_1^2 + a_2^2 + b_1^2 + b_2^2 + 2\surd(a_1^2 + a_2^2) \cdot \surd(b_1^2 + b_2^2)$$

So $AB^2 \leqslant (OA + OB)^2 \Leftrightarrow$ qn 55.

57 Consider the inequality $0 \leqslant \sum_{i=1}^{i=n}(a_ix + b_i)^2.$

58 Follows from qn 14.

59 Consider the three cases $0 < a$, $0 = a$ and $a < 0$.

60 Consider the three cases $0 < a$, $0 = a$ and $a < 0$.

61 Consider the nine cases $0 < a$, $0 = a$ and $a < 0$, and $0 < b$, $0 = b$ and $b < 0$.

62 Put $b = 1/a$ in qn 61.

63 Use qn 62 in qn 61.

64 $|a| \leqslant b \Leftrightarrow -b \leqslant -|a| \Rightarrow -b \leqslant -|a| \leqslant a \leqslant |a| \leqslant b$, from qn 59.
If $-b \leqslant a \leqslant b$, then $-b \leqslant -a \leqslant b$, from qn 16, so $|a| \leqslant b$.

65 Since $-|a| \leqslant a \leqslant |a|$ and $-|b| \leqslant b \leqslant |b|$ from qn 59, we obtain the first inequality from qn 20. The second inequality then comes from qn 64.

66 Put $-b$ for b in qn 65 and use qn 60.

3

Sequences
A first bite at infinity

Preliminary reading: Hemmings and Tahta, O'Brien.
Concurrent reading: Hight.
Further reading: Knopp.

The notion of a sequence is a familiar and intuitive one. The following examples of sequences of numbers illustrate this notion.

1, 2, 3, 4, 5, ...

2, 4, 6, 8, 10, ...

$\frac{1}{2}, \frac{1}{4}, \frac{1}{8}, \frac{1}{16}, \frac{1}{32}, \ldots$

3, 3.1, 3.14, 3.141, 3.1415, ...

$-1, 1, -1, 1, -1, \ldots$

$\frac{1}{2}, \frac{2}{3}, \frac{3}{4}, \frac{4}{5}, \frac{5}{6}, \ldots\ldots$

The general way of writing down a sequence is

$$a_1, a_2, a_3, \ldots, a_n, \ldots$$

so that, for each counting number n, there is an element of the sequence, a_n. Such a sequence is denoted by (a_n).

1 For each of the sequences listed above, except the fourth, suggest a formula for the nth term, a_n.

2 It is possible to draw a graph of the sequence (a_n) by plotting the points (n, a_n) on a conventional cartesian graph of $\mathbb{R} \times \mathbb{R}$.
Draw the graphs of the first five terms of the six sequences given above.

There is one disadvantage of this method of graph drawing.

Although it provides a useful picture of the first few terms of the sequence, it does not illustrate what happens in the long run. Most of the important questions about sequences, which go on for ever, are not decided by the first few terms but depend on the behaviour of the nth term a_n as n gets larger and larger, and these terms are not illustrated on the graphs we have drawn. Another method of drawing a graph of a sequence of numbers, which *can* illustrate what happens for large values of n, was used by Hight in his book. Hight plots the points $(-1/n, a_n)$ on a conventional cartesian graph, and page 146 of his book contains some excellent illustrations.

3 Illustrate the formulae you have found in qn 1, by plotting the points $(-1/n, a_n)$.

Although the central property of sequences that we will be exploring is that of convergence, there are other properties of sequences which are less difficult to define and which we consider now in order to have some descriptive language available for the discussion of sequences.

Monotonic sequences

4 Test each of the sequences (i)–(vii), defined below, to determine whether any one or more of the following four properties, (a)–(d), holds for all values of n.

(a) $a_n < a_{n+1}$.

(b) $a_n \leqslant a_{n+1}$.

(c) $a_n > a_{n+1}$.

(d) $a_n \geqslant a_{n+1}$.

(i) $a_n = n$, (ii) $a_n = 1/n$, (iii) $a_n = (-1)^n$,

(iv) $a_{2n-1} = n$, $a_{2n} = n$, (v) $a_n = -1/n$

(vi) $a_n = 1$, (vii) $a_n = 2^{-n}$.

When property (a) in qn 4 holds, the sequence is said to be *strictly monotonic increasing* or just *strictly increasing*.

When property (b) holds, the sequence is said to be *monotonic increasing* or just *increasing*.

When property (c) holds, the sequence is said to be *strictly monotonic decreasing* or *strictly decreasing*.

When property (d) holds, the sequence is said to be *monotonic decreasing* or *decreasing*.

When any one of (a), (b), (c) or (d) holds, the sequence is said to be monotonic.

The rather bizarre status of sequence 4(vi) is a consequence of wanting simple and practically useful definitions. The fact that constant sequences satisfy conditions (b) and (d) simultaneously is not a sufficient reason to revise our definitions.

Bounded sequences

5   Test each of the sequences (i)–(iv) defined here to determine whether either or both of the following properties applies.
   (a)   There is a number U such that $a_n \leqslant U$ for all n.
   (b)   There is a number L such that $a_n \geqslant L$ for all n.

   (i) $a_n = (-2)^n$,

   (ii) $a_n = (-1)^n/n$,

   (iii) $a_n = \sin n$,

   (iv) $a_n = \sqrt{n}$.

When property (a) in qn 5 holds, the sequence (a_n) is said to be *bounded above*, and the number U is called an *upper bound* of the sequence.

When property (b) holds, the sequence is said to be *bounded below*, and the number L is called a *lower bound* of the sequence.

When both properties (a) and (b) hold, so that there are numbers L and U such that $L \leqslant a_n \leqslant U$, for all values of n, the sequence (a_n) is said to be *bounded*.

A sequence which is neither bounded above nor bounded below is said to be *unbounded*. [Some authors say that a sequence is unbounded when it is either not bounded above, or not bounded below, but differing conventions in this matter will not affect the formulation of any of the theorems in this book.] Notice that upper bounds and lower bounds, if they exist, are not unique: for if U is an upper bound, then so is $U + 1$, and if L is a lower bound, then so is $L - 1$.

Beware of the language here. The sequence $1, 2, 3, \ldots, n, \ldots$ is neither bounded nor unbounded!

6 A sequence (a_n) is known to be monotonic increasing.
  (i) Might it have an upper bound?
  (ii) Might it have a lower bound?
  (iii) Must it have an upper bound?
  (iv) Must it have a lower bound?
  Give a numerical example to illustrate each possibility or impossibility.
  How must your answers be revised if the same questions were asked about an arbitrary monotonic *decreasing* sequence?

7 If a sequence is not bounded above, must it contain
  (i) a positive term,
  (ii) an infinity of positive terms?

Subsequences

The sequences

$2, 4, 6, 8, \ldots, 2n, \ldots,$
$1, 3, 5, 7, \ldots, 2n - 1, \ldots,$
$1, 4, 9, 16, \ldots, n^2, \ldots,$
$2, 4, 8, 16, \ldots, 2^n, \ldots,$
$1, 3, 6, 10, \ldots, \frac{1}{2}n(n + 1), \ldots,$
$5, 6, 7, 8, \ldots, n + 4, \ldots,$

all provide examples of infinite subsets of the counting numbers, $\mathbb{N}$.

From any such infinite subset, a subsequence of a sequence may be constructed.

From the sequence

$a_1, a_2, a_3, \ldots, a_n, \ldots$

the subsequences

$a_2, a_4, a_6, \ldots, a_{2n}, \ldots$
$a_1, a_4, a_9, \ldots, a_{n^2}, \ldots$
$a_5, a_6, a_7, \ldots, a_{n+4}, \ldots$

may be constructed, for example.

In general, if $M = \{n_1, n_2, n_3, \ldots\}$, with $n_1 < n_2 < n_3 < \ldots$, is any infinite subset of $\mathbb{N}$ then the sequence (a_{n_i}) is called a *subsequence* of the sequence (a_n).

8 A sequence (a_n) is known to be monotonic increasing, but not strictly monotonic increasing.

  (i) Might there be a strictly monotonic increasing subsequence of (a_n)?

  (ii) Must there be a strictly monotonic increasing subsequence of (a_n)?

9 If a sequence is bounded, must every subsequence be bounded?

10 (i) If the subsequence $a_2, a_3, \ldots, a_{n+1}, \ldots$ is bounded, does it follow that the sequence (a_n) is bounded?

  (ii) If the subsequence $a_3, a_4, \ldots, a_{n+2}, \ldots$ is bounded does it follow that the sequence (a_n) is bounded?

  (iii) If the subsequence $a_{N+1}, a_{N+2}, \ldots, a_{N+n}, \ldots$ is bounded does it follow that the sequence (a_n) is bounded?

We describe the result of qn 10 by saying that if a sequence is *eventually bounded*, then it is *bounded*.

11 A sequence (a_n) is known to be unbounded.
  (i) Might it contain a bounded subsequence?
  (ii) Must it contain a bounded subsequence?
Look back to the definition of *bounded* and *unbounded*.

Rather unexpectedly, it is possible to show that *any* sequence has a monotonic subsequence. The proof is a little tricky and you may find it helpful to give the next question just a brief scrutiny now and come back to it again later. The result will be used in chapter 4. Bryant gives it as a theorem about Spanish hotels!

12 Let (a_n) be any sequence whatsoever. We will call a term a_f a 'floor term' when none of its successors is less: that is if $a_f \le a_n$ when $f \le n$. A floor term is a lower bound for the rest of the sequence.
  (i) If there is an infinity of floor terms, show that they form a monotonic increasing subsequence.
  (ii) If there is a finite number of floor terms, and the last one is a_F, form a monotonic decreasing subsequence with a_{F+1} as its first term.
  (iii) If there are no floor terms at all, form a monotonic decreasing subsequence with a_1 as its first term.

Sequences tending to infinity

13 Describe the following four sequences saying whether or not they are monotonic and whether or not they are bounded above or below.

(i) 1, 1, 2, 1, 3, 1, ..., n, 1, ...

(ii) −1, 2, −3, 4, ..., $(-1)^n n$, ...

(iii) 2, 1, 4, 3, ..., $2n$, $2n - 1$, ...

(iv) 11, 12, 11, 12, ..., 11, 12, ...

In each case decide whether there is a stage in each sequence beyond which all terms in the sequence are

(a) greater than 1, in which case we say the sequence is eventually greater than 1,

(b) greater than 10, so that the sequence is eventually greater than 10,

(c) greater than 100, so that the sequence is eventually greater than 100.

Only sequence (iii) is said to tend to $+\infty$ (plus infinity).

We say that a sequence tends to $+\infty$ when any number, however chosen, is eventually a lower bound for the sequence. This means that to be said to tend to $+\infty$ a sequence must pass a potential infinity of tests – because the chosen number can be *any* number whatever, and the sequence must eventually exceed each of them.

If the chosen number is C, we are looking for a value N of n sufficiently great to make $a_n \geq C$, for all $n > N$.

14 Select values of C to demonstrate that qn 13(i), (ii) and (iv) do not tend to $+\infty$.

In relation to a sequence (a_n), the term 'eventually', the phrase 'provided n is large enough', and its equivalent 'for sufficiently large n', simply mean '**for all $n >$ some number N**'.

We can now gather the discussion into a formal definition and say that **the sequence (a_n) tends to $+\infty$, when, given any number C, there is a number N such that $n > N \Rightarrow a_n \geq C$.**

Yet another way to express this definition is to say that $a_n \geq C$ for all but a finite number of terms of the sequence.

'(a_n) tends to infinity' is often written $(a_n) \to +\infty$ as $n \to \infty$.

Axiom of Archimedean order

A formal definition is quite useless if it does not help us to formulate any theorems or conjectures. It may come as a shock to learn that the axioms for an ordered field are not sufficient to prove that the sequence 1, 2, 3, ..., n, ... of counting numbers tends to

$+\infty$. Some ordered fields contain other kinds of numbers which cannot be reached, let alone passed, by any progression along the counting numbers, so these numbers are not, even eventually, lower bounds for any sequence of counting numbers. In order to proceed with the study of convergence within the real number system we need an additional axiom:

given any number C, there is an integer n, which is greater than C.

This is the so-called Axiom of Archimedean order and an ordered field in which it holds is called an Archimedean ordered field. It is this axiom which makes the familiar 'number line' a possible model for the real numbers. Notice, too, that in an Archimedean ordered field there cannot be an infinite element, nor, because of the existence of multiplicative inverses (or reciprocals), can there be infinitesimals. A positive infinitesimal would have to be less than every positive rational but still greater than 0, so its reciprocal would be greater than any positive integer.

It follows from this axiom that the sequence of counting numbers tends to infinity because if C is any number we can choose a positive integer n such that $C < n$; then $n < n + 1$, so $C < n + 1$ and by induction all subsequent counting numbers are also greater than C, so C is an eventual lower bound for the sequence of counting numbers.

It also follows that there is an integer *less than* any given number C: if we apply the Axiom of Archimedean order to $-C$ there is an integer n such that $-C < n$, and so $-n < C$. So any number C divides the set of integers into two sets, those $\leqslant C$ and those $> C$. If, for a given number C, m and n are integers such that $m < C < n$, we need only scrutinise the $n - m$ integers $m, m + 1, \ldots, n - 1$ to determine the largest integer $\leqslant C$ which we denote by $[C]$ or in BASIC by $\text{INT}(C)$.

$[\pi] = 3$, $[2] = 2$ and $[-2 \cdot 1] = -3$. $[C] \leqslant C < [C] + 1$.

15 Prove that the sequence $(\sqrt{n}) \to +\infty$ as $n \to \infty$.

16 If $1 < x$ so that, for some positive y, $1 + y = x$, use qn 2.33 to show that $1 + ny \leqslant x^n$, and deduce that $(x^n) \to +\infty$.

17 Construct a definition for '(a_n) tends to $-\infty$ as n tends to ∞'. Check that your definition guarantees that the sequence defined by $a_n = -n$ tends to $-\infty$.

Null sequences

Even to the untutored eye the sequence

$$1, 1/2, 1/3, 1/4, \ldots, 1/n, \ldots$$

will be seen to 'tend to 0'. We work towards a precise definition of this phrase and its synonym 'converges to 0'.

18 (a) Name some lower bounds for the sequence $(1/n)$.
 Are all possible lower bounds negative?
 (b) Name some upper bounds for the sequence $(1/n)$.
 Are all possible upper bounds positive?
 Are all positive numbers upper bounds for the sequence?
 Is each positive number *eventually* an upper bound for the sequence?

As the notion of convergence was evolving during the eighteenth century, the contexts in which the notion developed were ones in which the limit of the sequence was not reached. For example, when the sum of a series of positive terms such as $\frac{1}{2} + \frac{1}{4} + \frac{1}{8} + \ldots$ tends to a limit, the limit is not one of the intermediate values. The ordinary usage of the terms 'converges to' or 'tends to' also excludes reaching the destination. But if our interest lies in the bounds, and the eventual bounds, of a sequence, this condition becomes irrelevant.

19 Answer the questions of qn 18 in relation to the two sequences

$$1, 0, 1/2, 0, 1/3, 0, 1/4, 0, \ldots, 1/n, 0, \ldots$$

and

$$0, 0, 0, \ldots, 0, \ldots.$$

The similarity of answers in qns 18 and 19 will make us accept the sequences in qn 19 as 'tending to 0'.

20 Exhibit the inadequacy of the following attempts (a)–(e) to define a null sequence, that is, a sequence which tends to 0, by reference to the succeeding set of sequences, (i)–(v), none of which is null.
 (a) A sequence in which each term is strictly less than its predecessor.

(b)   A sequence in which each term is strictly less than its predecessor while remaining positive.

(c)   A sequence in which, for sufficiently large n, each term is less than some small positive number.

(d)   A sequence in which, for sufficiently large n, the absolute value of each term is less than some small positive number.

(e)   A sequence with arbitrarily small terms.

   (i) $2, 1, 0, -1, -2, -3, -4, \ldots, -n, \ldots$

   (ii) $2, 3/2, 4/3, 5/4, 6/5, \ldots, (n+1)/n, \ldots$

   (iii) $2, 1, 0, -.1, -.1, -.1, \ldots, -.1, \ldots$

   (iv) $2, 1, 0, -.1, .01, -.001, .01, -.001, \ldots, .01, -.001, \ldots$

   (v) $1, \frac{1}{2}, 1, \frac{1}{4}, 1, \frac{1}{8}, \ldots$

21 Examine the sequence

$$-1, 1/2, -1/3, 1/4, -1/5, \ldots, (-1)^n/n, \ldots$$

   (i) Beyond what stage in the sequence are the terms between -0.1 and 0.1?

   (ii) Beyond what stage in the sequence are the terms between -0.01 and 0.01?

   (iii) Beyond what stage in the sequence are the terms between -0.001 and 0.001?

   (iv) Beyond what stage in the sequence are the terms between $-\varepsilon$ and ε, where ε is a given positive number? The choice of ε, the Greek e, here, is to stand for 'error', where the terms of the sequence are thought of as successive attempts to hit the target 0.

   A sequence (a_n) is said to be a null sequence, or to tend to 0, or to converge to 0, when,

for any positive number ε, however small, there is a stage in the sequence beyond which the terms of the sequence are between $-\varepsilon$ and ε;

or, equivalently,

any positive ε is eventually an upper bound for the sequence and $-\varepsilon$ eventually a lower bound for the sequence;

or, equivalently

all but a finite number of terms of the sequence lie between $\pm\varepsilon$, or satisfy $|a_n| < \varepsilon$.

Since this must hold for every possible choice of ε, the sequence must satisfy an infinity of conditions.

Definition of a null sequence

> **Formally, we write $(a_n) \to 0$ as $n \to \infty$ when, given $\varepsilon > 0$, there is an n_0 such that $n > n_0 \Rightarrow |a_n| < \varepsilon$.**

This definition could also be given by saying that the subsequence of all terms after a_{n_0} is bounded by $\pm\varepsilon$.

22 Write down a sequence which is not null, and prove that it is not null by naming an ε for which no related n_0 exists.

23 Prove that the sequence $(1/n)$ is a null sequence.
You must use the Axiom of Archimedean order.

24 Prove that the sequence $(1/\sqrt{n})$ is a null sequence.

25 If $(a_n) \to +\infty$, prove that $(1/a_n)$ is a null sequence.
Give an example to show that the converse is false.

26 If (a_n) is a null sequence and c is a constant number, prove that $(c \cdot a_n)$ is a null sequence.
Consider the cases $c \neq 0$ and $c = 0$, in turn.
Deduce that $(10/\sqrt{n})$ is a null sequence.

27 If (a_n) is a null sequence, prove that $(|a_n|)$ is a null sequence, and conversely that if $(|a_n|)$ is a null sequence then so is (a_n).

28 (*A sandwich theorem*) If (a_n) is a null sequence and $0 \leqslant b_n \leqslant a_n$, prove that (b_n) is a null sequence.

29 Prove that each of the following sequences is null:

(i) $(1/(n + 1))$, (ii) $(10/(n + 1))$, (iii) $(20/(7n + 3))$,

(iv) $(1/(n^2 + 1))$, (v) $(\sqrt{(n + 1)} - \sqrt{n})$, (vi) $((\sin n)/n)$.

(vii) $(\sin n\pi)$, (viii) $(\sin(n!\pi \cdot c))$, where c is a rational number.
The standard properties of the sine function should be used: they will be proved in chapter 11.

30 Must every subsequence of a null sequence be null?

31 Prove that each of the following sequences is null:

(i) $(1/(2n - 1))$, (ii) $(1/(3n - 100))$, (iii) $(1/(2n^2 - 1))$.

32 The most important family of null sequences is formed by
geometric progressions. Suppose $0 < x < 1$.
- (i) Prove that $1 < 1/x$, using qn 2.27, and let $1/x = 1 + y$, so
  $y > 0$.
- (ii) Show that $(1/x)^n \geqslant 1 + ny$, using qn 2.33.
- (iii) Prove that $0 < x^n \leqslant 1/(yn + 1)$.
- (iv) Prove that (x^n) is a null sequence.
- (v) Prove that $(c \cdot x^n)$ is a null sequence.

33 (a)   Use a calculator or a computer to evaluate terms of the
   sequence $(n^2/2^n)$.
   Calculate the values for $n = 1, 2, 5, 10, 20, 50$.
   Would you conjecture that the sequence is null?
- (b)   Verify that the ratio $a_{n+1}/a_n = \frac{1}{2}(1 + 1/n)^2$.
- (c)   Find an integer N such that, if $n \geqslant N$, then the ratio
   $a_{n+1}/a_n < \frac{3}{4}$.
- (d)   By considering that $a_{N+1} < \frac{3}{4}a_N$ and $a_{N+2} < \frac{3}{4}a_{N+1}$, etc.
   prove that $a_{N+n} < (\frac{3}{4})^n a_N$
- (e)   Why is $((\frac{3}{4})^n a_N)$ a null sequence?
- (f)   Use the sandwich theorem, qn 28, to show that $(n^2/2^n)$ is a
   null sequence.
- (g)   What numbers might have been used in place of $\frac{3}{4}$ in part
   (c) (with perhaps a different N) which would still have led
   to a proof that the sequence was null?

34 Use the method of qn 33 to establish that the following sequences
are null:

(i) $(n^4/2^n)$, (ii) $(2^n/n!)$, (iii) $(n!/n^n)$, using qn 2.51,

(iv) $(n^3(0.9)^n)$.

35 Use qn 27 to extend qn 32 and establish that, when $-1 < x < 1$,
$(c \cdot x^n)$ is a null sequence.

36 Suppose (a_n) and (b_n) are both null sequences, and $\varepsilon > 0$ is
given.
- (i) Must there be an n_1 such that $|a_n| < \varepsilon$ when $n > n_1$?
- (ii) Must there be an n_2 such that $|b_n| < 1$ when $n > n_2$?
- (iii) Is there an n_0 such that when $n > n_0$, then both $n > n_1$ and
  $n > n_2$?
- (iv) If $n > n_0$, must $|a_n b_n| < \varepsilon$?

You have now proved that the termwise *product* of two null
sequences is null.
- (v) If the sequence (c_n) is also null, what about $(a_n b_n c_n)$? And
  the termwise product of k null sequences?

37 Suppose (a_n) and (b_n) are both null sequences, and $\varepsilon > 0$ is given.

    (i) Must there be an n_1 such that $|a_n| < \frac{1}{2}\varepsilon$ when $n > n_1$?

    (ii) Must there be an n_2 such that $|b_n| < \frac{1}{2}\varepsilon$ when $n > n_2$?

    (iii) Is there an n_0 such that when $n > n_0$, then both $n > n_1$ and $n > n_2$?

    (iv) If $n > n_0$, must $|a_n + b_n| < \varepsilon$?

You have proved that the termwise *sum* of two null sequences is null.

    (v) If the sequence (c_n) is also null, what about $(a_n + b_n + c_n)$? And the sum of k null sequences?

38 If (a_n) and (b_n) are both null sequences, show that $(c \cdot a_n + d \cdot b_n)$ is a null sequence, where c and d are constants.

39 If (a_n) and (c_n) are both null sequences, and $a_n \leqslant b_n \leqslant c_n$,

    (i) say why $(c_n - a_n)$ is a null sequence,

    (ii) and why $(b_n - a_n)$ is a null sequence,

    (iii) and finally why (b_n) is a null sequence.

This is another *sandwich theorem*.

40 Let $A_n = a_1 + a_2 + \ldots + a_n$.

Give an example of a sequence (a_n) which is not null, but for which the sequence (A_n/n) is null.

Convergent sequences and their limits

41 Examine the terms of the sequence given by

$$a_n = \frac{n}{n+1}.$$

To what limit do you think this sequence tends?

What can you say about the sequence $(a_n - 1)$?

Up to this point we have avoided the term 'limit' because in non-mathematical usage (e.g. a speed limit) it is synonymous with 'bound'. However, for mathematicians, bounds need not be approached, whereas limits are numbers to which the terms of sequences and the values of functions get close.

Definition of convergence
We will say that

a sequence (a_n) tends to the limit a, or converges to a, when $(a_n - a)$ is a null sequence.

This is expressed symbolically by writing

either $(a_n) \to a$ as $n \to \infty$ *or* $\lim\limits_{n \to \infty} a_n = a$.

We read each of the arrows here as 'tends to'.
A sequence of numbers with a number as limit is said to be convergent.

42 If $(a_n - a)$ is a null sequence and $(a_n - b)$ is a null sequence, prove that $a = b$ by considering the difference of these two null sequences and using qn 38. This shows that limits are unique when they exist.

43 Write down the first four terms of the sequence $([10^n a]/10^n)$
(i) when $a = 6\frac{1}{4}$, (ii) when $a = \frac{1}{3}$, (iii) when $a = \sqrt{2}$.
Prove that, for *any a*,

$$0 \leqslant a - \frac{[10^n a]}{10^n} < \frac{1}{10^n}.$$

Deduce that there is a sequence of rational numbers tending to any number a in an Archimedean ordered field.

This question shows both the power and the limitations of decimal expressions. If the sequence $([10^n a]/10^n)$ is eventually constant, then the number a is a *terminating decimal*. Terminating decimals provide an excellent way of approximating to any real number, but many real numbers may only be expressed as the *limit* of a sequence of terminating decimals and not by a single terminating decimal. An infinite decimal is really a sequence of terminating decimals whose convergence can be examined. When the sequence of terminating decimals is convergent (as in the case of 1, 1.4, 1.41, 1.414, ... with the limit $\sqrt{2}$) the infinite decimal determines a unique number – the limit of the sequence.
Many of the theorems we have established for null sequences can be extended to give theorems about convergent sequences in general.

44 If $(a_n) \to a$ and $(b_n) \to b$,
    (i) use qn 26 to show that $(c \cdot a_n) \to c \cdot a$,
    (ii) use qn 37 to show that $(a_n + b_n) \to a + b$,
    (iii) use qn 38 to show that $(c \cdot a_n + d \cdot b_n) \to c \cdot a + d \cdot b$,

(iv) use qn 36 and the equation

$$a_n b_n - ab = (a_n - a)(b_n - b) + a(b_n - b) + b(a_n - a)$$

to show that $(a_n b_n) \to ab$,

(v) use qn 2.67 and qn 27 to show that $(|a_n|) \to |a|$,

(vi) use qn 39 when $a_n \leqslant c_n \leqslant b_n$ and $a = b$ to show that $(c_n) \to a$.

45 (i) Extend qn 44(ii) to a sum of three convergent sequences and then to a sum of k convergent sequences.

(ii) Extend qn 44(iv) to a product of three convergent sequences, and then to a product of k convergent sequences.

(iii) If $(a_n) \to a$, what is the limit of the sequence $(1 + 2a_n + 3a_n^2 + 4a_n^3 + 5a_n^4)$?

46 Use a calculator to explore the sequences $(\sqrt[n]{2})$, $(\sqrt[n]{10})$ and $(\sqrt[n]{1000})$. Repeated use of the square root button gives a subsequence in each case.

47 (i) If $1 < b$, use qn 2.35 to show that $1 < \sqrt[n]{b}$.

(ii) If $1 + a_n = \sqrt[n]{b}$, use qn 2.33 to show that $1 + na_n \leqslant b$.

(iii) Prove that (a_n) is a null sequence.

(iv) Deduce that $(\sqrt[n]{b}) \to 1$ as $n \to \infty$.

48 (i) Use a calculator to explore the sequence $(\sqrt[n]{(2^n + 3^n)})$.

(ii) Find the limit of the sequence $(\sqrt[n]{(a^n + b^n)})$ if $0 \leqslant a \leqslant b$.

49 Give a definition of what it means to say that $(a_n) \to a$ as $n \to \infty$, without mentioning null sequences as such but using ε and n_0 as in the definition of a null sequence.

Boundedness of convergent sequences

50 For what values of L and U is the inequality $L < x < U$ equivalent to the inequality $|x - a| < \varepsilon$? For these values of L and U the interval $\{x | L < x < U\}$ is called an *ε-neighbourhood* of a.

51 Prove that, if $(a_n) \to a$, then the sequence is eventually bounded above by $a + 1$ and eventually bounded below by $a - 1$. (*Hint:* take $\varepsilon = 1$.) Deduce from qn 10 that every convergent sequence is bounded.

Give an example to show that a bounded sequence need not be convergent.

52 If $(a_n) \to a$ and $a > 0$, identify a positive number which is eventually a lower bound for the sequence.

Quotients of convergent sequences

53 What is the relationship between the limits of the sequences $(2n/(n + 1))$ and $((n + 1)/2n)$?

54 If $(a_n) \to a$ and $a > 0$, show that, for sufficiently large n, $a/2 < a_n < 3a/2$, and hence $2/3a < 1/a_n < 2/a$.
Multiply this inequality by $|a_n - a|/a$ and use qns 27, 26 and 39 to prove that $(1/a_n) \to 1/a$.

55 Revise the argument of qn 54 to obtain the same result in the case when $a < 0$.

56 If $(a_n) \to a$ and $(b_n) \to b \neq 0$, prove that $(a_n/b_n) \to a/b$.

57 Find the limits of the sequences with nth terms given here.

(a) $\dfrac{1 + 1/n}{2 + 1/n}$, (b) $\dfrac{3n + 2}{4n - 3}$, (c) $\dfrac{n^2 - 1}{3n^2 + n}$, (d) $\dfrac{(\frac{1}{2})^n - 1}{(\frac{1}{2})^n + 1}$,

(e) $\dfrac{2^n - 1}{2^n + 1}$, (f) $\dfrac{n^3 + n^2 + n + 1}{3n^4 + 5}$, (g) $\dfrac{(\sqrt{n} + 3)(\sqrt{n} - 1)}{3\sqrt{n} - 5n}$,

(h) $\dfrac{a^n - 1}{a^n + 1}$, $0 < a$, (i) $\dfrac{1 + 2 + \ldots + n}{n^2}$,

(j) $\sqrt[n]{b}$, where $0 < b < 1$.

d'Alembert's ratio test

58 Give examples of sequences (a_n) of positive terms for which

$$\left(\frac{a_{n+1}}{a_n}\right) \to \frac{1}{2}.$$

Are all the sequences which you have found convergent?
To what limit?
By choosing $\varepsilon = \frac{1}{4}$ and using the limit of ratios, show that the method of qn 33 can be applied here to prove that any such sequence is null.

59 If (a_n) is a sequence of positive terms and

$$\left(\frac{a_{n+1}}{a_n}\right) \to l < 1,$$

show that, for sufficiently large n, the sequence (a_n) is sandwiched between 0 and an appropriately chosen null geometric progression.
Deduce that (a_n) is a null sequence.

60 If (a_n) is a sequence of positive terms and

$$\left(\frac{a_{n+1}}{a_n}\right) \to l > 1,$$

show that, for sufficiently large n, the terms of the sequence are greater than the corresponding terms of an appropriately chosen geometric progression which, from qn 16, is tending to $+\infty$.
Deduce that $(a_n) \to +\infty$.

61 Give an example of a null sequence of positive terms (a_n) for which

$$\left(\frac{a_{n+1}}{a_n}\right) \to 1.$$

Also give an example of a non-convergent sequence of positive terms (a_n) for which

$$\left(\frac{a_{n+1}}{a_n}\right) \to 1.$$

This means that the condition that the ratio of successive terms tend to 1 does not determine the convergence of the original sequence, one way or the other.

62 Let $k \in \mathbb{Z}$. Determine for what values of x the sequences with nth terms

(i) $n^k x^n$ and   (ii) $\dfrac{x^n}{n!}$ are null.

Can the value of k affect the answer?

Convergent sequences in closed intervals

63 If $(a_n) \to a$, $0 \le a_n$ and $a \le 0$, use the inequality
$0 \le a_n \le a_n - a$ to prove that $a = 0$.
Deduce that, if $(a_n) \to a$ and $0 \le a_n$, then necessarily $0 \le a$.

64 If $(a_n) \to a$, $(b_n) \to b$ and $a_n \leqslant b_n$ for all n, by considering the sequence $(b_n - a_n)$, prove that $a \leqslant b$.

65 By considering the sequences defined by $a_n = -1/n$ and $b_n = 1/n$, show that even if $a_n < b_n$ for all n, $(a_n) \to a$ and $(b_n) \to b$ does not imply $a < b$.

66 If $(a_n) \to a$ and $A \leqslant a_n \leqslant B$ for all n, prove that $A \leqslant a \leqslant B$.
This is the reason for calling the interval $\{x \mid A \leqslant x \leqslant B\}$ a *closed* interval, since no convergent sequence in the interval can have a limit which escapes from the interval.
Give an example to show that if $(a_n) \to a$ and $A < a_n < B$ for all n, it does not necessarily follow that $A < a < B$.
The interval $\{x \mid A < x < B\}$ is called an *open* interval, partly because a convergent sequence in the interval may have a limit outside the interval.

67 If $(a_n) \to a$ and (a_n) is a sequence of positive terms such that

$$a_{n+1}^2 = a_n + 2$$

prove that $a = 2$. Use qn 44 (ii), (iv) and the closed interval property.

68 If (a_n) is a monotonic sequence with a convergent subsequence $(a_{n_i}) \to A$, show that $(a_n) \to A$.
Suppose that (a_n) is monotonic *increasing*.
  (i) First show that $a_n \leqslant A$ for all n, by considering that, if this were false, then, for some particular n, $A < a_n$; then find a term of the convergent subsequence which is greater than A, and deduce that all succeeding terms of the subsequence are at least as far from A, thus contradicting its convergence to A.
  (ii) Then show that, given $\varepsilon > 0$, for some k, $A - \varepsilon < a_{n_k} \leqslant A$. Now name an N such that $A - \varepsilon < a_n \leqslant A$ when $n > N$.
  (iii) Having completed the proof for (a_n) monotonic increasing, show how to apply the result when (a_n) is monotonic decreasing.

Intuition and convergence

69 An isosceles right-angled triangle is constructed with a given line segment l as hypoteneuse. It has perimeter p_1 and area A_1.

Two isosceles right-angled triangles are constructed each with hypoteneuse on half of the line segment l, but with the two hypoteneuses occupying the whole of l. The two triangles, which only overlap at one vertex, have combined perimeter p_2 and combined area A_2.

This replacement of a triangle by two smaller triangles is repeated indefinitely, so that at each stage of replacement there are 1, 2, 4, 8, ..., 2^n, ... congruent triangles with hypoteneuses filling the line segment l, and at the nth stage, the combined perimeter of these triangles is p_n and their combined area is A_n.

(i) What is the relation between p_n and p_{n+1}?

(ii) What is the relation between A_n and A_{n+1}?

(iii) What can you say about the two sequences (p_n) and (A_n)?

70 (*Koch*, 1904) An equilateral triangle is given with perimeter p_1 and area A_1. This is stage 1.

Now an equivalent triangle is added to each side on the outside of the original figure, transforming

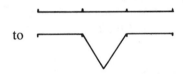

to

where the four sides in the new figure are of equal length.

The result is a polygon with 12 sides, perimeter p_2 and area A_2. This is stage 2.

The same kind of extension is now made on each side of the new polygon, resulting in a further polygon with 48 sides, perimeter p_3 and area A_3. This is stage 3.

If this process is repeated indefinitely

(i) What is the number of sides of the nth polygon?

(ii) Show that $p_{n+1} = (4/3) \cdot p_n$.

(iii) Show that, in making the $(n + 1)$th polygon, each little triangle being adjoined has area $(1/9)^n A_1$.

(iv) Show that $A_{n+1} = A_n + (3/9)(4/9)^{n-1} A_1$.

(v) Prove, from qn 1.3(vi), that

$$1 + 4/9 + (4/9)^2 + \ldots + (4/9)^{n-1} = \frac{(4/9)^n - 1}{4/9 - 1}$$

(vi) Prove that $(A_n) \to (8/5) A_1$ as $n \to \infty$.

(vii) Prove that $(p_n) \to +\infty$ as $n \to \infty$.

Summary

Definition   A sequence (a_n) is said to be *monotonic increasing*
qn 4   when $a_n \leqslant a_{n+1}$ for all n, and is said to be
            monotonic decreasing when $a_{n+1} \leqslant a_n$ for all n.

Definition   A sequence (a_n) is said to be *bounded above* when
qn 5   there is a number U such that $a_n \leqslant U$ for all n.
            A sequence (a_n) is said to be *bounded below* when
            there is a number L such that $L \leqslant a_n$ for all n.
            A sequence which is bounded above and bounded
            below is said to be *bounded*.

Definition   If M is an infinite subset of $\mathbb{N}$, then $\{a_{n_i} | n_i \in M\}$ is a
qn 8   *subsequence* of the sequence (a_n) provided that, for
            the subsequence (a_{n_i}), we have $n_i < n_j$ when $i < j$.

Theorem   Every sequence has a monotonic subsequence.
qn 12

Definition   A sequence (a_n) is said to tend to infinity when, for
qn 14   any number C, there is an N such that $C \leqslant a_n$, for
            all $n > N$.
            This is written $(a_n) \to +\infty$ as $n \to \infty$.

Axiom of Archimedean order
qn 15   There is an integer greater than any given number.

Definition   A sequence (a_n) is said to tend to zero, or to be
qn 21   null, when, given $\varepsilon > 0$, there exists an n_0 such that
            $|a_n| < \varepsilon$ when $n > n_0$.
            This is written $(a_n) \to 0$ as $n \to \infty$.

Theorems   (i) $(1/n)$ is a null sequence.
qns 23, 32   (ii) (x^n) is a null sequence when $|x| < 1$.

Theorems   If (a_n) and (b_n) are null sequences, then
qns 27,   (i) $(|a_n|)$ is a null sequence;
38,   (ii) $(c \cdot a_n + d \cdot b_n)$ is a null sequence;
36,   (iii) $(a_n b_n)$ is a null sequence;
39   (iv) if $a_n \leqslant c_n \leqslant b_n$, (c_n) is a null sequence.

Definition   A sequence (a_n) is said to tend to the limit a, or to
            converge to a, when $(a_n - a)$ is a null sequence.
            This is written $(a_n) \to a$ as $n \to \infty$, or $\lim a_n = a$.
            A sequence is said to be convergent when it has a
            limit.

Theorem   Every element of an Archimedean ordered field is
qn 43   the limit of an infinite decimal.

Theorems If $(a_n) \to a$ and $(b_n) \to b$ as $n \to \infty$, then

$\qquad$ (i) $(|a_n|) \to |a|$;

qns 44 $\quad$ (ii) $(c \cdot a_n + d \cdot b_n) \to c \cdot a + d \cdot b$;

$\qquad$ (iii) $(a_n b_n) \to ab$;

$\qquad$ (iv) if $a = b$ and $a_n \leqslant c_n \leqslant b_n$, then $(c_n) \to a$;

56, $\qquad$ (v) if $b \neq 0$, then $(a_n/b_n) \to a/b$;

64, $\qquad$ (vi) if $a_n \leqslant b_n$, then $a \leqslant b$;

66 $\qquad$ (vii) if $A \leqslant a_n \leqslant B$, then $A \leqslant a \leqslant B$.

Theorem $\quad$ If $1 < a$, then $(\sqrt[n]{a}) \to 1$ as $n \to \infty$.

qn 47

Theorem $\quad$ Every convergent sequence is bounded.

qn 51

Historical note

The Archimedean axiom appears in Euclid in the form of a definition of ratio (Book V, def. 4), and thus predates Archimedes (250 B.C.). It was important to Euclid because it precluded comparing a measurement of length with a measurement of area or a measurement of area with a measurement of volume. Its importance in modern times dates from 1891, when G. Veronese recognised that it eliminated constant infinitesimals. In Hilbert's *Foundations of Geometry* (1899) it is referred to as the 'axiom of continuity'.

During the eighteenth century a great deal of work was done on particular sequences and series, and sophisticated methods for approximating and finding limits were developed. d'Alembert had studied the binomial expansion of $(1 + x)^m$ when m is rational (1768) and established the boundedness of the series by comparison with two geometric series when $|x| < 1$. There are indications of how to prove that the limit of a product is equal to the product of the limits in d'Alembert and de la Chapelle (1789) and of how to prove that the limit of a quotient is equal to the quotient of the limits (without attention to the possibility of a zero denominator) in L'Huilier (1795). But formal proofs, as in this chapter, had to await a formal definition of convergence and this came in Cauchy's *Cours d'Analyse* (1821) in his discussion of series. 'There exists an n_0 such that when $n > n_0 \ldots$' and the abbreviation 'for sufficiently large n' are both expressions which we owe to Cauchy. The suffix notation for sequences stems from Lagrange (1759) who actually used exponents and not suffices. The suffix notation proper is due to S. F. Lacroix (1800), but it was in Cauchy's hands that it became the powerful tool we know today. Most of the limits which were studied during the eighteenth century were one-sided, as was mentioned in the section preceding qn 19, and there was also a presumption that limits were not to be reached.

The notation for absolute value was not available to Cauchy, though he sometimes wrote $\sqrt{(x^2)}$ where we would write $|x|$. Where absolute value was pertinent, there could be a reference to an inequality holding 'without regard to sign'. The symbolism of absolute value and the notion of neighbourhood were introduced by K. Weierstrass in his lectures in 1859.

The symbol ∞ was introduced by J. Wallis in 1655, but he manipulated it like a number. The 'lim' notation was due to L'Huilier in 1786, and mid-nineteenth century mathematicians, including Weierstrass, were content with expressions such as

$$\lim_{n=\infty} n^2 = \infty$$

This only became unacceptable with the greater precision about what is and what is not a real number in the latter part of the nineteenth century, and it was in 1905 that the Cambridge mathematician J. G. Leathem proposed the use of '$n \to \infty$'. G. H. Hardy (1908) vehemently exhorts his readers to forgo writing $n = \infty$, though even he allowed the statement $\lim a_n = \infty$!

The language of bounds and boundedness, introduced by Pasch (1882), was adopted at the beginning of this century. For most of the nineteenth century a bound was called a limit and the boundedness of a set described by saying that its elements were finite.

The symbol $[x]$ for the integral part of x was used for certain special cases by Dirichlet in 1849, and in the sense that we now use it by F. Mertens in 1874. G. Peano wrote Ex for $[x]$ in 1899 (E being the first letter of *entier*, the word for *whole number* in French).

Answers

1   n, $2n$, $1/2^n$, π to $n-1$ places of decimals, $(-1)^n$, $n/(n+1)$.

4

	(a)	(b)	(c)	(d)
(i)	✓	✓	✗	✗
(ii)	✗	✗	✓	✓
(iii)	✗	✗	✗	✗
(iv)	✗	✓	✗	✗
(v)	✓	✓	✗	✗
(vi)	✗	✓	✗	✓
(vii)	✗	✗	✓	✓

5

	(a)	(b)
(i)	✗	✗
(ii)	✓	✓
(iii)	✓	✓
(iv)	✗	✓

6   (i) yes $(-1/n)$; (ii) yes; (iii) no (n); (iv) yes, always, a_1.
    Monotonic decreasing: (i) yes; (ii) yes $(1/n)$; (iii) yes, always, a_1;
    (iv) no $(-n)$.

7   (i) yes; (ii) yes, because, if there were only a finite number, the greatest
    would be an upper bound.

8   (i) yes, see qn 4(iv); (ii) not for a constant sequence.

9   Yes, the bounds for the sequence are certainly bounds for any
    subsequence.

10   (i) If the subsequence is bounded above by U and bounded below by
     L, then the sequence has $\max(a_1, U)$ as an upper bound and
     $\min(a_1, L)$ as a lower bound. 'max' is formally defined at qn 6.44.
     (ii) As (i) with upper bound $\max(a_1, a_2, U)$, and lower bound
     $\min(a_1, a_2, L)$.
     (iii) As (i) with upper bound $\max(a_1, a_2, \ldots, a_N, U)$, and lower bound
     $\min(a_1, a_2, \ldots, a_N, L)$.

11   (i) yes, $a_{2n-1} = 0$, $a_{2n} = (-2)^n$; (ii) no, $(-2)^n$.

12   (i) If the floor term a_{f_j} succeeds the floor term a_{f_i}, then by definition
     $a_{f_i} \le a_{f_j}$, and the subsequence (a_{f_i}) is monotonic increasing.
     (ii) Since no term after a_F is a floor term, every term thereafter is
     eventually succeeded by a lesser term, giving a decreasing
     subsequence.

     (iii) Since there are no floor terms, every term is eventually succeeded by a lesser term, again giving a decreasing subsequence.

13   (i) Not monotonic; bounded below; all terms $\geqslant 1$.
     (ii) Not monotonic; unbounded.
     (iii) Not monotonic; bounded below; all terms $\geqslant 1$, all terms past the 10th $\geqslant 10$, all terms past the 100th $\geqslant 100$.
     (iv) Not monotonic; bounded; all terms $\geqslant 10$.

14 (i) $1 < C$, (ii) any C, (iv) $11 < C$.

15 Take $N \geqslant C^2$.

16 Take $N \geqslant C/y$.

17 $(a_n) \to -\infty$ when, given any number C, there exists an N such that $a_n \leqslant C$ when $n > N$.

18 (a)   Any negative number is a lower bound; also 0.
     (b)   Any positive number $\geqslant 1$ is an upper bound; any positive number is eventually an upper bound.

19 First sequence: answers as in qn 18.
Second sequence: any positive number is an upper bound; any negative number is a lower bound. 0 is both.

20   (i) satisfies (a) and (c) but tends to $-\infty$.
     (ii) satisfies (a) and (b) but tends to 1.
     (iii) satisfies (c) and perhaps (d) but keeps its distance from 0.
     (iv) possibly satisfies (c) and (d) but still keeps a definite distance from 0.
     (v) satisfies (e) but oscillates between 0 and 1.

21 (i) $n > 10$, (ii) $n > 100$, (iii) $n > 1000$, (iv) $n > 1/\varepsilon$.

22 For example, take $\varepsilon = 1$ for the sequence with nth term $1 + 1/n$.

23 Given $\varepsilon > 0$, let n_0 be an integer greater than $1/\varepsilon$, (using the Axiom of Archimedean order). Then $n > n_0$ implies that $n > 1/\varepsilon$, so $1/n < \varepsilon$.

24 Given $\varepsilon > 0$, let n_0 be an integer greater than $1/\varepsilon^2$. Then $n > n_0$ implies that $n > 1/\varepsilon^2$, so $1/\sqrt{n} < \varepsilon$.

25 Given $\varepsilon > 0$, there exists an n_0 such that $a_n > 1/\varepsilon$ when $n > n_0$, so $0 < 1/a_n < \varepsilon$. Converse false, put $a_n = (-1)^n/n$.

26 Case (i), $c \neq 0$. Given $\varepsilon > 0$, choose n_0 so that $|a_n| < \varepsilon/|c|$ when $n > n_0$. Then $|c \cdot a_n| < \varepsilon$.
Case (ii), $c = 0$. Given $\varepsilon > 0$, $|c \cdot a_n| = 0 < \varepsilon$, for all n.
Use qn 24 and put $c = 10$.

27 Since $\|a_n\| = |a_n|$, for given $\varepsilon > 0$, the same n_0 is sufficient for either sequence.

28 If $|a_n| < \varepsilon$ and $0 \leqslant b_n \leqslant a_n$, then $0 \leqslant b_n < \varepsilon$, so $|b_n| < \varepsilon$. So the n_0 which is sufficient for (a_n) is sufficient for (b_n).

29   (i) Use qns 28 and 23. (ii) Use qn 26 and (i).
     (iii) Show that $1/(7n + 3) < 1/n$. Then use qns 23, 28 and 26.
     (iv) Show that $1/(n^2 + 1) < 1/n$. Then use qns 23 and 28.
     (v) nth term $= 1/(\sqrt{(n + 1)} + \sqrt{n}) < 1/(2\sqrt{n})$. Use qns 24, 26, and 28.
     (vi) $|(\sin n)/n| \leqslant 1/n$. Use qns 23, 28 and 27.
     (vii) All terms are 0. (viii) If $c = p/q$, where p and q are integers, then for $n \geqslant q$ the terms are 0.

30 If $|a_n| < \varepsilon$ when $n > n_0$, then $|a_{n_i}| < \varepsilon$ when $n_i > n_0$.

31 (i) and (iii) are subsequences of $(1/n)$. For $n \geqslant 34$, so is (ii).

32 (iv) Use qn 23, then qn 26 with $c = 1/y$, and then qn 28 twice.
    (v) From qn 26.

33 (c) $N = 5$ or greater. (d) For example $a_{N+2} < (\tfrac{3}{4})^2 a_N$
   (e) $((\tfrac{3}{4})^n)$ is a null sequence from qn 32. Then use qn 26. Don't forget that N is constant.
   (f)   Part (d) provides the sandwich for all terms of the original sequence after the 5th. All terms are positive.
   (g)   A number between 0 and 1 must be used to be able to appeal to qn 32. Provided the number is between $\tfrac{1}{2}$ and 1, a serviceable N can be found for this sequence. To use $3/5$, we must take $N = 11$ or more.

34   (i) $a_{n+1}/a_n = \tfrac{1}{2}(1 + 1/n)^4 < \tfrac{3}{4}$, when $n > 10$.

     (ii) $a_{n+1}/a_n = 2/(n + 1) < \tfrac{3}{4}$, when $n > 2$.

     (iii) $a_{n+1}/a_n = 1/(1 + 1/n)^n < \tfrac{1}{2}$, for all n.

     (iv) $a_{n+1}/a_n = (0.9)(1 + 1/n)^3 < 0.95$, when $n > 55$.

35 $|x^n| = |x|^n$.

36 (iii) Take $n_0 = \max\{n_1, n_2\}$.
    (iv) $|a_n b_n| = |a_n| \cdot |b_n| < |a_n| < \varepsilon$, from qns 2.61 and 2.21.
    (v) $a_n b_n c_n$ is the product of $a_n b_n$ and c_n. k terms by induction.

37 (iii) Take $n_0 = \max\{n_1, n_2\}$.
    (iv) $|a_n + b_n| \leqslant |a_n| + |b_n| < \tfrac{1}{2}\varepsilon + \tfrac{1}{2}\varepsilon = \varepsilon$, from qns 2.65 and 2.20.
    (v) $a_n + b_n + c_n$ is the sum of $a_n + b_n$ and c_n. k terms by induction.

38 $(c \cdot a_n)$ and $(d \cdot b_n)$ are both null sequences, from qn 26. Their sum is a null sequence, from qn 37.

39   (i) Choice of scalars, c, $d = +1$, -1 in qn 38.
   (ii) $0 \le b_n - a_n \le c_n - a_n$ and qn 28.
   (iii) $b_n = (b_n - a_n) + a_n$. Use qn 37.

40 If $a_n = 1$ when $n = 2^k$ and $a_n = 0$ otherwise, then $A_{2^k} = k + 1$, and (A_n/n) is null by comparison with qn 33.

41 Terms tend to 1. $(a_n - 1)$ is null, from qns 23 and 28.

42 In qn 38 use $c = 1$, $d = -1$. Then the constant sequence $(b - a)$ is null. If not null, the possibility that a positive ε might equal $\frac{1}{2}|b - a|$ gives a contradiction.

43   (i) 6.2, 6.25, 6.25, 6.25. (ii) 0.3, 0.33, 0.333, 0.3333,
   (iii) 1.4, 1.41, 1.414, 1.4142.
   $[10^n a] \le 10^n a < [10^n a] + 1$ gives the inequality.
   $(1/10^n)$ is null by qn 32, so $([10^n a]/10^n) \to a$ by qn 28. Every term of this sequence is rational.

44   (iv) Also use qns 26 and 37.
   (v) $(a_n - a)$ null implies $(|a_n - a|)$ is null. Now use qn 28 with
   $0 \le ||a_n| - |a|| \le |a_n - a|$ from qn 2.67, and then use qn 27 again.
   (vi) $a_n - a \le c_n - a \le b_n - a$.

45   (i) $a_n + b_n + c_n$ is the sum of $a_n + b_n$ and c_n. k terms by induction.
   (ii) $a_n b_n c_n$ is the product of $a_n b_n$ and c_n. k terms by induction.
   (iii) Repeated use of (ii), qn 44(i) and part (i). $1 + 2a + 3a^2 + 4a^3 + 5a^4$.

46 Each tends to 1.

47   (iii) $0 < a_n \le (b - 1)/n$. Use qns 23, 26 and 28.

48   (ii) $\sqrt[n]{(a^n + b^n)} = b \cdot \sqrt[n]{(1 + (a/b)^n)}$, so
   $b < \sqrt[n]{(a^n + b^n)} \le b \cdot \sqrt[n]{(1 + a/b)}$. Now use qn 47, 44(i) and 44(vi).

49 Given $\varepsilon > 0$, there exists an n_0, such that $|a_n - a| < \varepsilon$ when $n > n_0$.

50 $L = a - \varepsilon$, $U = a + \varepsilon$.

51 From the definition there is an n_0 such that $n > n_0$ implies $|a_n - a| < 1$. This shows that a convergent sequence is eventually bounded. $((-1)^n)$ is bounded but not convergent.

52 $\frac{1}{2}a$, by taking $\varepsilon = \frac{1}{2}a$.

53 2 and $\frac{1}{2}$ are reciprocal.

54 Take $\varepsilon = \frac{1}{2}a$, then use qn 2.27. $2|a_n - a|/3a^2 < |1/a - 1/a_n| < 2|a_n - a|/a^2$.

55 Take $\varepsilon = -\frac{1}{2}a$. Then, for sufficiently large n, $|a_n - a| < -\frac{1}{2}a$, so $3a/2 < a_n < a/2 < 0$. Now $2/a < 1/a_n < 2/3a$. Since a is negative, multiplying through by $|a_n - a|/a$ reverses the inequalities again but, with both a and a_n negative, $aa_n = |aa_n|$.

56 Use qns 54, 55 and 44(iv).

57 (a) $\frac{1}{2}$ by qn 56. (b) nth term $= \dfrac{3 + 2/n}{4 - 3/n}$, limit $\frac{3}{4}$.

(c) nth term $= \dfrac{1 - 1/n^2}{3 + 1/n}$, limit $= \frac{1}{3}$.

(d) Limit -1. (e) Limit 1. (f) Limit 0, divide top and bottom by n^4.
(g) Limit $-1/5$. (h) Limit -1 when $a < 1$, limit 0 when $a = 1$, limit 1 when $1 < a$. (i) Limit $\frac{1}{2}$, use qn 1.3(i). (j) nth term $= 1/\sqrt[n]{(1/b)}$, limit $= 1$.

58 $(n^k/2^n)$ for any positive integer k. All null.
Take $\varepsilon = \frac{1}{4}$, then for sufficiently large n, $a_{n+1}/a_n < \frac{3}{4}$.

59 Take $\varepsilon = \frac{1}{2}(1 - l)$, then $0 \leqslant a_{N+n} < \left(\frac{1}{2}(1 + l)\right)^n a_N$.

60 Take $\varepsilon = \frac{1}{2}(l - 1)$, then $\left(\frac{1}{2}(1 + l)\right)^n a_N < a_{N+n}$.

61 $(1/n)$, (n).

62 (i) null when $|x| < 1$, or when $|x| = 1$ and $k < 0$.
(ii) null for all values of x.

63 $(a_n - a)$ is a null sequence, so (a_n) is a null sequence by qn 28, and so $a = 0$. This result bars $a < 0$.

64 Since $0 \leqslant b_n - a_n$, from qns 63 and 44(iii), $0 \leqslant b - a$.

66 Since $0 \leqslant a_n - A$, $0 \leqslant a - A$, from qn 63.
Since $0 \leqslant B - a_n$, $0 \leqslant B - a$, from qn 63. $0 < 1/n < 2$.

68 (i) $A < a_n \Rightarrow A < a_n \leqslant a_{n_k} \leqslant a_{n_i}$, for $n_k > n$ and $i \geqslant k$.
(ii) Let $N = n_k$. $n > N \Rightarrow A - \varepsilon < a_{n_k} \leqslant a_n \leqslant A$. So $(a_n) \to A$.
(iii) If (a_n) is monotonic decreasing, then $(-a_n)$ is monotonic increasing and $(-a_n) \to -A \Rightarrow (a_n) \to A$.

69 (i) $p_{n+1} = p_n$. (ii) $A_{n+1} = \frac{1}{2}A_n$. (iii) (p_n) is constant, (A_n) is null.

70 (i) $3 \cdot 4^{n-1}$. (iv) from (i) and (iii). (v) note that the limit of the right hand side as $n \to \infty$ is 9/5.
(vi) $A_{n+1} = A_1 + (3/9)A_1\left(1 + 4/9 + \ldots + (4/9)^{n-1}\right)$. (vii) from (i) and qn 16.

4

Completeness
What the rational numbers lack

Preliminary reading: Hemmings and Tahta, Niven, Burn (1990), Gardiner part II, Lieber, Yarnelle, Zippin.
Concurrent reading: Baylis and Haggarty, Wheeler.
Further reading: Armitage and Griffiths, part II ch. 1, Cohen and Ehrlich chs 4 and 5, Artmann chs 1 to 3.

The first sixteen questions of this chapter are concerned with integers and their quotients.

The Fundamental Theorem of Arithmetic

2 prime	$21 = 3 \cdot 7$	41 prime	61 prime	$81 = 3^4$
3 prime	$22 = 2 \cdot 11$	$42 = 2 \cdot 3 \cdot 7$	$62 = 2 \cdot 31$	$82 = 2 \cdot 41$
$4 = 2^2$	23 prime	43 prime	$63 = 3^2 \cdot 7$	83 prime
5 prime	$24 = 2^3 \cdot 3$	$44 = 2^2 \cdot 11$	$64 = 2^6$	$84 = 2^2 \cdot 3 \cdot 7$
$6 = 2 \cdot 3$	$25 = 5^2$	$45 = 3^2 \cdot 5$	$65 = 5 \cdot 13$	$85 = 5 \cdot 17$
7 prime	$26 = 2 \cdot 13$	$46 = 2 \cdot 23$	$66 = 2 \cdot 3 \cdot 11$	$86 = 2 \cdot 43$
$8 = 2^3$	$27 = 3^3$	47 prime	67 prime	$87 = 3 \cdot 29$
$9 = 3^2$	$28 = 2^2 \cdot 7$	$48 = 2^4 \cdot 3$	$68 = 2^2 \cdot 17$	$88 = 2^3 \cdot 11$
$10 = 2 \cdot 5$	29 prime	$49 = 7^2$	$69 = 3 \cdot 23$	89 prime
11 prime	$30 = 2 \cdot 3 \cdot 5$	$50 = 2 \cdot 5^2$	$70 = 2 \cdot 5 \cdot 7$	$90 = 2 \cdot 3^2 \cdot 5$
$12 = 2^2 \cdot 3$	31 prime	$51 = 3 \cdot 17$	71 prime	$91 = 7 \cdot 13$
13 prime	$32 = 2^5$	$52 = 2^2 \cdot 13$	$72 = 2^3 \cdot 3^2$	$92 = 2^2 \cdot 23$
$14 = 2 \cdot 7$	$33 = 3 \cdot 11$	53 prime	73 prime	$93 = 3 \cdot 31$
$15 = 3 \cdot 5$	$34 = 2 \cdot 17$	$54 = 2 \cdot 3^3$	$74 = 2 \cdot 37$	$94 = 2 \cdot 47$
$16 = 2^4$	$35 = 5 \cdot 7$	$55 = 5 \cdot 11$	$75 = 3 \cdot 5^2$	$95 = 5 \cdot 19$
17 prime	$36 = 2^2 \cdot 3^2$	$56 = 2^3 \cdot 7$	$76 = 2^2 \cdot 19$	$96 = 2^5 \cdot 3$
$18 = 2 \cdot 3^2$	37 prime	$57 = 3 \cdot 19$	$77 = 7 \cdot 11$	97 prime
19 prime	$38 = 2 \cdot 19$	$58 = 2 \cdot 29$	$78 = 2 \cdot 3 \cdot 13$	$98 = 2 \cdot 7^2$
$20 = 2^2 \cdot 5$	$39 = 3 \cdot 13$	59 prime	79 prime	$99 = 3^2 \cdot 11$
	$40 = 2^3 \cdot 5$	$60 = 2^2 \cdot 3 \cdot 5$	$80 = 2^4 \cdot 5$	$100 = 2^2 \cdot 5^2$

1 Find non-negative integers a, b, c, d, e, f, g such that
$3\,089\,988 = 2^a \cdot 3^b \cdot 5^c \cdot 7^d \cdot 11^e \cdot 13^f \cdot 17^g$.

The *Fundamental Theorem of Arithmetic* states that each counting number greater than 1 may be expressed in a unique way as a product of prime numbers. The proof of this theorem can be found in most undergraduate textbooks on number theory. See for example the first chapter of Burn (1982), or of Hardy and Wright. For the simple enunciation of this theorem it is important that the number 1 should *not* be called prime.

2 Find integers a, b, c, d, e, f, g such that
$2775/999\,999 = 2^a \cdot 3^b \cdot 5^c \cdot 7^d \cdot 11^e \cdot 13^f \cdot 17^g$.

3 Use the Fundamental Theorem of Arithmetic to enunciate a unique factorisation theorem for *rational* numbers.
(The set of rational numbers, $\mathbb{Q}$ was defined at the end of chapter 1 and again at the start of chapter 2.)

Dense sets of rational numbers on the number line

4 Illustrate on a number line those portions of the sets
$\{m \mid m \in \mathbb{Z}\}$, $\{m/2 \mid m \in \mathbb{Z}\}$, $\{m/4 \mid m \in \mathbb{Z}\}$, $\{m/8 \mid m \in \mathbb{Z}\}$
which lie between ± 3. Is each set contained in the set which follows in this list?
What would an illustration of the set $\{m/2^n \mid m \in \mathbb{Z}\}$ look like for some large positive integer n?

5 Find a rational number which lies between $57/65$ and $64/73$ and may be written in the form $m/2^n$, where m is an integer and n is a non-negative integer.

6 To show that there is always a rational number of the form $m/2^n$, where m is an integer and n is a non-negative integer, between the numbers a and b, where $a < b$, let k be a non-negative integer such that $1/2^k < b - a$. Use qn 3.32 to explain why such a k must exist.
Use the inequality $[a \cdot 2^k] \leqslant a \cdot 2^k < [a \cdot 2^k] + 1$ to show that $m = [a \cdot 2^k] + 1$, and $n = k$, satisfy the required conditions, using the notation preceding qn 3.15.

7 By repeatedly bisecting the interval $\{x \mid a < x < b\}$, show that between any two distinct numbers on the number line there is an infinity of numbers of the form $m/2^n$, where m is an integer and n is a non-negative integer.

A set of numbers containing an element in every interval on the number line is said to be *dense* on the line. The argument of qn 7 can be used to show that there is an *infinity* of elements of a dense set in every interval on the number line. We have shown that the rational numbers are dense on the number line, even though the proof only uses rationals with denominator a power of 2.

8 Is there a smallest positive number in the set
   $\{m/2^n \mid m \in \mathbb{Z}, n \in \mathbb{Z}^+\}$?
   Is there a smallest positive rational number?

9 Let $T = \{m/2^n \mid m \in \mathbb{Z}, n \in \mathbb{Z}^+\}$ and
   $D = \{m/10^n \mid m \in \mathbb{Z}, n \in \mathbb{Z}^+\}$.
   Show that $1/5 \in D$ and use the Fundamental Theorem of Arithmetic to prove that $1/5 \notin T$.
   Show that every element of T is an element of D, that is, $T \subset D$. What is the usual name for an element of the set D? See the paragraph following qn 3.43.
   Is the set T closed under
      (i) addition,
      (ii) subtraction,
      (iii) multiplication,
      (iv) division except by zero?
   Would the answers have been the same if the questions had been asked about the set D?

10 By considering the sequence
   $a_1 = \frac{1}{4}, a_2 = \frac{1}{4} - (\frac{1}{4})^2, a_3 = \frac{1}{4} - (\frac{1}{4})^2 + (\frac{1}{4})^3, \ldots,$
   $a_n = -((-\frac{1}{4}) + (-\frac{1}{4})^2 + (-\frac{1}{4})^3 + \ldots + (-\frac{1}{4})^n), \ldots$

   show that a sequence of fractions in T (as defined in qn 9) may converge to a number which is not in T. Use qns 1.3(vi) and 3.35.

Decimals

11 Use the Fundamental Theorem of Arithmetic to prove that $\frac{1}{3}$ does not belong either to the set T or to the set D as defined in qn 9.
   By considering the sequence
   $a_1 = 3/10, a_2 = 3/10 + 3/100, a_3 = 3/10 + 3/100 + 3/1000, \ldots,$
   $a_n = 3/10 + 3/100 + \ldots + 3/10^n, \ldots,$

show that there is a sequence of numbers in D which converges to $\frac{1}{3}$. Compare this sequence with qn 3.43(ii). This sequence is the infinite decimal $0.33\dot{3}\ldots$ (point three recurring), with the rational number $\frac{1}{3}$ as limit. The dot above the last 3 indicates that that digit recurs.

12 Use the idea of qn 11 to give a sequence in D for $0.99\dot{9}\ldots$ (point nine recurring) and find its limit.
See the note which follows qn 3.43.

13 Suppose that a and b are rational numbers. Use qn 1.3(vi) and qn 3.32 to prove that the limit of the sequence with nth term

$$a + b\left(\frac{1}{10^3} + \frac{1}{10^6} + \frac{1}{10^9} + \ldots + \frac{1}{10^{3n}}\right)$$

is $a + b/999$.
Find terminating decimals a and b such that the limit of $123.456\,789\,789\,789\,7\dot{8}\dot{9}\ldots$ (with 789 recurring) $= a + b/999$.
Must every recurring decimal converge to a rational number?

14 Show that every rational number, not belonging to the set D of qn 9, is the limit of a recurring decimal. (*Hint*: analyse the remainders on dividing the integer A by the integer B.)

The *infinite decimal* $d_0.d_1d_2d_3\ldots d_n\ldots$ is the sequence with nth term

$$d_0 + \frac{d_1}{10} + \frac{d_2}{10^2} + \ldots + \frac{d_n}{10^n},$$

where d_0 is an integer, and, when $1 \leqslant i$, d_i is either 0, 1, 2, $\ldots$, or 9. In qn 3.43 we showed that every number in an Archimedean ordered field is the limit of an infinite decimal. Now we ask when two infinite decimals may have the same limit.

15 If two recognisably different infinite decimals have the same limit, show that one of them is a terminating decimal (and eventually has recurring 0s) and the other eventually has recurring 9s.

Although the sequence for an infinite decimal may be constructed in any ordered field, its convergence to a limit in that field is not guaranteed.

16 An infinite decimal $0.101\,001\,000\,100\,001\ldots$ has 1s in the $\frac{1}{2}n(n + 1)$th positions and zeros elsewhere. Can its limit be a rational number?

Irrational numbers

17 If p and q are non-zero integers, prove that an equation of the form $p^2 = 2q^2$ would contradict the Fundamental Theorem of Arithmetic, and so there can be no rational number equal to $\sqrt{2}$. List some other square roots and cube roots which cannot equal rational numbers for similar reasons.

18 If a and b are rational numbers with $b \neq 0$, is it possible for $a + b\sqrt{2}$ to be a rational number?

(19) Prove that the set $\mathbb{Q}[\sqrt{2}] = \{a + b\sqrt{2} \mid a, b \in \mathbb{Q}\}$ forms a field. Most of the axioms are easily checked. Make sure you can find a multiplicative inverse in $\mathbb{Q}[\sqrt{2}]$ for each non-zero element.

20 Let a and b be rational numbers, with $a < b$. Show that $(a + b\sqrt{2})/(1 + \sqrt{2})$ is not rational. Show also that $a < (a + b\sqrt{2})/(1 + \sqrt{2}) < b$. Deduce that there is an irrational number between any two distinct rationals. This, with qn 7, shows that the irrationals are dense on the number line.

21 Show that there can be no rational number equal to $\log_{10} 2$.

Infinity: countability

Galileo noted in his *Discourses on Two New Sciences* (1638) that the mapping $n \mapsto n^2$ matched the set $\mathbb{N}$, one-to-one, with a proper subset of $\mathbb{N}$. He also noted that the mapping $x \mapsto \frac{1}{2}x$ matched the set $\{x \mid 0 \leqslant x \leqslant 2\}$, one-to-one, with the set $\{x \mid 0 \leqslant x \leqslant 1\}$, another one-to-one matching of a set with a proper subset. Either one of these mappings provides a counter-example to Euclid's axiom: the whole is always greater than the part. It was Bolzano (1851) who first recognised that this was always a possibility with infinite sets.

22 By running through the integers in an order beginning $0, +1, -1, \ldots$ exhibit a one-to-one matching of $\mathbb{Z}$ and $\mathbb{N}$. If, for $a \in \mathbb{Z}^+$, we map $a \mapsto 2^a$ and $-a \mapsto 2^a \cdot 3$, show that $\mathbb{Z}$ can be matched with a, rather small(!) though still infinite, proper subset of $\mathbb{N}$.

23 If a and b are positive integers, show that $a/b \mapsto 2^a \cdot 3^b$, gives a one-to-one matching of $\mathbb{Q}^+$ with a subset of $\mathbb{N}$, provided fractions are always taken in their lowest terms. By also mapping $-a/b \mapsto 2^a \cdot 3^b \cdot 5$ establish a one-to-one matching of $\mathbb{Q}$ with a subset of $\mathbb{N}$.

24 If a, b, c and d are positive integers, show that
$(a/b) + (c/d)\sqrt{2} \mapsto 2^a \cdot 3^b \cdot 5^c \cdot 7^d$ gives a one-to-one matching,
provided fractions are always taken in their lowest terms.
Use this to construct a one-to-one matching of the set
$\mathbb{Q}[\sqrt{2}] = \{x + y\sqrt{2} \mid x, y \in \mathbb{Q}\}$ with a subset of $\mathbb{N}$.

When there is a one-to-one matching of the elements of a set A
with the elements of $\mathbb{N}$, the set A is said to be *countably infinite*.

25 If a set A can be matched, one-to-one, with an infinite subset of
$\mathbb{N}$, show that it can be matched, one-to-one, with $\mathbb{N}$ itself, and is
thus countably infinite. Hence T and D of qn 9, $\mathbb{Z}$, $\mathbb{Q}$ and $\mathbb{Q}[\sqrt{2}]$
are all countably infinite. Note that T, D, $\mathbb{Q}$ and $\mathbb{Q}[\sqrt{2}]$ are
dense sets on the number line.

26 Explain why a set A is countably infinite if and only if there is a
sequence of distinct terms taken from A which uses up all the
elements of A. The point here is to see how the word 'sequence'
catches hold of the crux of the definition of 'countably infinite'.

We can clearly write down infinitely many infinite decimals with 0
before the decimal point (so that, if they exist as numbers, the limits
are between 0 and 1). There is a classic proof (Cantor's second proof,
1890) that, though this set is infinite, it is **not countably** infinite. It is
a proof by contradiction. The proof starts by presuming that this set
of decimals **is** countably infinite. We can therefore write all of these
decimals in one sequence. We then construct an infinite decimal, with
0 to the left of the decimal point, which is different from every
member of the sequence, and so does not occur as a term of the
sequence.
Let the sequence of decimals be:

$x_1 = 0.a_{11}a_{12}a_{13}a_{14}a_{15} \ldots$
$x_2 = 0.a_{21}a_{22}a_{23}a_{24}a_{25} \ldots$
$x_3 = 0.a_{31}a_{32}a_{33}a_{34}a_{35} \ldots$
$x_4 = 0.a_{41}a_{42}a_{43}a_{44}a_{45} \ldots$
$\vdots$
$x_n = 0.a_{n1}a_{n2}a_{n3}a_{n4}a_{n5} \ldots$

and so on.
To construct a decimal $0.b_1b_2b_3 \ldots b_n \ldots$ which is not in the
list, make the entry b_1, in the first decimal place, different from a_{11}
and 0 and 9 (for example if $0 \leqslant a_{11} \leqslant 7$, let $b_1 = a_{11} + 1$, and if
$a_{11} = 8$ or 9, let $b_1 = a_{11} - 1$); make the entry, b_2, in the second
decimal place, different from a_{22} and 0 and 9; and, in general, make

the entry, b_n, in the nth decimal place, different from a_{nn} and 0 and 9. (The reason for avoiding 0s and 9s lies in qn 15.) This shows that any attempt to construct a sequence which contains all the decimals between 0 and 1 is bound to fail. So the decimals with 0 to the left of the decimal point are *not* countably infinite. Such an infinite set is said to be *uncountably infinite*.

After postulating that the limit of each infinite decimal is a real number, it will follow that the set of all real numbers is uncountably infinite. The contrast between the countably infinite rationals and the uncountably infinite real numbers is not only itself counter-intuitive but has further counter-intuitive consequences.

Any countably infinite set of numbers can be shown to occupy only a tiny portion of the real line.

27 If, for each $n \in \mathbb{N}$, the interval

$$\{x \mid n - 1/2^n \leqslant x \leqslant n + 1/2^n\}$$

is deleted from the number line, is every element of $\mathbb{N}$ deleted? What is the total length deleted?

28 Let ε be a fixed positive number. If, for each $n \in \mathbb{N}$, the interval

$$\{x \mid n - \varepsilon/2^n \leqslant x \leqslant n + \varepsilon/2^n\}$$

is deleted from the number line, is every element of $\mathbb{N}$ deleted? What is the total length deleted? Can this length be as small as we like? How much length should we say the set $\mathbb{N}$ occupies?

29 Let (a_n) be any sequence and let ε be any positive number. If, for each $n \in \mathbb{N}$, the interval

$$\{x \mid a_n - \varepsilon/2^n \leqslant x \leqslant a_n + \varepsilon/2^n\}$$

is deleted from the number line, is every element of the sequence (a_n) deleted? What is the total length deleted? Deduce that a countably infinite set of numbers does not occupy any length on the number line.

When we remember that the set of rational numbers is countably infinite (qn 23) and the set of rational numbers is dense on the line (qn 7), it is particularly strange to realise that qn 29 implies that, when every rational is removed, together with a small interval around it, most of the real line remains untouched. Despite appearances, a real line on which only points corresponding to rational numbers can be identified is mostly empty!

The construction of a new infinite decimal, in the uncountability proof, could be done even if all the original infinite decimals were rational. Can you see why this does *not* establish that the rationals are uncountable?!

Axiom of Completeness: infinite decimals are convergent

The axioms for an Archimedean ordered field are satisfied by the set of rational numbers, $\mathbb{Q}$. From qn 29 it follows that these axioms are not sufficient to guarantee that there are no gaps on the number line. However, we found in qn 2.31 that every ordered field (whether Archimedean or not) contains a copy of the rational numbers, and therefore contains all terminating decimals. So in any ordered field we can form the sequence which gives any infinite decimal. From qn 3.43 every element of an Archimedean ordered field is the limit of an infinite decimal. But, from qn 16, there are infinite decimals which have no limit in $\mathbb{Q}$.

So we will adopt a further axiom:

every infinite decimal is convergent.

This axiom is called the **Axiom of Completeness**. An Archimedean ordered field in which this axiom holds is called a complete Archimedean ordered field. The elements of such a field are called real numbers, and the set is denoted by $\mathbb{R}$.

30 The infinite decimal 0.123 456 789 101 112 . . . is the sequence of terminating decimals which starts

0, 0.1, 0.12, 0.123, 0.1234, 0.123 45, 0.123 456, 0.123 4567, . . .

State an upper bound for this sequence. Can you find an upper bound which is smaller than this?
Is the sequence monotonic increasing?
Is every term in this sequence rational?
Explain how a negative real number, such as -3.1, may be expressed as a decimal with only the integer part negative.
[This is a useful convention when constructing algebraic proofs with decimals as it ensures that *every* infinite decimal is a monotonic increasing sequence].

31 Explain how the limits of infinite decimals may be added, subtracted, multiplied and divided by one another.

The remainder of this chapter will be devoted to establishing six properties of the real numbers, each of which is in fact equivalent to the Axiom of Completeness in an Archimedean ordered field. These are

I.   every bounded monotonic sequence of real numbers is convergent;

II. the intersection of a set of *nested* closed intervals is not empty (the term 'nested' is defined in qn 42);

III. every bounded sequence of real numbers has a convergent subsequence;

IV. every bounded infinite set of real numbers has a *cluster point* (this term is defined before qn 48);

V.   every *Cauchy sequence* of real numbers is convergent (the term 'Cauchy sequence' is defined before qn 56);

VI. every non-empty set of real numbers which is bounded above has a least upper bound.

Bounded monotonic sequences

32 Explain why every monotonic sequence is either bounded above or bounded below. Deduce that an increasing sequence which is bounded above is bounded, and that a decreasing sequence which is bounded below is bounded.

33 (After *Osgood*, 1907) We suppose that (a_n) is a monotonic decreasing sequence which is bounded below, by L, and we seek to prove that (a_n) is convergent.
   (i)   Name a lower bound for the sequence (a_n) which is an integer.
   (ii) Name an upper bound for the sequence (a_n) which is an integer.
   (iii) Must there be consecutive integers c, $c + 1$ for which c is a lower bound for (a_n) and $c + 1$ is not?
         For such integers, let $t_0 = d_0 = c$.
   (iv) Consider the eleven numbers:

   $$t_0, \quad t_0 + \tfrac{1}{10}, \quad t_0 + \tfrac{2}{10}, \quad \ldots, \quad t_0 + \tfrac{10}{10}.$$

   Is there a $c = 0, 1, 2, \ldots,$ or 9 such that

   $$t_0 + \frac{c}{10} \text{ is a lower bound for } (a_n) \text{ and } t_0 + \frac{c + 1}{10} \text{ is not?}$$

   Let $d_1 = c$ and $t_1 = d_0.d_1$.

(v) Proceed inductively to define a sequence (t_n) for an infinite decimal, in such a way that t_n is a lower bound for the sequence (a_n) while $t_n + 1/10^n$ is not a lower bound. Define

$$t_{n+1} = t_n + \frac{c}{10^{n+1}}, \quad \text{with } c = 0, 1, 2, \ldots \text{ or } 9,$$

where t_{n+1} is a lower bound for (a_n) while $t_{n+1} + 1/10^{n+1}$ is not.

$c = d_{n+1}$, and $t_n = d_0.d_1 d_2 d_3 \ldots d_n$.

(vi) Why is the sequence (t_n) convergent? Let its limit be D.
(vii) For a constant k, say why $(a_k - t_n) \to a_k - D \geq 0$.
(viii) How do you know that, for sufficiently large n,

$$D \leq a_n < D + 1/10^i \quad \text{for a given positive integer } i?$$

(ix) Prove that $(a_n) \to D$.

34 Use qn 33 to prove that a monotonic increasing sequence which is bounded must be convergent.

Now we know that monotonic sequences which are bounded are convergent, and it follows easily that monotonic sequences which are not bounded tend to $\pm\infty$.

35 Use qn 2.51 to prove that the sequence with nth term $(1 + 1/n)^n$ has a limit which lies between 2 and 3.

nth roots of positive real numbers

36 If a and x are positive real numbers with $a \leq x^2$, prove that $a \leq (\frac{1}{2}(x + a/x))^2 \leq x^2$. Deduce that, if x_1 is positive and $a < x_1^2$, the sequence (x_n) defined by

$$x_{n+1} = \frac{1}{2}\left(x_n + \frac{a}{x_n}\right)$$

is convergent. Find its limit. Deduce that every positive real number has a unique positive real number as square root. Compare with qn 2.42.

37 Prove that the sequences (a_n) and (b_n) of qn 2.43 are both convergent to the same limit.

38 Use qn 36 to show that positive real numbers have real 4th, 8th and 16th roots, etc.

39 If we know that positive real numbers have real kth roots, for a positive integer k, and a and x are positive real numbers such that $x^{k-1} \leqslant a$, prove that $x^{k-1} \leqslant \left(\sqrt[k]{(ax)}\right)^{k-1} \leqslant a$.
Deduce that if $0 < x_1$ and $x_1^{k-1} < a$, then the sequence given by $x_{n+1} = \sqrt[k]{(ax_n)}$ is convergent, and has limit $\sqrt[k-1]{a}$.
With the help of qn 38, deduce that positive real numbers have real kth roots for every positive integer k.

40 (i) Prove that the sequence $\left(n(\sqrt[n]{a} - 1)\right)$ of qn 2.48 is convergent when $1 < a$.
(ii) Prove that $\lim n\left(\sqrt[n]{(1/a)} - 1\right) = -\lim n(\sqrt[n]{a} - 1)$ and that
(iii) $\lim n\left(\sqrt[n]{(ab)} - 1\right) = \left(\lim n(\sqrt[n]{a} - 1)\right) + \left(\lim n(\sqrt[n]{b} - 1)\right)$
for any two positive numbers a and b.

41 Use qns 2.52 and 3.66 to prove that the sequence $\left(\sqrt[n]{n}\right)$ has a limit $l \geqslant 1$. Since $\left(\sqrt[2n]{(2n)}\right) \to l$, deduce from qn 3.44(iv) that $\left(\sqrt[n]{(2n)}\right) \to l^2$, and hence from qns 3.47 and 3.44(iv) that $l = l^2$. Deduce that $l = 1$.

Nested closed intervals

42 *The Chinese Box Theorem*
$\left(a_n\right)$ and $\left(b_n\right)$ are sequences such that

$$a_n \leqslant a_{n+1} < b_{n+1} \leqslant b_n \quad \text{for all values of } n.$$

Let $[a, b]$ denote the closed interval $\{x \mid a \leqslant x \leqslant b\}$.
   (i) Say why $[a_1, b_1] \cap [a_2, b_2] \cap \ldots \cap [a_n, b_n] = [a_n, b_n]$.
       We describe this by saying that these intervals are *nested*.
   (ii) Why must each of the two sequences be convergent?
   (iii) By considering the limits of the two sequences prove that at least one number lies in all the intervals $[a_n, b_n]$.
   (iv) By considering intervals of the type $\{x \mid 0 < x < 1/n\}$, show that the intersection of nested *open* intervals may be empty.

Convergent subsequences of bounded sequences

43 Find an upper bound and a lower bound for the sequences with nth term (i) $(-1)^n$,   (ii) $(-1)^n(1 + 1/n)$.

Is either sequence convergent? In each case find a convergent subsequence. Is the convergent subsequence monotonic?

44 Look back to your proof in qn 3.51 that every convergent sequence is bounded. Is it true that every bounded sequence is convergent?

45 Let A be a point on the circumference of a circle with unit radius. Let A_1 be the image of A under a rotation about the centre of the circle through 1 radian, and likewise let A_n be the image of A under a rotation through n radians. The perpendicular distance from A_n to the diameter through A is $\sin n$.
[The sine function will be formally defined in chapter 11 with a view to supporting this geometrical description.]
Clearly $-1 \leqslant \sin n \leqslant 1$. We consider whether the sequence $(\sin n)$ may have a convergent subsequence.

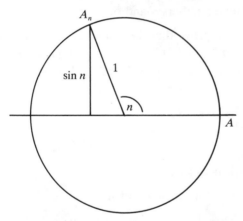

Figure 4.1

   (i) Assuming that π is irrational, prove that there is no repetition in the sequence, so that it contains an infinity of distinct terms. [π will also be formally defined in chapter 11: there is a proof of its irrationality at the end of Burkill ch. 7.]
  (ii) Use a calculator to find what might be the first four terms of an increasing subsequence.
 (iii) By appealing to qn 3.12, say why the sequence $(\sin n)$ must contain a monotonic subsequence.
  (iv) Use qns 33 and 34 to show that $(\sin n)$ has a convergent subsequence. (In fact, $(\sin n)$ has subsequences which converge to *every* point in the interval $[-1, 1]$ and you may wish to investigate this.)

46 Prove that **every bounded sequence has a convergent subsequence**, using qn 3.12 and qns 33 and 34. [Without paying proper regard to history, this theorem is sometimes called the Bolzano–Weierstrass Theorem, the name being transferred from qn 53, to which it is closely related. The theorem will be used repeatedly in our study of real functions.]

47 By considering the infinite decimal for $\sqrt{2}$, show that if only rational numbers are considered, a bounded sequence need not have a convergent subsequence.

Cluster points

Because of the matching of real numbers with the points of a number line, it is sometimes helpful to refer to real numbers as 'points'. For any given positive ε, there is an infinity of $n \in \mathbb{N}$ such that $-\varepsilon < 1/n < \varepsilon$, and we know this because $(1/n)$ is a null sequence. This can also be expressed by saying that in any neighbourhood of 0 (see qn 3.50) there is an infinity of points of the set $\{1/n \mid n \in \mathbb{N}\}$. This makes 0 a *cluster point* or an *accumulation point* for the set.

The point $\frac{1}{4}$ is *not* a cluster point for the set because we can find a positive value of ε such that $\frac{1}{4} - \varepsilon < 1/n < \frac{1}{4} + \varepsilon$ is only possible for the single counting number $n = 4$.

48 State a value of ε such that $\frac{1}{4} - \varepsilon < 1/n < \frac{1}{4} + \varepsilon$ only holds for a unique counting number n.

49 State a positive value of ε such that

$$2/7 - \varepsilon < 1/n < 2/7 + \varepsilon$$

does not hold for any counting number n. This means that $2/7$ is not a cluster point for the set $\{1/n \mid n \in \mathbb{N}\}$.

50 Name three cluster points for the set

$$\left\{ \frac{1}{2^m} + \frac{1}{3^n} \,\middle|\, m, n \in \mathbb{N} \right\}.$$

Name two points between 0 and 1 which are not cluster points for the set.

51 Can a finite set of points on the number line have a cluster point?

52 Describe an infinite set of points that has no cluster point.

A set A of real numbers has an *upper bound* U if $a \leqslant U$ for all $a \in A$, and has a *lower bound* L if $L \leqslant a$ for all $a \in A$. A set A is said to be *bounded* if it has an upper bound and a lower bound. This language is like that for the upper bound and lower bound of a sequence introduced in qn 3.5.

53 Let A be an infinite set of real numbers with an upper bound U and a lower bound L.
  (i) Construct a sequence (a_n) of distinct points in A.
  (ii) Is the sequence (a_n) bounded?
  (iii) Why must it contain a convergent subsequence?
  (iv) If the convergent subsequence $(a_{n_i}) \rightarrow a$, must $a \in A$?
     Must $L \leqslant a \leqslant U$?
  (v) Whether $a \in A$ or not, show that a is a cluster point of A.

[This theorem is called the **Bolzano–Weierstrass Theorem** following Cantor (1870) and Heine (1872). It was formulated by Weierstrass about 1867. He proved it using a method due to Bolzano (1817) of repeated bisections of the bounded interval containing the set A, choosing always a half containing an infinity of points of A. Cantor said that this theorem was the basis of all the more important mathematical truths.]

So **if an infinite set of real numbers is bounded, then it has a cluster point**. To understand the significance of this it is necessary to recognise that this result would be false if the rational numbers were the only ones available.

54 Let A be the set of terminating decimals which make up the infinite decimal for $\sqrt{2}$. State integer bounds for this set. Prove that no rational number can be a cluster point for A.

Cauchy sequences

The great virtue of the theorem that a bounded monotonic sequence is convergent is that it provides a method of determining the fact of convergence without knowing the value of the limit.

Is there a condition guaranteeing the convergence of a non-monotonic sequence which does not require knowledge of the value of the limit?

A reasonable conjecture would be to suggest that a sequence is convergent if the differences between consecutive terms form a null sequence. But there are counter-examples to such a conjecture. One is provided by the harmonic series, which we will examine in the next chapter. Another derives from the sequence $(\surd n)$.

We have shown that the differences between consecutive terms $(\surd(n+1) - \surd n)$ gives a null sequence in qn 3.29(v), and that the sequence itself tends to infinity in qn 3.15.

Examining differences of the form $(a_{n+k} - a_n)$ for some constant k is no more effective for a large value of k than for $k = 1$ since the sequence $(\surd(n/k))$ also tends to infinity. However it is worth examining the hypothesis that the difference $(a_{n+k} - a_n)$ is sandwiched between two null sequences for *all* vaues of k. If, for example,

$$-1/n < a_{n+k} - a_n < 1/n, \quad \text{for all positive integers } k,$$

is this sufficient for the sequence (a_n) to be convergent?

55 If $(a_n) \to a$, use the fact that

$$|a_{n+k} - a_n| \leqslant |a_{n+k} - a| + |a - a_n| \text{ (from qn 2.65)}$$

to show that $|a_{n+k} - a_n|$ may be made arbitrarily small, irrespective of the value of k, provided only that n is large enough, or, to use more precise language, that, given $\varepsilon > 0$, there exists an n_0 such that $|a_{n+k} - a_n| < \varepsilon$, when $n > n_0$. This shows that the condition we are investigating is a necessary consequence of convergence, and will always hold for a convergent sequence. The argument shows that this claim holds whether the ordered field in which we are working is complete or not.

The condition we are investigating is called the *Cauchy criterion*. A sequence (a_n) is said to satisfy the Cauchy criterion, when,

given $\varepsilon > 0$, there exists an n_0 such that $|a_{n+k} - a_n| < \varepsilon$ when $n > n_0$, for all positive integers k.

Such a sequence is called a **Cauchy sequence**.

In qn 55 we proved that every convergent sequence is a Cauchy sequence. We now explore the converse.

56 By considering the infinite decimal for $\surd 2$, show that a sequence of rational numbers which satisfies the Cauchy criterion need not have a rational limit. This suggests that if we can construct a proof that a Cauchy sequence is convergent it is likely to depend upon the Axiom of Completeness.

57 Let (a_n) be a Cauchy sequence of real numbers.
By putting $\varepsilon = 1$ in the Cauchy criterion prove that every Cauchy sequence is bounded.
Using qn 46, we let $(a_{n_i}) \to a$ be a convergent subsequence.
Now use the inequality $|a_n - a| \leqslant |a_n - a_{n_i}| + |a_{n_i} - a|$ to prove that $(a_n) \to a$. This establishes the *General Principle of Convergence* (so-called by *Du Bois Reymond*, 1882), that every Cauchy sequence of real numbers converges to a real limit.

58 Locate the continued fractions

$$1,\ 1 + \tfrac{1}{2},\ 1 + \cfrac{1}{2 + \tfrac{1}{2}},\ 1 + \cfrac{1}{2 + \cfrac{1}{2 + \tfrac{1}{2}}},\ 1 + \cfrac{1}{2 + \cfrac{1}{2 + \cfrac{1}{2 + \tfrac{1}{2}}}}.$$

on a number line.
If, for all n, the terms of a sequence (a_n) satisfy the inequalities $a_{2n-1} \leqslant a_{2n+1} \leqslant a_{2n+2} \leqslant a_{2n}$, and $|a_{n+1} - a_n| < 1/n$, prove that the sequence is convergent.

Least upper bounds and greatest lower bounds

Remind yourself of the definitions which follow qn 52 of an upper bound and a lower bound of a set of real numbers.

59 Check that 0 is a lower bound and that 2 is an upper bound of each of the sets

(a) $\{x \mid 0 \leqslant x \leqslant 1\}$, (b) $\{x \mid 0 < x < 1\}$,

(c) $\{1 + 1/n \mid n \in \mathbb{N}\}$, (d) $\{2 - 1/n \mid n \in \mathbb{N}\}$,

(e) $\{1 + (-1)^n/n \mid n \in \mathbb{N}\}$. (f) $\{q \mid q^2 < 2,\ q \in \mathbb{Q}^+\}$.

For which of these sets can you find a lower bound greater than 0 and/or an upper bound less than 2?
Identify the greatest lower bound (g.l.b.) and the least upper bound (l.u.b.) for each set.
Can a least upper bound or a greatest lower bound for a set A belong to the set?
Must a least upper bound or a greatest lower bound for a set A belong to the set?

A lower bound of a set of real numbers, A, is called the *greatest lower bound*, or *infimum* of A, and denoted by **inf** A, when any number greater than $\inf A$ is not a lower bound of A.

60 (i) If a given set A has an infimum, prove that it has only one infimum.

(ii) Given $\varepsilon > 0$, prove that there is an $a \in A$ such that $\inf A \leqslant a < \inf A + \varepsilon$.

(iii) If L is a lower bound of a set A, and, for any $\varepsilon > 0$, there is an $a \in A$ such that $L \leqslant a < L + \varepsilon$, show that $L = \inf A$.

61 Does the set $\{q \mid q^2 < 2, q \in \mathbb{Q}\}$ have a rational infimum?
The result here indicates that the existence of a supremum or of an infimum of a bounded set of real numbers may depend upon the Axiom of Completeness.

62 Let A be a non-empty set of real numbers with a lower bound L.

(i) Name an integer which is a lower bound for A.

(ii) If $a \in A$, name an integer which is not a lower bound for A.

(iii) Must there be consecutive integers c and $c + 1$ such that c is a lower bound for A and $c + 1$ is not? Let $t_0 = c$.

(iv) Construct a sequence of terminating decimals (t_n) as in qn 33 such that t_n is a lower bound for A, and $t_n + 1/10^n$ is not.

(v) Why is (t_n) convergent? Let its limit be D.

(vi) For a given $a \in A$, prove that $(a - t_n) \rightarrow a - D \geqslant 0$.

(vii) Since t_n is a lower bound for A and $t_n + 1/10^n$ is not deduce that for some $a \in A$,

$$D \leqslant a < t_n + 1/10^n \leqslant D + 1/10^n.$$

(viii) Deduce that $D = \inf A$.

We have proved that a **non-empty set of real numbers which is bounded below has a greatest lower bound.**

63 Fomulate theorems analogous to qns 60 and 62 about the existence of a least upper bound, or *supremum*, of a set of real numbers.
If a set A of real numbers has a supremum, it is denoted by **sup A**.

64 If A and B denote bounded sets of real numbers, how do the numbers sup A, inf A, sup B, inf B relate if $B \subseteq A$?
Give examples of unequal sets for which sup A = sup B and inf A = inf B.

lim sup and lim inf

The ideas in qns 65, 66 and 67 will only be used hereafter in qns 5.97 and 5.102, and qn 5.102 will be used at the end of chapter 12. They have been bracketed, not because they are incidental in the development of analysis, but because they may most profitably be studied at a second reading.

(65) For any bounded sequence (a_n) with an upper bound U and a lower bound L,
  (i) how do you know that (a_n) contains a convergent subsequence;
  (ii) if a is the limit of a convergent subsequence, how do you know that $L \leqslant a \leqslant U$;
  (iii) how do you know that the set of limits of convergent subsequences has a supremum and an infimum?
  The supremum is called **lim sup a_n**, and the infimum is called **lim inf a_n**.

Question 65 shows the power of the theorems we have developed by establishing the existence of numbers about which we know very little. A concrete and constructive approach to the same concepts is given in the following question.

(66) (i) Illustrate on a graph the first few terms of the sequence with nth term $a_n = (-1)^n(1 + 1/n)$.
  (ii) Is the sequence bounded?
  (iii) Is the sequence convergent?
  (iv) Identify a convergent subsequence.
  (v) Find $\sup \{a_n | \ n \in \mathbb{N}\}$ and $\inf \{a_n | \ n \in \mathbb{N}\}$.
  (vi) Let $u_k = \sup \{a_n | \ k \leqslant n\}$ and let $l_k = \inf \{a_n | \ k \leqslant n\}$. Find the first four terms of the sequence (u_n) and the first four terms of the sequence (l_n).
  (vii) Explain why $u_{k+1} \leqslant u_k$ for all k, and why $l_k \leqslant l_{k+1}$ for all k.
  (viii) Are both (u_n) and (l_n) bounded monotonic sequences?
  (ix) The limits of these sequences are lim sup a_n and lim inf a_n respectively. Find these limits.

The method used in qn 66 applies quite generally and provides a more illuminating way of describing lim sup a_n and lim inf a_n. This description enables us to prove, for example, that a sequence (a_n) is convergent if and only if lim sup a_n = lim inf a_n.

(67) To recognise how the existence of lim sup or lim inf may fail in

the absence of completeness, consider the set of rational numbers with squares less than 2.

   (i) Is this set bounded?

  (ii) Is this set countably infinite?

 (iii) May a sequence be constructed which contains each of the elements of this set among its terms?

 (iv) Is this sequence bounded?

  (v) To what rational numbers may subsequences of this sequence converge?

 (vi) Is there a greatest or a least among these rational limits?

Summary

Definition qn 7	In an Archimedean ordered field, a subset, A, is said to be *dense* when there is an infinity of elements of A belonging to any interval $\{x \mid a < x < b\}$.
Theorem qn 7	The rational numbers are dense.
Theorem qn 13, 14	Every rational number is the limit of a recurring decimal or is equal to a terminating decimal. Every recurring decimal has a rational limit.
Theorem qn 15	If two distinct infinite decimals have the same limit, then one terminates and the other eventually has recurring 9s.
Theorem qn 17	There is no rational number with square equal to 2.
Theorem qn 20	If an Archimedean ordered field contains irrational numbers, they are dense.
Definition	A set A is said to be *countably infinite* when its elements may be matched in one-to-one correspondence with $\mathbb{N}$.
Theorem qn 23, 25	The set, $\mathbb{Q}$, of rational numbers is countably infinite.
Theorem	The set of infinite decimals between 0 and 1 is not countably infinite. Thus $\mathbb{R}$ is not countably infinite.
Theorem qn 29	A countably infinite subset of points of $\mathbb{R}$ does not occupy length on the line.
Axiom of Completeness	Every infinite decimal is convergent. A complete Archimedean ordered field is called the set of *real numbers*, and is denoted by $\mathbb{R}$.
Theorem qn 33, 34	Every bounded monotonic sequence of real numbers is convergent.

Theorem qn 35	The sequence with nth term $(1 + 1/n)^n$ has a limit between 2 and 3.
Theorem qn 39	Every positive real number has a real nth root.
Theorem qn 41	$\left(\sqrt[n]{n}\right) \to 1$, as $n \to \infty$.

Chinese Box Theorem

qn 42	A set of nested closed intervals has non-empty intersection.
Theorem qn 26	Every bounded sequence of real numbers has a convergent subsequence.
Definition	When every neighbourhood of a point (i.e. of a number), c, contains an infinity of points of a set A, then c is called a *cluster point* of A.
Definition	If there is a number U such that $a \leqslant U$ for every element $a \in A$, then U is called an *upper bound* of A, and A is said to be *bounded above*. Likewise, if there is a number L such that $L \leqslant a$ for every element $a \in A$, then L is called a *lower bound* of A and A is said to be *bounded below*. A set is *bounded* when it is bounded above and below.

Bolzano–Weierstrass Theorem

qn 53	Every infinite bounded set of real numbers has a cluster point.
Definition	A sequence (a_n) is called a *Cauchy sequence*, if, given $\varepsilon > 0$, there exists an n_0 such that $|a_{n+k} - a_n| < \varepsilon$, when $n > n_0$, for all values of k.

General Principle of Convergence

qns 55, 57	A sequence of real numbers is convergent if and only if it is a Cauchy sequence.
Theorem qns 62, 63	A non-empty set of real numbers which is bounded above has a least upper bound. The least upper bound of a bounded set A is denoted $\sup A$. A non-empty set of real numbers which is bounded below has a greatest lower bound. The greatest lower bound of a bounded set A is denoted by $\inf A$.
Definition	If (a_n) is a bounded sequence of real numbers and $$u_k = \sup\{a_n \mid k \leqslant n\} \quad \text{and} \quad l_k = \inf\{a_n \mid k \leqslant n\},$$ we say that $\lim u_n = \limsup a_n$ and $\lim l_n = \liminf a_n$.
Theorem qns 65, 66	If (a_n) is a bounded sequence of real numbers, (a_n) has a subsequence tending to $\limsup a_n$ and a subsequence tending to $\liminf a_n$, and if a is the limit

of any convergent subsequence of (a_n) then
$\liminf a_n \le a \le \limsup a_n$.

Historical note

About 500 B.C. the Pythagoreans discovered the irrationality of
$\sqrt{2}$. A multiplicity of irrationals are discussed in Euclid, Book X.
The treatment of ratios given in Euclid, Book V (c. 300 B.C.) includes
a careful description (in definition 5) of how to compare two
incommensurable lengths. Until the nineteenth century, European
mathematicians took their notions of number from a geometric view
of measurements.

Eighteenth-century writers implicitly assumed that bounded
monotonic sequences were convergent in their discussion of series and
also assumed that lines which crossed on a graph necessarily had a
point of intersection.

The description of a least upper bound which is not attained
appears in the thesis of C. F. Gauss (1799), and the notion of lim sup
and lim inf, geometrically defined, in an unpublished notebook of his
about 1800.

In 1817 B. Bolzano proved that the convergence of Cauchy
sequences implies the least upper bound property, though his proof
that Cauchy sequences converge was defective, having no Axiom of
Completeness.

In 1821, A. L. Cauchy proved that the partial sums of a
convergent series (see chapter 5) satisfy the Cauchy criterion and
claimed that partial sums satisfying the Cauchy criterion were those of
a convergent series, but without proof. Cauchy also presumed that
bounded monotonic sequences were convergent, both in his work on
series and in his work on continuity. He used lim sup (*la plus grande
des limites*) repeatedly, in tests for convergence of series, though
without formal definition. Cauchy stated that irrational numbers were
limits of sequences of rational numbers, but did not argue from this
proposition.

Until the 1860s there was no public discussion to clarify the status
of irrational numbers. In his lectures in Berlin in 1865, K. Weierstrass
gave an elaborate construction of irrational numbers as bounded
infinite sums of rational numbers, and the Axiom of Completeness we
have adopted is a restricted and simplified form of his ideas.
Weierstrass insisted on the fundamental importance of the theorem
that an infinite bounded set has a cluster point, which he proved in
his lectures from about 1867.

In attempting to simplify the treatment of Weierstrass, both E. Heine and G. Cantor defined real numbers as the limits of rational Cauchy sequences which they called *fundamental sequences* (1872). After reading Heine's and Cantor's treatments, Dedekind decided to publish work that he had done in 1858, when he acknowledged the unprovability of the convergence of bounded monotonic sequences from a geometric standpoint. Dedekind described points on the number line (whether rational or irrational) as producing a cut in the rational numbers, separating those above the cut from those below. This description is similar to that in Euclid, Book V, but Dedekind's new definition of numbers as cuts in the rationals led to formal proofs of their algebraic properties and of completeness, in that every cut of this new system is a cut of the rationals.

Independently of the German tradition, C. Méray had proved that the convergence of bounded monotonic sequences implied the convergence of Cauchy sequences by considering their eventual upper bounds and eventual lower bounds respectively. In the same paper (1869) Méray also defined irrational numbers as fictitious limits of rational Cauchy sequences.

All of this work presumed that the rational numbers were well founded and that it was only irrationals that were not precisely defined. But any attempt to define irrationals by means of monotonic sequences or Cauchy sequences of rationals is plagued by the fact that many rational sequences converge to the same irrational number and that a seemingly convergent sequence without a known limit is not itself a number. It was not until 1907 that the British mathematician Hobson recognised the need to redefine all real numbers (rationals as well as irrationals) as equivalence classes of Cauchy sequences of rationals, with two Cauchy sequences being regarded as equivalent when their term-wise difference is null. Dedekind's cuts did not suffer from the ambiguity of Cauchy sequences but they also required a new definition of real numbers based on rational numbers. Today's axiomatic approach avoids any notion that some numbers are more 'real' than others. But the fact that the real numbers can be constructed from the rational numbers establishes that there is nothing inherently contradictory in introducing an Axiom of Completeness.

Almost at the same time that Weierstrass was successfully banishing infinitesimals and the infinite from definitions of limits (these two notions had been the main tools for discussing limits in the seventeenth and eighteenth centuries), Cantor and Dedekind were in correspondence about infinite sets and the possibility of transfinite cardinals. Cantor's first paper on the subject was published in 1874 and contained a proof that the set of algebraic numbers (solutions of polynomial equations with integer coefficients) was countably infinite while the set of real numbers was not. Modestly, he claimed that this provided a new proof that transcendental numbers were dense, which had been shown by Liouville in 1851. In 1879 Cantor introduced the notion of a dense set of numbers. It was in 1885 that A. Harnack showed that a countable infinity of points did not occupy length on the line.

4 Completeness: what the rational numbers lack   88

Answers

1 $3\,089\,988 = 2^2 \cdot 3^5 \cdot 11 \cdot 17^2$.

2 $2775/999\,999 = 3^{-2} \cdot 5^2 \cdot 7^{-1} \cdot 11^{-1} \cdot 13^{-1}$.

3 Every positive rational number may be expressed in a unique way as a product of prime numbers raised to integer powers.

4 Points evenly spaced at a distance of $1/2^n$.

5 Must take $2^n > 65 \times 73 = 4745$. $7183/2^{13}$.

6 $(1/2^n)$ is a null sequence. Take $\varepsilon = b - a$.
$a < ([a \cdot 2^k] + 1)/2^k$ for all k. $[a \cdot 2^k]/2^k \leqslant a$ and $a + 1/2^k < b$ implies $([a \cdot 2^k] + 1)/2^k < b$.

7 From qn 6, there is at least one element in $(a, \frac{1}{2}(a + b))$ and one element in $(\frac{1}{2}(a + b), b)$ so at least two in (a, b). When each subinterval is bisected the number of elements doubles, and so will eventually exceed any predetermined bound.

8 Were there to be a smallest q in the set, $\frac{1}{2}q$ would be smaller still, but still in the set. Contradiction.

9 $1/5 = 2/10$. $1/5 = m/2^n \Rightarrow 2^n = 5 \cdot m$. $m/2^n = m \cdot 5^n/10^n$. The elements of D are terminating decimals. See qn 3.43. Closed under (i), (ii), (iii). Not (iv), e.g. $1/3$. Same for D.

10 From qn 9(i) and (iii) every term of the sequence $\in T$.
$a_n = (1/5)(1 - (-\frac{1}{4})^n) \to 1/5$ as $n \to \infty$, using qn 3.44.

11 $1/3 = m/2^n \Rightarrow 2^n = 3 \cdot m$. $1/3 = m/10^n \Rightarrow 2^n \cdot 5^n = 3 \cdot m$.
$a_n = \frac{1}{3}(1 - 1/10^n) \to \frac{1}{3}$ as $n \to \infty$.
$0.3, 0.33, 0.333, 0.3333, \ldots$ or $0.\dot{3}$.

12 $0.9, 0.99, 0.999, 0.9999, \ldots$. The nth term of the sequence is $1 - 1/10^n$.
$(1 - 1/10^n) \to 1$ as $n \to \infty$.

13 $a = 123\,456/1000$, $b = 789/1000$. An infinite decimal with a recurring block of k digits may be written in the form

$$a + b(1/10^k + 1/10^{2k} + \ldots + 1/10^{nk} + \ldots)$$

where a and b are rational numbers. From qns 1.3(vi), 3.32 and 3.44 this has a rational limit $= a + b/(10^k - 1)$.

14 If A/B is not a terminating decimal, long division of A by B always gives a non-zero remainder less than B. There are only $B - 1$ distinct remainders, so they will repeat and therefore recur, giving a recurring decimal. A more detailed proof is given in Hardy and Wright, ch. 9.

15 Each sequence of terminating decimals must be monotonic increasing. If the two sequences of terminating decimals are (a_n) and (b_n) with the suffix denoting the number of digits after the decimal place, and they first differ when $n = N$, with $a_N < b_N$, then for all $n \geqslant N$, $a_n < b_N \leqslant b_n$. Now, from qns 3.64 and 3.66, both sequences converge to b_N. Thus (a_n) has recurring 9s and (b_n) is constant for $n \geqslant N$.

16 Since this decimal is not terminating and does not have recurring 9s it cannot have the same limit as any other decimal. Any rational number is the limit of a terminating or recurring infinite decimal, so this decimal cannot have a rational limit.

17 Consider the exponents of 2 on either side of the equation. One is odd, one is even. $\sqrt{3}, \sqrt{5}, \sqrt{6}, \sqrt[3]{2}$ are not rational. If n is a natural number, $\sqrt{n}$ is not rational unless n is a square.

18 $a + b\sqrt{2} = r \Rightarrow \sqrt{2} = (r - a)/b$. So r rational $\Rightarrow \sqrt{2}$ rational.

19 $1/(a + b\sqrt{2}) = (a - b\sqrt{2})/(a^2 - 2b^2)$

$$= \left(a/(a^2 - 2b^2)\right) + \left(-b/(a^2 - 2b^2)\right)\sqrt{2}$$

20 $(a + b\sqrt{2})/(1 + \sqrt{2}) = r \Rightarrow \sqrt{2} = (r - a)/(b - r)$.

So r rational $\Rightarrow \sqrt{2}$ rational.

21 If $\log_{10} 2 = a/b$ with a and b integers, $2 = 10^{a/b}$, so $2^b = 10^a = 2^a \cdot 5^a$.

22 $0, +1, -1, +2, -2, +3, -3, \ldots$. With $a \mapsto 2^a$, map $0 \mapsto 3$.

23 Map $0 \mapsto 2$.

24 If $x < 0$, multiply image by 11. If $y < 0$, multiply image by 13. Map $0 \mapsto 10$.

25 If $A \leftrightarrow M \subseteq \mathbb{N}$, then, for each $a \in A$, let $a \leftrightarrow m(a) \in M$, let $m(a_1) = \min\{m(a)|\, a \in A\}$, and let $m(a_{n+1}) = \min\{m(a)|\, m(a) > m(a_n), a \in A\}$. These elements exist by the Well-ordering Principle. Now match $a_i \leftrightarrow i$, and we have $A \leftrightarrow \mathbb{N}$.

26 If A is countably infinite, denote the element of A which is matched with n by a_n, then the elements of A form the sequence (a_n). Conversely, if A consists of the terms of the sequence (a_n), $a_n \leftrightarrow n$ gives the required matching.

27 All of $\mathbb{N}$ is deleted. Intervals removed have length $1, \frac{1}{2}, \frac{1}{4}, \ldots$, overall length $= 2$.

28 Overall length deleted $\leqslant 2\varepsilon$ ($= 2\varepsilon$, but for the possibility of overlap) which may be arbitrarily small. So $\mathbb{N}$ does not occupy any length on the line.

29 Every term of the sequence (a_n) has been deleted along with a length $\leqslant 2\varepsilon$ ($= 2\varepsilon$, but for the possibility of overlap). Again, ε may be arbitrarily small, so there is no positive minimum to the length deleted.

The infinite decimals with rational limits are terminating or recurring. The newly constructed infinite decimal will therefore be neither terminating nor recurring. Being outside the original set, no contradiction will arise.

30 1 is an upper bound, so is 0.2. The sequence is a monotonic increasing sequence of rational numbers. The sequence need not be strictly increasing. This argument applies to all positive infinite decimals. For negative numbers, take just the integer part to be negative. E.g. $\sqrt{2} - 2 = (-1) + 0.4142.\ldots$.

31 Work with the sequences of terminating decimals and use qn 3.44(ii) for addition, qn 3.44(iii) for subtraction, qn 3.44(iv) for multiplication and qn 3.56 for division.

32 An increasing sequence is bounded below by its first term. A decreasing sequence is bounded above by its first term.

33   (i) If L is a lower bound, then $[L]$ is an integer lower bound.
    (ii) $[a_1] + 1$ is an upper bound.
    (iii) An integer less than a lower bound is a lower bound. An integer greater than a non lower bound is a non lower bound, so the property partitions the integers into {lower bounds} and {non lower bounds}. c is the greatest in the first set and $c + 1$ is the least in the second.
    (iv) Since t_0 is a lower bound and $t_0 + 1$ is not a lower bound we can argue as in (iii).
    (v) Argue as in (iii).
    (vi) By the Axiom of Completeness.
    (vii) t_n is a lower bound for (a_n), so $a_k - t_n \geqslant 0$. $(a_k - t_n) \to a_k - D$ by qn 3.44(iii) and $a_k - D \geqslant 0$ by qn 3.63.
    (viii) From (vii) $D \leqslant a_n$ for all n. Now t_i is a lower bound for (a_n) and $t_i + 1/10^i$ is not a lower bound, so there are terms of the sequence in the interval between these numbers.
$t_i \leqslant D \leqslant a_n < t_i + 1/10^i \leqslant D + 1/10^i$, for some value of n, and because (a_n) is decreasing, for all subsequent n.
    (ix) $D \leqslant a_n < D + 1/10^i$ for all sufficiently large $n \Rightarrow$
$0 \leqslant a_n - D < 1/10^i$. Given $\varepsilon > 0$, $1/10^i < \varepsilon$ for some i. So $(a_n) \to D$ by definition.

34 If (a_n) is monotonic increasing and bounded above, then $(-a_n)$ is monotonic decreasing and bounded below. From qn 33, $(-a_n) \to A$, say, so $(a_n) \to -A$.

35 Sequence strictly increasing and bounded. Use qn 3.66. In qn 11.32 we will see that the limit is e.

36 $a \leqslant x^2 \Rightarrow a/x \leqslant x \Rightarrow a/x \leqslant \frac{1}{2}(x + a/x) \leqslant x \Rightarrow (\frac{1}{2}(x + a/x))^2 \leqslant x^2$ and $0 \leqslant (x - a/x)^2 \Rightarrow a \leqslant (\frac{1}{2}(x + a/x))^2$. (x_n) is monotonic decreasing and bounded. Let $(x_n) \to l$. Apply qns 3.44 and 3.56 to the defining equation to obtain $l = \frac{1}{2}(l + a/l)$, so $l^2 = a$ and, by qn 3.63, $l = \sqrt{a}$.

37 (a_n) is increasing, (b_n) is decreasing. Both are bounded, so both are convergent by qns 33 and 34. Suppose $(a_n) \to a$ and $(b_n) \to b$ then, from $b_{n+1} = \frac{1}{2}(a_n + b_n)$, we have $b = \frac{1}{2}(a + b)$, so $a = b$.

38 If $\sqrt{a}$ exists for all positive real a, then $\sqrt{\sqrt{a}}$, etc. exist.

39 $x^{k-1} \leqslant a \Rightarrow x^k \leqslant ax \Rightarrow x \leqslant \sqrt[k]{(ax)}$. Also $x^{k-1} \leqslant a \Rightarrow (ax)^{k-1} \leqslant a^k$. Use qn 2.35 in each case. The sequence is increasing and bounded above, and so convergent by qn 33. The limit is given by qn 3.44(iv).
To show that real nth roots exist for positive real numbers, identify a power of 2, say, $2^k = m$, which is greater than n. Then mth roots exist by qn 38, and so $(m-1)$th roots, etc. by the argument here. $n = m - (m - n)$.

40 (i) Sequence is decreasing and bounded below by 0, and so convergent, by qn 34.
(ii) $n(\sqrt[n]{(1/a)} - 1)) = n(1 - \sqrt[n]{a})/\sqrt[n]{a}$. Now use qns 3.47 and 3.56. Deduce that (i) holds even when $0 < a < 1$.
(iii) $\sqrt[n]{(ab)} - 1 = (\sqrt[n]{a} - 1) \cdot \sqrt[n]{b} + (\sqrt[n]{b} - 1)$.

41 Sequence is strictly decreasing for $n \geqslant 3$, and bounded below by 1, and so convergent to $l \geqslant 1$ by qn 3.66. Every subsequence of a convergent sequence has the same limit. $(\sqrt[n]{2} \cdot \sqrt[n]{n}) \to 1 \cdot l$.

42 (i) $a_1 \leqslant a_2 \leqslant \ldots \leqslant a_n < b_n \leqslant \ldots \leqslant b_2 \leqslant b_1$.
(ii) (a_n) is monotonic increasing and bounded above by b_1.
(b_n) is monotonic decreasing and bounded below by a_1.
(iii) If $(a_n) \to a$ and $(b_n) \to b$, then $a \leqslant b$ by qn 3.64. For fixed k and $n > k$, $a_k \leqslant a_n$, so $0 \leqslant a_n - a_k$. Therefore $(a_n - a_k) \to a - a_k \geqslant 0$, by qn 3.64, so $a_k \leqslant a$, for all k. Likewise $b \leqslant b_k$. So, any c such that $a \leqslant c \leqslant b$, lies in all the intervals.
(iv) Suppose $c > 0$ was in all the intervals. Consider $n > 1/c$.

43 (i) $[-1, +1]$, even terms constant and so convergent.

(ii) $[-2, +2]$, even terms $(1 + 1/2n) \to 1$.

44 Convergent $\Rightarrow$ bounded, but not conversely, from qn 43.

45 (i) $\sin n = \sin m \Rightarrow n \pm m = k\pi$. Since k, m and n are integers and π is irrational, this is impossible unless $k = 0$.

(ii) $\sin 1, \sin 2, \sin 8, \sin 14$.

(iii) From qn 3.12, every sequence has a monotonic subsequence.

(iv) $-1 \leqslant \sin n \leqslant 1$, so the sequence $(\sin n)$ and hence every subsequence is bounded. A sequence which is bounded and monotonic is convergent by qns 33 and 34.

46 Any sequence contains a monotonic subsequence, from qn 3.12, so any bounded sequence contains a bounded monotonic subsequence, and this is convergent, from qns 33 and 34.

47 The sequence of terminating decimals which gives the infinite decimal for $\sqrt{2}$ is convergent to $\sqrt{2}$, so every subsequence is convergent to $\sqrt{2}$. Thus when the field of discussion is $\mathbb{Q}$, the sequence is bounded but contains no convergent subsequences.

48 $1/5 < 1/4 < 1/3$. $\varepsilon \leqslant 1/20$.

49 $1/4 < 2/7 < 1/3$. Take $\varepsilon \leqslant 1/28$.

50 $1/2$, $1/3$ and $1/4$ are cluster points. $3/5$ and $2/5$ are not.

51 No. Can take $\varepsilon <$ shortest distance between two points.

52 $\mathbb{Z}$

53 (i) Let a_1 be any point of A. Let a_2 be a point of A which is different from a_1. Proceed by choosing a new point for each successive term.

(ii) $L \leqslant a_n \leqslant U$, since each term of the sequence belongs to A.

(iii) By qn 46.

(iv) a need not be an element of A, but $L \leqslant a \leqslant U$ by qn 3.66.

(v) Given $\varepsilon > 0$, there is an n_k such that $|a_{n_i} - a| < \varepsilon$ when $n_i > n_k$. So there are an infinity of elements of A in any neighbourhood of a.

54 A is bounded by 1 and 2. If q is any rational number, $1 \leqslant q \leqslant 2$, take $0 < \varepsilon < |q - \sqrt{2}|$, then at most a finite number of terms of A lie in an ε-neighbourhood of q.

55 Choose n_0 such that $|a_n - a| < \frac{1}{2}\varepsilon$ when $n > n_0$.

56 The infinite decimal for $\sqrt{2}$ satisfies the Cauchy criterion by qn 55, since it is convergent in $\mathbb{R}$, but all terminating decimals are rational.

57 If $n \geq n_0$ implies $|a_{n+k} - a_n| < 1$, then the sequence is bounded above by $\max\{a_1, a_2, \ldots, a_{n_0-1}, a_{n_0} + 1\}$ and bounded below by

$\min\{a_1, a_2, \ldots, a_{n_0-1}, a_{n_0} - 1\}$.

If $n_i > N_1 \Rightarrow |a_{n_i+k} - a_{n_i}| < \frac{1}{2}\varepsilon$, and $n_i > N_2 \Rightarrow |a_{n_i} - a| < \frac{1}{2}\varepsilon$, then for any $n_i > \max\{N_1, N_2\}$, $n > n_i \Rightarrow |a_n - a| < \varepsilon$.

58 In such a case $0 \leq |a_{n+k} - a_n| \leq |a_{n+1} - a_n| < 1/n$ for all k, so the sequence satisfies the Cauchy criterion.

59

	(a)	(b)	(c)	(d)	(e)	(f)
l.u.b.	1	1	2	2	$1\frac{1}{2}$	$\sqrt{2}$
g.l.b.	0	0	1	1	0	0

l.u.b. and g.l.b. may belong to set as in (a).
Neither l.u.b. nor g.l.b. need belong to set as in (b).

60 (i) If a set has two infima, unless they are equal, one is less than the other and then the lesser is not a greatest lower bound.
(ii) Since $\inf A < \inf A + \varepsilon$, $\inf A + \varepsilon$ is not a lower bound, so such an a exists.
(iii) $L \leq \inf A$. If $L < \inf A$, take $\varepsilon = \inf A - L$ to obtain a contradiction.

61 If $q > -\sqrt{2}$, then q is not a lower bound. If $q < -\sqrt{2}$, then q is not the *greatest* lower bound.

62 (i) $[L]$. (ii) $[a] + 1$. (iii) As in qn 33.
(v) By the Axiom of Completeness. (vi) $(a - t_n) \to a - D$ by qn 3.44(iii). $a - D \geq 0$ by qn 3.63.
(vii) By (vi), $D \leq a$ for all $a \in A$. (viii) Follows from qn 60(iii).

63 If a set A has a supremum it is unique.
There is an $a \in A$ such that $\sup A - \varepsilon < a \leq \sup A$.
If U is an upper bound of A and for any $\varepsilon > 0$, there is an $a \in A$ such that $U - \varepsilon < a \leq U$, then $U = \sup A$.
Every non-empty set of real numbers which is bounded above has a least upper bound, or supremum.

64 $\inf A \leq \inf B \leq \sup B \leq \sup A$. Qn 59 (a) and (b).

65 (i) By qn 46. (ii) By qn 3.66. (iii) By qns 62 and 63.

66 (ii) $-2 \leq a_n \leq 1\frac{1}{2}$. (iii) Not convergent. (iv) Even terms converge to 1, odd terms converge to -1. (v) $1\frac{1}{2}$ and -2. (vi) $1\frac{1}{2}, 1\frac{1}{2}, 1\frac{1}{4}, 1\frac{1}{4}$: $-2, -1\frac{1}{3}, -1\frac{1}{3}, -1\frac{1}{5}$. (vii) $\{a_n| k + 1 \leq n\} \subseteq \{a_n| k \leq n\}$, see qn 64.
(viii) Monotonic from (vii), bounded by the bounds of the sequence.
(ix) ± 1.

67 (i) Bounded between ±2, so ±2 are rational bounds. (ii) An infinite subset of $\mathbb{Q}$; see qn 23. (iii) See qn 26. (iv) As (i). (v) Any rational number between ±√2, (vi) This set has no rational least upper bound, nor rational greatest lower bound, so a sequence in this set, though bounded, has neither lim sup nor lim inf in $\mathbb{Q}$.

5

Series
Infinite sums

Preliminary reading: Cohen, D., Northrop ch. 7.
Concurrent reading: Ferrar.
Further reading: Rudin ch. 3, Knopp.

Sequences of partial sums

1 Criticise the following argument:

if $S = 1 + x + x^2 + \ldots$,
then $xS = x + x^2 + x^3 + \ldots$,
so $S - xS = 1$,
and therefore $S = \dfrac{1}{1 - x}$.

Try putting $x = 2$!

If the argument in qn 1 were sound we could put $x = -1$ and obtain the sum of the series

$$1 - 1 + 1 - 1 + 1 - \ldots$$

to be $\frac{1}{2}$. But the same series could plausibly be thought to have a sum of 0:

$$(1 - 1) + (1 - 1) + (1 - 1) + \ldots,$$

or a sum of 1:

$$1 + (-1 + 1) + (-1 + 1) + (-1 + 1) + \ldots.$$

These paradoxical conclusions show that great care must be used in arguing with infinite sums.

2 For $x \neq 1$, let $s_n = 1 + x + x^2 + \ldots + x^{n-1}$, a sum of only n
terms.
By considering $x \cdot s_n - s_n$, prove that $s_n = (x^n - 1)/(x - 1)$.
Compare with qn 1.3(vi). It is also conventional to write s_n, as
defined in the first line, in the form

$$\sum_{r=1}^{r=n} x^{r-1} \quad \text{or} \quad \sum_{r=0}^{r=n-1} x^r.$$

3 By decomposing $1/r(r + 1)$ into partial fractions, or by induction,
prove that

$$\frac{1}{1 \cdot 2} + \frac{1}{2 \cdot 3} + \ldots + \frac{1}{n(n + 1)} = 1 - \frac{1}{n + 1}.$$

Express this result using the $\sum$ notation.

4 If $s_n = \displaystyle\sum_{r=1}^{r=n} \frac{1}{r(r + 1)}$ prove that $(s_n) \to 1$ as $n \to \infty$.

This result is also written $\displaystyle\sum_{r=1}^{\infty} \frac{1}{r(r + 1)} = 1$.

When discussing whether the series

$$a_1 + a_2 + a_3 + \ldots + a_n + \ldots$$

has a sum, we construct the sequence (s_n) thus:

$s_1 = a_1$

$s_2 = a_1 + a_2$

$s_3 = a_1 + a_2 + a_3$

$\vdots$

$s_n = a_1 + a_2 + a_3 + \ldots + a_n$

$\vdots$

and so on.
   The sequence (s_n) is called the *sequence of partial sums* of the
series. When the sequence (s_n) is convergent to s we say that *the
series is convergent to s*, or has the sum s.
Symbolically, the series $\sum a_r$ is said to be *convergent* when *the
sequence of partial sums* (s_n), defined by

$$s_n = \sum_{r=1}^{r=n} a_r$$

is convergent. When $(s_n) \to s$ as $n \to \infty$, we write

$$\sum_{r=1}^{\infty} a_r = s,$$

and say that the series has the sum s.

5 Write down a formula for $\sum_{r=1}^{r=n} (\frac{1}{2})^{r-1}$ and find the sum $\sum_{r=1}^{\infty} (\frac{1}{2})^{r-1}$.

6 Find the sum $\sum_{r=1}^{\infty} 1/10^r$.
   What recurring decimal have you evaluated?

7 Construct an infinite series for which the sum is $1/11$.

8 Find the sum $\sum_{r=0}^{\infty} (9/10)^r$.

9 If $s_n = \sum_{r=1}^{r=n} r/2^r$, show that $s_n - \frac{1}{2}s_n = (\sum_{r=1}^{r=n} 1/2^r) - n/2^{n+1}$
   and use qn 3.62 to deduce that $(s_n) \to 2$ as $n \to \infty$. Compare with
   qn 1.3(vii).

10 Use qn 2 to determine whether the series $\sum x^n$ is convergent
   when

   (i) $|x| < 1$,

   (ii) $|x| > 1$.

   Also examine the two cases $x = 1$ and $x = -1$.

   Any series which is *not* convergent is said to be *divergent*. The
partial sums of a divergent series need not 'diverge' to $\pm\infty$ but may
oscillate, as with $\sum (-1)^n$. When all the terms of a series are
positive, the sequence of partial sums is monotonic increasing: so, if
the sequence is bounded, the series is convergent; and, if it is not
bounded, it tends to $+\infty$. See qn 22.

Convergence and null sequence of terms

11 *The null sequence test*
   Suppose that the series $\sum a_n$ is convergent, and its sequence of
   partial sums $(s_n) \to s$ as $n \to \infty$. Use the equation $a_n = s_{n+1} - s_n$
   and qn 3.44(iii) to prove that (a_n) is a null sequence.

   The importance of this result lies in its contrapositive which
enables us to recognise many series as being divergent: *when the
sequence of terms (a_n) is not a null sequence, then the series $\sum a_n$
must be divergent*.

12 By considering when the sequence (x^n) is *not* null, use qn 11 to check your claims of divergence in qn 10.

13 Does there exist a real number $\alpha > 0$ for which the series $\sum n^\alpha$ is convergent?

Simple consequences of convergence

14 Prove that $\sum_{r=1}^{\infty} a_r$ is convergent if and only if $\sum_{r=k}^{\infty} a_r$ is convergent for any fixed positive integer k.
This shows that the symbol $\sum a_n$ is not ambiguous if it is only convergence which is at stake. It *is* an ambiguous symbol if the sum is to be found.

15 If $\sum a_n$ is convergent, show that $\left(\sum_{r=n}^{\infty} a_r \right) \to 0$ as $n \to \infty$.

16 Write down the claim that a sequence of partial sums (s_n) is a Cauchy sequence. Interpret this claim for the infinite series from which the partial sums were formed.

17 For a fixed real number $c \neq 0$, prove that $\sum c \cdot a_n$ is convergent if and only if $\sum a_n$ is convergent.

18 Use qns 10 and 17 to determine exactly which geometric series are convergent and which are divergent.

19 If $\sum a_n$ and $\sum b_n$ are convergent, prove that $\sum (a_n + b_n)$ is convergent.

Questions 17 and 19 show that the set of convergent series have the structure of a real vector space.

Series of positive terms

First comparison test

20 Let $e_n = 1 + 1 + 1/2! + 1/3! + \ldots + 1/n!$,
and let $s_n = 1 + 1 + 1/2 + 1/4 + \ldots + 1/2^{n-1}$, with $n \geqslant 1$.
Prove that $e_n \leqslant s_n = 3 - 1/2^{n-1}$.
Deduce that the sequence (e_n) is monotonic increasing and bounded above, and so convergent. We say $(e_n) \to e$.

21 Prove that e is irrational by supposing that e is rational, letting $e = p/q$ in lowest terms, showing that $e - e_q = k/q!$ for some integer k and also $e - e_q < (1/q!) \cdot \frac{1}{2}$.

22 If $a_n \geq 0$ for all n, prove that the sequence of partial sums of the series $\sum a_n$ is monotonic increasing. Deduce that $\sum a_n$ is convergent if and only if its partial sums are bounded.

23 The result of qn 22 provides a sandwich theorem (the *first comparison test*) for series of positive terms.
If $0 \leq a_n \leq b_n$, and $\sum b_n$ is convergent, show that the partial sums of $\sum a_n$ are bounded and deduce that $\sum a_n$ is convergent. Contrapositively, if $\sum a_n$ is divergent, show that the partial sums of $\sum b_n$ are unbounded so $\sum b_n$ is divergent.

24 Use the fact that

$$0 < \frac{1}{(n + 1)^2} < \frac{1}{n(n + 1)},$$

and qns 4 and 23, to prove that $\sum 1/(n + 1)^2$ is convergent. Deduce from qn 14 that $\sum 1/n^2$ is convergent.

25 Let $a_n = \sqrt{(n + 1)} - \sqrt{n}$, and let $b_n = 1/\sqrt{n}$.
What is the partial sum of the first hundred terms of the series $\sum a_n$?
  (i) Use qn 22 to prove $\sum a_n$ is divergent.
  (ii) Prove that $a_n = 1/(\sqrt{(n + 1)} + \sqrt{n})$.
  (iii) Prove that $0 < a_n < \frac{1}{2}b_n$.
  (iv) Use qns 23 and 17 to prove that $\sum b_n$ is divergent.

So, although (b_n) is a null sequence, $\sum b_n$ is divergent. Rather disconcertingly, this establishes that the converse of qn 11 is false. So do not fall into the trap of thinking that $\sum b_n$ must be convergent when (b_n) is null.

26 Use qns 23, 24 and 25 to show that $\sum 1/n^\alpha$ is convergent when $\alpha \geq 2$ and divergent when $0 \leq \alpha \leq \frac{1}{2}$.

The harmonic series

27 For the series

$$1 + \frac{1}{2} + \frac{1}{3} + \frac{1}{4} + \ldots + \frac{1}{n} + \ldots,$$

let s_n denote the sum of the first n terms. For each n, show that

$$s_{2n} - s_n \geq \frac{n}{2n} = \frac{1}{2}.$$

Write down these inequalities for $n = 1, 2, 4, 8, \ldots, 2^{k-1}$, and add them to prove that

$$\sum_{r=1}^{r=2^n} \frac{1}{r} \geqslant 1 + \tfrac{1}{2}n.$$

Is $\sum 1/n$ convergent or divergent?

Convergence of $\sum 1/n^\alpha$

28 For what values of the positive real number α can you be sure that the series $\sum 1/n^\alpha$ is divergent? Just use the results of qns 23 and 27 at this stage.

29 In qn 26 we proved that the series $\sum 1/n^\alpha$ was convergent when $\alpha \geqslant 2$, and in qn 28 we saw that this series is divergent when $\alpha \leqslant 1$. We have not yet examined the convergence of $\sum 1/n^\alpha$ for values of α between 1 and 2. Let $1 < \alpha < 2$. Let s_n denote the sum of the first n terms of the series

$$1 + \frac{1}{2^\alpha} + \frac{1}{3^\alpha} + \frac{1}{4^\alpha} + \ldots + \frac{1}{n^\alpha} + \ldots.$$

Verify that

$$s_{2n-1} - s_{n-1} < \frac{n}{n^\alpha}.$$

Write down these inequalities for $n = 2, 4, 8, \ldots, 2^k$ and add them to show that

$$\sum_{r=1}^{r=2^{n+1}-1} \frac{1}{r^\alpha} < 1 + \frac{2}{2^\alpha} + \frac{4}{4^\alpha} + \ldots + \frac{2^n}{2^{n\alpha}} = \frac{1 - (1/2^{\alpha-1})^{n+1}}{1 - (1/2^{\alpha-1})},$$

using qn 2 for the equality.
Show that $\alpha > 1$ implies $1/2^{\alpha-1} < 1$.
Deduce that the partial sums of $\sum 1/r^\alpha$ are bounded and so the series is convergent. This proof holds for all real $\alpha > 1$.

Cauchy's nth root test

30 Use the fact that $n/(2n + 1) < \tfrac{1}{2}$ and the sandwich theorem in qn 23 to prove that the series $\sum \left(n/(2n + 1)\right)^n$ is convergent.

31 If $n > 3$, prove that

$$\frac{4n^2}{5n^2 - 2} < \frac{36}{43},$$

and using qns 14 and 23 prove that the series $\sum \left(4n^2/(5n^2 - 2) \right)^n$ is convergent.

32 To generalise the results of qns 30 and 31, we suppose that we have a series of positive terms, $\sum a_n$, for which the sequence $\left(\sqrt[n]{a_n} \right) \to k$ as $n \to \infty$, and that $0 \leqslant k < 1$. By choosing $\varepsilon = \frac{1}{2}(1 - k)$, show that, for sufficiently large n, $a_n < \rho^n$ where $\rho = k + \varepsilon = \frac{1}{2}(1 + k) < 1$. Deduce from questions 14 and 23 that $\sum a_n$ is convergent.

This method of proving that a series $\sum a_n$ converges is called *Cauchy's nth root test*.

33 Apply Cauchy's nth root test to the series $\sum n/2^n$, and to the series $\sum n^2(0.8)^n$.

34 If $\sqrt[n]{a_n} \geqslant 1$ for all n, prove that $\sum a_n$ is divergent.

35 If we have a series of positive terms $\sum a_n$, for which the sequence $\left(\sqrt[n]{a_n} \right) \to k$ as $n \to \infty$, where $1 < k$. Prove that the series is divergent.

36 By considering the two series $\sum 1/n$ and $\sum 1/n^2$, show that Cauchy's nth root test gives no information about the convergence of the series when $k = 1$.

d'Alembert's ratio test

37 Let $a_n = n^2/2^n$. Prove that, **if $n \geqslant 3$**, then $a_{n+1}/a_n \leqslant 8/9$. By using this inequality when $n = 3, 4, 5, \ldots$ prove that $a_{n+3} \leqslant (8/9)^n a_3$. Use qns 18 and 23 to show that $\sum a_{n+3}$ is convergent. Use qn 14 to prove $\sum a_n$ is convergent.

38 Let $a_n = n^3(0.8)^n$. Prove that if $n \geqslant 14$, then $a_{n+1}/a_n < 0.99$. Use arguments like those in qn 37 to prove $a_{n+14} < (0.99)^n a_{14}$. Deduce that $\sum a_n$ is convergent.

39 To generalise the results of qns 37 and 38, we suppose that we have a series $\sum a_n$ of positive terms for which

$$0 < a_{n+1}/a_n < \rho < 1, \quad \text{for all } n.$$

Show that $a_{n+1} < \rho^n \cdot a_1$. Use qns 18 and 23 to prove $\sum a_n$ is convergent.

40 Suppose that $\sum a_n$ is a series of positive terms and the sequence of ratios $\left(a_{n+1}/a_n \right) \to k < 1$. By a suitable choice of $\varepsilon > 0$ (what ε?), show that there exists an n_0 such that

$a_{n+1}/a_n < \frac{1}{2}(k + 1) < 1$,   for $n > n_0$.

Let $\rho = \frac{1}{2}(k + 1)$, and apply qn 39 and qn 14 to prove that $\sum a_n$ is convergent.

41 Prove that $\sum 2^n/n!$ is convergent.

42 Prove that $\sum n!/n^n$ is convergent.

43 Suppose that we have a series $\sum a_n$ of positive terms for which $1 \leqslant a_{n+1}/a_n$, for all n. Prove that $a_1 \leqslant a_n$ for all n. Deduce that (a_n) is not a null sequence and so $\sum a_n$ is divergent.

44 Suppose that we have a series of positive terms $\sum a_n$ for which the sequence of ratios $(a_{n+1}/a_n) \to k > 1$. By a suitable choice of $\varepsilon > 0$ (what ε?), show that there exists an n_0 such that $1 < \frac{1}{2}(1 + k) < a_{n+1}/a_n$, for $n > n_0$.
Apply qns 43 and 14 to show that $\sum a_n$ is divergent.

45 For which positive real numbers x is the series $\sum (2x)^n n^x$ convergent and for which is it divergent? Use both d'Alembert's ratio test and Cauchy's nth root test.

Questions 40 and 44 give *d'Alembert's ratio test*. It is important to note that this test gives no information about the convergence of $\sum a_n$ when $(a_{n+1}/a_n) \to 1$.
This means that d'Alembert's test gives no information about the convergence or divergence of $\sum 1/\sqrt{n}$, $\sum 1/n$ or $\sum 1/n^2$.

46 Give an example to show that the condition $0 < a_{n+1}/a_n < 1$ is not sufficient to ensure the convergence of the series $\sum a_n$.

47 By considering the sequence defined by $a_{2n-1} = 1/(2n)^3$ and $a_{2n} = 1/(2n)^2$, show that a series $\sum a_n$ may be convergent even when the sequence (a_{n+1}/a_n) is unbounded.

Second comparison test

48 If $a_n = \dfrac{\sqrt{n} + 1}{n^2 + 2}$ and $b_n = \dfrac{1}{n^{3/2}}$,

prove that $(a_n/b_n) \to 1$ as $n \to \infty$. Deduce that for some n_0, $1/2 < a_n/b_n < 3/2$ when $n > n_0$. Now use the convergence of $\sum b_n$ to establish the convergence of $\sum (3/2)b_n$ and thus the convergence of $\sum a_n$.

49 If

$$a_n = \frac{\sqrt{n} + 1}{n + 2} \text{ and } b_n = \frac{1}{\sqrt{n}},$$

prove that $(a_n/b_n) \to 1$ as $n \to \infty$. Deduce that for some n_0, $1/2 < a_n/b_n < 3/2$ when $n > n_0$. Now use the divergence of $\sum b_n$ to establish the divergence of $\sum (1/2)b_n$ and deduce the divergence of $\sum a_n$.

50 Let $0 < m < a_n/b_n < M$, for all n.
If $\sum a_n$ is convergent, prove that $\sum m \cdot b_n$ is convergent and deduce that $\sum b_n$ is convergent.
If $\sum b_n$ is convergent, prove that $\sum M \cdot b_n$ is convergent and deduce that $\sum a_n$ is convergent.
Thus $\sum a_n$ and $\sum b_n$ are both convergent or both divergent.

This result is called the *second comparison test*.

51 If $a_n = 1/n^2$ and $b_n = 1/n$, prove that $0 < a_n/b_n < 1$. Why does the convergence of $\sum a_n$ and the divergence of $\sum b_n$ not contradict the second comparison test?

52 Suppose that $\sum a_n$ and $\sum b_n$ are both series of positive terms and that $(a_n/b_n) \to l \neq 0$ as $n \to \infty$. By a suitable choice of $\varepsilon > 0$ (what ε?) show that there exists an n_0 such that $0 < \frac{1}{2}l < a_n/b_n < \frac{3}{2}l$, for all $n > n_0$. By using qn 14 and the second comparison test prove that $\sum a_n$ and $\sum b_n$ are both convergent or both divergent.

Integral test

This section uses the notion of integral which is discussed formally in chapter 10 and the logarithm function which is discussed formally in chapter 11. Sixth-form notions of integrals and logarithms are sufficiently reliable to answer qns 53–58.

53 Sketch the graph of $f(x) = 1/x$ for positive x. Why is

$$\frac{1}{n + 1} < \int_n^{n+1} \frac{dx}{x} < \frac{1}{n}?$$

Prove that

$$\sum_{r=2}^{r=n} \frac{1}{r} < \int_1^n \frac{dx}{x} < \sum_{r=1}^{r=n-1} \frac{1}{r}.$$

If $s_n = \sum\limits_{r=1}^{r=n} \dfrac{1}{r}$, prove that $\dfrac{1}{n} < s_n - \log_e n < 1$.

Let $D_n = s_n - \log_e n$.
Calculate some values of D_n, using a computer if possible. Show that

$$D_n - D_{n+1} = \int_n^{n+1} \frac{dx}{x} - \frac{1}{n+1}.$$

Deduce that (D_n) is a monotonic decreasing sequence of positive terms and so has a limit. The limit in this case is usually denoted by γ, and is known as Euler's constant. Its value is slightly greater than 0.577.

54 If $f(x) \geqslant 0$ and f is a decreasing function for $x \geqslant 1$, what may be said about the sequence $(f(n))$? Is it necessarily convergent? Since f is monotonic, f is integrable, see qns 10.7 and 10.8. Why is $f(n+1) \leqslant \int_n^{n+1} f(x)dx \leqslant f(n)$?
Deduce that $\sum_{r=2}^{r=n} f(r) \leqslant \int_1^n f(x)dx \leqslant \sum_{r=1}^{r=n-1} f(r)$.
Now let $s_n = \sum_{r=1}^{r=n} f(r)$.
Prove that $0 \leqslant f(n) \leqslant s_n - \int_1^n f(x)dx \leqslant f(1)$.
If $\lim_{n\to\infty} \int_1^n f(x)dx$ exists, prove that $\sum f(n)$ is convergent.
If $\sum f(n)$ is convergent, prove that $\lim_{n\to\infty} \int_1^n f(x)dx$ exists.

(55) Prove that $\sum_{n=1}^{\infty} 1/n^2 < 2$. Compare with qn 29.

(56) Prove that $2\sqrt{n} - 2 < \sum_{r=1}^{r=n} 1/\sqrt{r} < 2\sqrt{n} - 1$.

57 Prove that $\sum 1/(n \log n)$ is divergent for $n \geqslant 2$.

58 With the notation of qn 54, let $D_n = s_n - \int_1^n f(x)dx$.
Prove that $D_n - D_{n+1} = \int_n^{n+1} f(x)dx - f(n+1) \geqslant 0$.
Deduce that (D_n) is convergent. Notice that this result holds whether $\sum f(n)$ is convergent or divergent.

Series with positive and negative terms

Alternating series test

59 From our knowledge of $\sum 1/n$ we might expect the series

$$1 - \frac{1}{2} + \frac{1}{3} - \frac{1}{4} + \frac{1}{5} + \ldots + \frac{(-1)^{n+1}}{n} + \ldots$$

to be divergent.

However, if we let $s_n = \sum_{r=1}^{r=n} (-1)^{r+1}/r$ we can find $s_{2n+2} - s_{2n}$ and prove that it is positive.
This establishes that the sequence (s_{2n}) is monotonic increasing.
Now find $s_{2n+3} - s_{2n+1}$ and prove that it is negative.
This establishes that the sequence (s_{2n+1}) is monotonic decreasing.
From the proposition $s_2 < s_{2n} + 1/(2n + 1) = s_{2n+1} < s_1$, deduce that (s_{2n}) and (s_{2n+1}) are both convergent to the same limit and so (s_n) is convergent to this limit. We will find the limit in qn 61.

60 The argument of qn 59 is easily generalised to establish the convergence of any series of the form $\sum (-1)^{n+1} a_n$ when (a_n) is a monotonic decreasing null sequence of positive terms. This is called *the alternating series test*.
In this case, let $s_n = \sum_{r=1}^{r=n} (-1)^{r+1} a_r$.
Show that $s_{2n+2} - s_{2n} > 0$,
and that $s_{2n+3} - s_{2n+1} < 0$.
Deduce that $s_2 \leqslant s_{2n} + a_{2n+1} = s_{2n+1} \leqslant s_1$.
Show that the sequences (s_{2n}) and (s_{2n+1}) are both convergent and to the same limit.
Deduce that $\sum (-1)^{n+1} a_n$ is convergent.

61 We examine the series of qn 59, using the symbols D_n and γ from qn 53.

$$1 - \frac{1}{2} + \frac{1}{3} - \frac{1}{4} + \ldots - \frac{1}{2n}$$

$$= 1 + \frac{1}{2} + \frac{1}{3} + \frac{1}{4} + \ldots + \frac{1}{2n} - 2\left(\frac{1}{2} + \frac{1}{4} + \ldots + \frac{1}{2n}\right)$$

$$= \log 2n + D_{2n} - (\log n + D_n)$$

$$= \log 2 + D_{2n} - D_n.$$

Now use the fact that $(D_n) \rightarrow \gamma$ as $n \rightarrow \infty$, to prove that the sum of the series in qn 59 is $\log_e 2$.

Absolute convergence

62 For the series $\sum (-1)^n/n$ both the series of positive terms and the series of negative terms, taken separately, are divergent. In the case of the series $\sum (-1)^n/n^2$, the series of positive terms and the series of negative terms, taken separately, are convergent. For $\sum (-1)^n/n^2$, the alternating series test will establish convergence, but a more direct proof is available in such a case. We construct two new series thus.

$n =$	1	2	3	4	5	6	...
$a_n =$	-1	$+\frac{1}{4}$	$-\frac{1}{9}$	$+\frac{1}{16}$	$-\frac{1}{25}$	$+\frac{1}{36}$	...
$u_n =$	0	$\frac{1}{4}$	0	$\frac{1}{16}$	0	$\frac{1}{36}$	...
$v_n =$	1	0	$\frac{1}{9}$	0	$\frac{1}{25}$	0	...

In each case, the terms of the new sequences u_n, v_n are either equal to 0 or to $|a_n|$, so clearly $0 \leqslant u_n \leqslant |a_n|$, and $0 \leqslant v_n \leqslant |a_n|$. Now $|a_n| = 1/n^2$ and $\sum 1/n^2$ is convergent from qn 24, so

(i) $\sum u_n$ is convergent (why?),

(ii) $\sum v_n$ is convergent (why?),

(iii) $\sum (u_n - v_n)$ is convergent (why?),

(iv) $\sum a_n$ is convergent.

63 A series $\sum a_n$ for which $\sum |a_n|$ is convergent is said to be *absolutely convergent*. The argument used in qn 62 can be generalised to establish that a series *which is absolutely convergent is necessarily convergent*. The clue, as we have seen, is to construct the series formed by the positive terms and the series formed by the negative terms.
If we define $u_n = \frac{1}{2}(|a_n| + a_n)$, what values may u_n take?
If we define $v_n = \frac{1}{2}(|a_n| - a_n)$, what values may v_n take?
Now use the convergence of $\sum |a_n|$ to prove that $\sum u_n$ and $\sum v_n$ are convergent and hence that $\sum (u_n - v_n)$ is convergent. What is $u_n - v_n$ equal to?

64 Use d'Alembert's ratio test to devise a condition for a series to be absolutely convergent.

65 Does the condition $(|a_{n+1}/a_n|) \to k > 1$ as $n \to \infty$ imply that the series $\sum a_n$ is divergent?

66 Use Cauchy's nth root test to devise a condition for a series to be absolutely convergent.

Conditional convergence

A series $\sum a_n$ which is convergent, but not absolutely convergent is said to be *conditionally convergent*. The series $\sum (-1)^n/n$ is conditionally convergent.

67 With the notation of qn 63, use the equations $a_n = u_n - v_n$ and $|a_n| = u_n + v_n$ to prove that, if $\sum a_n$ is conditionally convergent, then both $\sum u_n$ and $\sum v_n$ are divergent, and the partial sums for both series tend to infinity.

Rearrangements

68 For the series $\sum a_n = 1 - 1 + \frac{1}{2} - \frac{1}{2} + \frac{1}{3} - \frac{1}{3} + \frac{1}{4} - \frac{1}{4} + \dots$ find the partial sum s_{2n} of the first $2n$ terms, and the partial sum s_{2n+1} of the first $2n + 1$ terms. Deduce that the series is convergent and find its sum. Find an expression for a_{2n-1} and for a_{2n}.

69 For the series

$$\sum b_n = 1 + \frac{1}{2} - 1 + \frac{1}{3} + \frac{1}{4} - \frac{1}{2} + \dots$$

$$+ 1/(2n - 1) + 1/2n - 1/n + \dots$$

compare the partial sum of the first $3n$ terms with the partial sum of the first $2n$ terms of the series in qn 59. Deduce with the help of qn 61 that this series has the sum $\log 2$. Find an expression for b_{3n}, b_{3n-1}, and for b_{3n-2}.

70 Prove that the set $\{a_n \mid n \in \mathbb{N}\}$ of qn 68 is identical to the set $\{b_n \mid n \in \mathbb{N}\}$ of qn 69. Deduce that a rearrangement of the terms of a conditionally convergent series may alter its sum.

71 (i) Write down the first nine terms of the series whose

$(3n - 2)$th term is $\dfrac{1}{\sqrt{(4n - 3)}}$,

$(3n - 1)$th term is $\dfrac{1}{\sqrt{(4n - 1)}}$

and $3n$th term is $\dfrac{-1}{\sqrt{(2n)}}$.

(ii) Check that this series is a rearrangement of

$$\sum (-1)^{n+1}/\sqrt{n}.$$

(iii) Is $\sum (-1)^{n+1}/\sqrt{n}$ convergent?

(iv) Prove that, if $n \geqslant 12$, then

$$\frac{1}{n} < \frac{1}{\sqrt{(4n-3)}} + \frac{1}{\sqrt{(4n-1)}} - \frac{1}{\sqrt{(2n)}}.$$

(v) Prove that the series in (i) is divergent.

So a conditionally convergent series may be rearranged to diverge!

With the result of qn 67, we can see how it is possible, starting with any conditionally convergent series, to rearrange the terms to produce another series which converges to any limit we please, say, l. First take *positive* terms until l is just exceeded, then take *negative* terms until the partial sum is just less than l, then positive terms until the sum is greater than l, then negative terms until the sum is less than l, and so on. The divergence of $\sum u_n$ and $\sum v_n$ means this is always possible, and the fact that both (u_n) and (v_n) are null sequences guarantees the convergence.

72 Let $\sum a_n$ be a convergent sequence of *positive* terms with sum A and let $\sum b_n$ be a rearrangement of the same series.

Let $s_n = \sum_{r=1}^{r=n} a_r$ and let $t_n = \sum_{r=1}^{r=n} b_r$.

Suppose that the first N terms of the sequence of bs is included within the first M terms of the sequence of as. Prove that $t_N \leqslant s_M \leqslant A$. Deduce that $\sum b_n$ is convergent to a sum B (say) and that $B \leqslant A$.

Now suppose that the first N terms of the sequence of as is included within the first L terms of the sequence of bs. Prove that $s_N \leqslant t_L \leqslant B$. Deduce that $A \leqslant B$, so that $A = B$.

This establishes that when a convergent series of *positive* terms is rearranged it is always convergent to the same sum.

73 Let $\sum a_n$ be an absolutely convergent series and let $\sum b_n$ be a rearrangement of the same series.

As in qn 63, we let

$$u_n = \tfrac{1}{2}(|a_n| + a_n), \quad v_n = \tfrac{1}{2}(|a_n| - a_n), \text{ and}$$

$$x_n = \tfrac{1}{2}(|b_n| + b_n), \quad y_n = \tfrac{1}{2}(|b_n| - b_n).$$

Why are the two series $\sum u_n$ and $\sum v_n$ necessarily convergent? Are they both sequences of positive terms?

How does $\sum u_n$ relate to $\sum x_n$, and $\sum v_n$ to $\sum y_n$?

Prove that $\sum a_n = \sum (u_n - v_n) = \sum (x_n - y_n) = \sum b_n$.

This establishes that when an *absolutely convergent* series is rearranged it is always convergent to the same sum.

Power series

Application of d'Alembert's ratio test and Cauchy's nth root test

74 For what values of x is $\sum x^n$ convergent? divergent? See qn 10.

75 By using d'Alembert's test for absolute convergence (qns 64 and 65) determine for what values of x the series $\sum nx^n$ is convergent and for what values it is divergent.

76 By using Cauchy's nth root test for absolute convergence (qn 66), determine the values of x for which the series $\sum nx^n$ is convergent and the values for which it is divergent.

77 For what values of x is the series $\sum n^2x^n$ convergent and for what values is it divergent?

78 For what values of x is the series $\sum x^n/n$ convergent and for what values is it divergent? This question is particularly illuminating because of the possibilities at the critical values when $|x| = 1$.

79 For what values of x is the series $\sum x^n/n^2$ convergent and for what values is it divergent?

80 Discuss the convergence of the series $\sum n^\alpha x^n$ for various values of x and α.

81 Prove that $\sum x^n/n!$ is absolutely convergent for each value of x.

82 Prove that $\sum (-1)^{n+1}x^{2n-1}/(2n - 1)!$ is absolutely convergent for each value of x.

83 Prove that $\sum n!x^n$ is only convergent when $x = 0$.

84 For what values of x is the series $\sum (2x)^n$ convergent and for what values is it divergent?

85 For what values of x is the series $\sum (2x)^n/n$ convergent and for what values is it divergent?

Radius of convergence

86 If $\sum a_n y^n$ is convergent, use qn 11 to prove that, for sufficiently large n, $|a_n y^n| < 1$.
For any real number x such that $|x| < |y|$, prove that, for these values of n, $|a_n x^n| \le |x/y|^n$, and deduce from the first comparison test that the series $\sum a_n x^n$ is absolutely convergent.

It is illuminating to illustrate this result on the number line. If a power series is convergent for a particular value of the variable other than 0, then it is absolutely convergent for any value of the variable nearer to the origin.

87 The result of qn 86 is available under weaker conditions. Suppose only that the terms of the series $\sum a_n y^n$ are bounded, say, $|a_n y^n| < K$. Use the argument of qn 86 to prove that $\sum a_n x^n$ is absolutely convergent when $|x| < |y|$.

88 Use the contrapositive of qn 86 to show that, if $\sum a_n y^n$ is divergent, then $\sum a_n x^n$ is divergent when $|x| > |y|$.

89 For a given power series $\sum a_n x^n$, we consider the values of x for which the series is convergent. Every power series is convergent for $x = 0$, so the set of such values is not empty.
Let $C = \{|x|: \sum a_n x^n \text{ is convergent}\}$.
The set C is a set of real numbers ≥ 0. Use qns 86 and 88 to show that if C is unbounded then $\sum a_n x^n$ is convergent for all values of x.
If C is bounded, let R denote its least upper bound,

$$R = \sup C.$$

For any x with $|x| < R$, use the properties of a least upper bound to say why there exists a y with $|x| < |y| < R$ and $\sum a_n y^n$ convergent, and deduce that $\sum a_n x^n$ is convergent.
If, for some x with $|x| > R$, the series $\sum a_n x^n$ were to be convergent, say why this would contradict the definition of R as an upper bound for C.

The work of question 89 gives the basis for the definition of the *radius of convergence, R,* of a power series $\sum a_n x^n$.
If the power series is only convergent when $x = 0$, we say that its radius of convergence $R = 0$.
If the power series is convergent for $|x| < R$ and divergent for $|x| > R$, for some positive real number R, then its radius of convergence is that real number R.
If the power series is convergent for all values of x, we say that its radius of convergence $R = \infty$.
When these results are extended to the plane of complex numbers we obtain a circle $|z| = R$ called the *circle of convergence*.
Question 89 establishes that every power series has a radius of convergence.

90 If $(|a_{n+1}/a_n|) \to k \neq 0$, prove that the radius of convergence of the power series $\sum a_n x^n$ is $1/k$.

91 If $\left(\sqrt[n]{|a_n|}\right) \to k \neq 0$, prove that the radius of convergence of the power series $\sum a_n x^n$ is $1/k$.

92 Find the radius of convergence of $\sum (nx)^n/n!$.

93 If a is a real number which is not a positive integer or 0, and

$$\binom{a}{n} = \frac{a(a-1)\ldots(a-n+1)}{n!},$$

the binomial coefficient, prove that the power series

$$\sum \binom{a}{n} x^n$$

has radius of convergence 1.

94 If the sequence $\left(\sqrt[n]{|a_n|}\right)$ is unbounded, show that, for any value of $x \neq 0$, infinitely many terms of the sequence are $> 1/|x|$, so the terms of the power series $\sum a_n x^n$ do not form a null sequence. Determine the radius of convergence.

95 If the sequence $\left(\sqrt[n]{|a_n|}\right) \to 0$ as $n \to \infty$, show that, for any value of $x \neq 0$, and for sufficiently large n, the terms of the sequence are $< \frac{1}{2}/|x|$ and hence show that the power series $\sum a_n x^n$ is absolutely convergent for all values of x. Determine the radius of convergence in this case.

96 Show that the series $\sum \left((\frac{1}{2}x)^{2n-1} + (\frac{1}{3}x)^{2n}\right)$ has radius of convergence 2. Notice that in this case the sequence $\left(\sqrt[n]{|a_n|}\right)$ is not convergent, but oscillates between $\frac{1}{2}$ and $\frac{1}{3}$.

Cauchy–Hadamard formula

Question 97 uses the notation of lim sup which was developed as an optional part of the book in qns 4.65 and 4.66.

97 If the sequence $\left(\sqrt[n]{|a_n|}\right)$ is bounded but not necessarily convergent, we let $\limsup \sqrt[n]{|a_n|} = A$. We further suppose that $A \neq 0$, since this case has been considered in qn 95. Show that if $|x| > 1/A$ then $\sqrt[n]{|a_n|} > 1/|x|$ for infinitely many n and so the terms of the power series $\sum a_n x^n$ do not form a null sequence and the series is divergent.
Show also that if $|x| < 1/A$, then, choosing r such that $|x| < r < 1/A$, for sufficiently large n, $\sqrt[n]{|a_n|} < 1/r$, so we have $\sqrt[n]{|a_n x^n|} < |x|/r < 1$, and so the series $\sum a_n x^n$ is convergent. Identify the radius of convergence.

The exploration in questions 94, 95 and 97 leads to the *Cauchy–Hadamard formula*

$$R = 1/(\limsup \sqrt[n]{|a_n|}),$$

which, with suitable interpretations, is a general statement of the radius of convergence of the power series $\sum a_n x^n$.

98 Give an example of a power series $\sum a_n x^n$, with radius of convergence 1, which is divergent at each point on the circle of convergence.

99 Give an example of a power series $\sum a_n x^n$, with radius of convergence 1, which is convergent at some points on the circle of convergence and divergent at others.

100 Give an example of a power series $\sum a_n x^n$, with radius of convergence 1, which is convergent at each point on the circle of convergence.

101 Compare regions of convergence of the three power series

$$\log(1 + x) = x - \frac{x^2}{2} + \frac{x^3}{3} - \frac{x^4}{4} + \dots,$$

$$\frac{1}{1 + x} = 1 - x + x^2 - x^3 + \dots,$$

$$\frac{-1}{(1 + x)^2} = -1 + 2x - 3x^2 + 4x^3 - \dots.$$

102 Use the Cauchy–Hadamard formula to show that the three series $\sum a_n x^n$, $\sum n a_n x^{n-1}$ and $\sum a_n x^{n+1}/(n + 1)$ have the same radius of convergence. This establishes that the radius of convergence is not changed when a series is differentiated or integrated term by term.

Products of series

Termwise products

103 If $\sum a_n$ and $\sum b_n$ are absolutely convergent, prove that $\sum a_n b_n$ is absolutely convergent.

104 Give examples of series $\sum a_n$ and $\sum b_n$ which are both conditionally convergent and for which $\sum a_n b_n$ is divergent.

105 Give an example of a divergent series $\sum a_n$ such that $\sum a_n^2$ is convergent.

106 Given an example of a convergent series $\sum a_n$ such that $\sum a_n^2$ is divergent.

107 Use the *Cauchy–Schwarz inequality*, qn 2.57, to prove that, if $\sum a_n^2$ is convergent and $a_n \geq 0$, then $\sum a_n/n$ is convergent. Why does the result hold without the condition $a_n \geq 0$?

108 The power series $\sum a_n x^n$ has finite radius of convergence $A \neq 0$, the power series $\sum b_n x^n$ has finite radius of convergence $B \neq 0$, and c is a real number such that $0 < c < AB$.
Show that $\sum a_n x^n$ is absolutely convergent when $x = \frac{1}{2}(A + c/B)$, and that $\sum b_n x^n$ is absolutely convergent when $x = \frac{1}{2}(B + c/A)$.
Prove that $c < \frac{1}{2}(A + c/B) \cdot \frac{1}{2}(B + c/A) < AB$, and deduce that $\sum a_n b_n x^n$ is absolutely convergent when $x = c$. Show that the radius of convergence of $\sum a_n b_n x^n \geq AB$.

109 Construct an example for which, with the notation of qn 108, the radius of convergence of $\sum a_n b_n x^n = AB$.

110 By using series like that in qn 96, construct an example for which, with the notation of qn 108, the radius of convergence of $\sum a_n b_n x^n > AB$.

Cauchy products

111 If $A_N(x) = \sum_{n=0}^{n=N} a_n x^n$ and $B_N(x) = \sum_{n=0}^{n=N} b_n x^n$, calculate the coefficients, c_n, in the product

$$A_N(x) \cdot B_N(x) = \sum_{n=0}^{n=N} c_n x^n + \sum_{n=N+1}^{n=2N} d_n x^n.$$

The series $\sum_{n=0}^{\infty} c_n$ is said to be the *Cauchy product* of the series $\sum_{n=0}^{\infty} a_n$ and $\sum_{n=0}^{\infty} b_n$.

112 By putting $a_n = b_n = (-1)^n/\sqrt{(n+1)}$, show that, even if the series $\sum a_n$ and $\sum b_n$ are convergent, their Cauchy product need not be convergent.

113 Evaluate the Cauchy product of the series $\sum a_n$ and the series $\sum b_n$ when $a_n = x^n/n!$ and $b_n = y^n/n!$, showing that $c_n = (x + y)^n/n!$.

With the result of qn 114, qn 113 illustrates that $\exp(x) \cdot \exp(y) = \exp(x + y)$, a result which we will establish by a different route in chapter 11. See qn 11.20(i) and 11.27$^+$.

114 The series $\sum a_n$ is absolutely convergent and $\sum_{n=0}^{\infty} a_n = A$. The series $\sum b_n$ is absolutely convergent and $\sum_{n=0}^{\infty} b_n = B$. The first array below is an array of products from these two series; the second array is a renaming of the first array, position by position.

a_0b_0	a_0b_1	a_0b_2	a_0b_3	$\ldots$	a_0b_n	$\ldots$
a_1b_0	a_1b_1	a_1b_2	a_1b_3	$\ldots$	a_1b_n	$\ldots$
a_2b_0	a_2b_1	a_2b_2	a_2b_3	$\ldots$	a_2b_n	$\ldots$
a_3b_0	a_3b_1	a_3b_2	a_3b_3	$\ldots$	a_3b_n	$\ldots$
$\vdots$	$\vdots$	$\vdots$	$\vdots$		$\vdots$	
a_nb_0	a_nb_1	a_nb_2	a_nb_3	$\ldots$	a_nb_n	$\ldots$
$\vdots$	$\vdots$	$\vdots$	$\vdots$		$\vdots$	

d_1	d_4	d_9	d_{16}	$\ldots$	$d_{(n+1)^2}$	$\ldots$
d_2	d_3	d_8	d_{15}	$\ldots$	$d_{(n+1)^2-1}$	$\ldots$
d_5	d_6	d_7	d_{14}	$\ldots$	$d_{(n+1)^2-2}$	$\ldots$
d_{10}	d_{11}	d_{12}	d_{13}	$\ldots$	$d_{(n+1)^2-3}$	$\ldots$
$\vdots$	$\vdots$	$\vdots$	$\vdots$		$\vdots$	
d_{n^2+1}	d_{n^2+2}	d_{n^2+3}	d_{n^2+4}	$\ldots$	d_{n^2+n+1}	$\ldots$
$\vdots$	$\vdots$	$\vdots$	$\vdots$		$\vdots$	

(i) Prove that
$$\sum_{n=1}^{n=(N+1)^2} d_n = \sum_{n=0}^{n=N} a_n \cdot \sum_{n=0}^{n=N} b_n.$$

(ii) Deduce that
$$\sum_{n=1}^{\infty} d_n = A \cdot B.$$

(iii) Prove that
$$\sum_{n=1}^{n=(N+1)^2} |d_n| = \sum_{n=0}^{n=N} |a_n| \cdot \sum_{n=0}^{n=N} |b_n|.$$

(iv) Prove that $\sum d_n$ is absolutely convergent.

(v) Show that the Cauchy product of $\sum_{n=0}^{\infty} a_n$ and $\sum_{n=0}^{\infty} b_n$ is a rearrangment of $\sum_{n=1}^{\infty} d_n$.

(vi) Prove that the Cauchy product of $\sum a_n$ and $\sum b_n$ is convergent to the sum $A \cdot B$.

115 By calculating the Cauchy product of $\sum x^n$ with itself when $-1 < x < 1$, prove that $1/(1 - x)^2 = \sum (n + 1)x^n$. The series begins with $n = 0$.

116 Prove the Binomial Theorem for negative integral index by induction using a Cauchy product, or, in other words, give a power series which converges to $1/(1 - x)^n$ when $-1 < x < 1$.

Summary

Definition   A series $\sum a_n$ is said to be *convergent* to the sum s, when its sequence of partial sums $\left(\sum_{r=1}^{r=n} a_r \right)$ is convergent to s as $n \to \infty$. A series which is not convergent is said to be *divergent*.

Theorem $\sum_{n=0}^{\infty} x^n = 1/(1 - x)$, provided $|x| < 1$.
qn 10

Theorem   If $\sum a_n$ is convergent, then (a_n) is a null sequence.
qn 11

Theorem   If $\sum a_n$ is a series of positive terms, then it is
qn 22   convergent if and only if its partial sums are bounded.

First comparison test
qn 23 $0 \leqslant a_n \leqslant b_n$. If $\sum b_n$ is convergent, then $\sum a_n$ is convergent. If $\sum a_n$ is divergent, then $\sum b_n$ is divergent.

Theorem $\sum 1/n^\alpha$ is convergent when $\alpha > 1$, and divergent
qns 28, 29   when $\alpha \leqslant 1$.

Cauchy's nth root test
qns 32, 35 $\left(\sqrt[n]{|a_n|} \right) \to k$. $\sum a_n$ is convergent when $k < 1$ and
and 66   divergent when $k > 1$.

d'Alembert's ratio test
qns 40, 44, $\left(|a_{n+1}/a_n| \right) \to k$. $\sum a_n$ is convergent when $k < 1$ and
64 and 65   divergent when $k > 1$.

Second comparison test
qn 50 $0 < a_n, b_n$ and $0 < m < a_n/b_n < M$.
$\sum a_n$ and $\sum b_n$ are both convergent or both divergent.

Integral test
qns 54, 58   If f is a real function which is positive and decreasing when $x \geqslant 1$, then $\left(\sum_{r=1}^{r=n} f(r) \right)$ and $\left(\int_1^n f(x)dx \right)$ are both convergent or both tend to infinity.
$\left(\sum_{r=1}^{r=n} f(r) - \int_1^n f(x)dx \right)$ is always convergent.

Alternating series test

qn 60   If (a_n) is a monotonic null sequence, then
$\sum (-1)^n a_n$ is convergent.

Definition   If $\sum |a_n|$ is convergent, then $\sum a_n$ is said to be *absolutely convergent*.

Theorem   If $\sum a_n$ is absolutely convergent, then $\sum a_n$ is
qn 63   convergent.

Definition   A series which is convergent but not absolutely convergent is said to be *conditionally convergent*.

Theorem   If the terms of a conditionally convergent series are
qn 70   rearranged, the sum may be altered.

Theorem   If the terms of an absolutely convergent series are
qn 73   rearranged, the sum is not changed.

Theorem   If the terms of $\sum a_n y^n$ are bounded then $\sum a_n x^n$ is
qn 87   absolutely convergent when $|x| < |y|$.

Definition   When $\{|x|: \sum a_n x^n$ is convergent$\}$ is bounded above, $\sup \{|x|: \sum a_n x^n$ is convergent$\}$ is called the *radius of convergence* of $\sum a_n x^n$.
When $\{|x|: \sum a_n x^n$ is convergent$\}$ is not bounded above, the radius of convergence of $\sum a_n x^n$ is said to be infinite.

Theorem   When $|x| <$ the radius of convergence, $\sum a_n x^n$ is
qn 89   absolutely convergent.

Cauchy–Hadamard formula

qn 97   The radius of convergence of $\sum a_n x^n$ is
$1/\limsup \sqrt[n]{|a_n|}$.

Theorem   If $\sum a_n$ and $\sum b_n$ are absolutely convergent, then
qn 102 $\sum a_n b_n$ is absolutely convergent.

Theorem   If A is the radius of convergence of $\sum a_n x^n$ and B is
qn 107   the radius of convergence of $\sum b_n x^n$, then the radius of convergence of $\sum a_n b_n x^n \geqslant AB$.

Definition   If $c_n = a_0 b_n + a_1 b_{n-1} + \ldots + a_i b_{n-i} + \ldots + a_n b_0$, then the series $\sum_{n=0}^{\infty} c_n$ is called the *Cauchy product* of the series $\sum_{n=0}^{\infty} a_n$ and $\sum_{n=0}^{\infty} b_n$.

Theorem   If $\sum a_n$ and $\sum b_n$ are both absolutely convergent,
qn 113   then their Cauchy product is convergent to the product of their sums.

Historical note

Archimedes (250 B.C.) could sum the series $\sum (\frac{1}{4})^n$. Oresme (1323–1382) was able to sum $\sum (\frac{3}{4})^n$ and $\sum n(\frac{1}{2})^n$ and he also showed the divergence of the harmonic series in the way we have done.

P. Mengoli summed the series $\sum 1/n(n + 1)$ in 1650. I. Newton in 1667 and N. Mercator in 1668 obtained the result $\sum (-1)^{n+1}/n = \log 2$ by integrating the power series for $(1 + x)^{-1}$. James Gregory in 1671 and Leibniz in 1673 obtained the result $\sum (-1)^{n+1}/(2n - 1) = \frac{1}{4}\pi$ by integrating the power series for $(1 + x^2)^{-1}$.

During the seventeenth and eighteenth centuries the word 'series' was equally used to describe a sequence or a series, and the word 'convergence' was used to refer to the sequence (a_n) or the series $\sum a_n$, even if it was the latter which appeared to be under discussion. In the eighteenth century, discussion of the sums of series was not confined to those cases where the sequence of partial sums was convergent. Insistence on at least the boundedness of partial sums for meaningful discussion of an overall sum is due to Gauss (1813). In 1742 C. Maclaurin used integrals to approximate to values of series and vice-versa, and illustrated his method with $\sum 1/n^2$. In 1748, by factorising the power series for $\sin x$ like a polynomial, L. Euler showed that $\sum 1/n^2 = \pi^2/6$. In 1785 E. Waring showed that $\sum 1/n^\alpha$ was convergent/divergent according as $\alpha >/\leqslant 1$. His proof used a rudimentary form of the integral test. The proof we have given is a special case of Cauchy's condensation test: that $\sum a_n$ is convergent if and only if $\sum (k^n)a_{k^n}$ is convergent when (a_n) is a decreasing sequence of positive terms.

The title of d'Alembert's ratio test derives from work which he published in 1768 establishing the convergence of the binomial series by comparison with geometric progressions with common ratio less than 1. The ratio test was used with modern precision in Gauss' paper on the hypergeometric series (1813). In 1815, Fourier proved that e, defined by the series $\sum 1/n!$, was irrational. The first comparison test, or sandwich theorem, is described in relation to geometric progressions by Bolzano (1816). Bolzano (1817) also explained why, for $\sum a_n$ to be convergent, it is necessary for (a_n) to be a null sequence, but not sufficient because of the harmonic series. Many of the theorems in this chapter were stated and proved for the first time by A. L. Cauchy in his *Cours d'Analyse* (1821). Cauchy formulated a modern definition of limit, and considered the convergence of a series only in terms of its sequence of partial sums. He proved that $\sum a_n$ was convergent/divergent when $\limsup_{n} \sqrt[n]{|a_n|} </> 1$ and also when $\lim |a_{n+1}/a_n| </> 1$. He used these results to give the radius of convergence of a power series, a result which was rediscovered by Hadamard (1892) in his thesis. C. Méray coined the terms 'radius' and 'circle of convergence' in 1872. In his *Cours d'Analyse* Cauchy argued that absolutely convergent series were convergent and he established the convergence of $\sum (-1)^n/n$ adding that 'the same reasoning can clearly be applied to every series of this kind'. He proved that the Cauchy product of two absolutely convergent series converges to the product of the two limits, and showed the necessity for the requirement of absolute convergence with the counter-example we have given. He used the Cauchy product to discuss Newton's binomial series. In 1827 Cauchy gave the integral test for convergence, though Cauchy's integrals were only for continuous functions.

In 1837 P. G. L. Dirichlet showed that a conditionally convergent series might be rearranged either to give a different sum or to diverge. The example of divergence that we have given is his. In the same paper, Dirichlet proved that any rearrangement of an absolutely convergent series converged to the same sum. In 1854, G. B. H. Riemann showed that a rearrangement of a conditionally convergent series could be devised to converge to any preassigned sum. The result was published posthumously in 1867.

Answers

1 Unless the sum is of a finite number of terms, S may not be defined.

3 $1/(r(r + 1)) = 1/r - 1/(r + 1)$.

$$s_n = (1 - \tfrac{1}{2}) + (\tfrac{1}{2} - \tfrac{1}{3}) + \ldots + \left(\frac{1}{n} - \frac{1}{n + 1}\right).$$

$$\sum_{r=1}^{r=n} \frac{1}{r(r + 1)} = 1 - \frac{1}{n + 1}.$$

5 $\left(1 - (\tfrac{1}{2})^n\right)/(1 - \tfrac{1}{2})$. Using qn 3.32, $1/(1 - \tfrac{1}{2}) = 2$.

6 $1/9 = 0.1111 \ldots$.

7 $\sum_{r=1}^{\infty} 9 \cdot (10^{-2})^r$.

8 10.

9 Put $k = 1$ and $x = \tfrac{1}{2}$ in qn 3.62(i).

10 (i) Convergent, from qn 3.32. (ii) Divergent, from qn 3.16. Partial sums tend to infinity when $x = 1$. Partial sums $= \tfrac{1}{2}(1 + (-1)^{n+1})$ when $x = -1$.

11 $\lim a_n = s - s = 0$, from qn 3.44(iii).

13 If $\alpha > 0$, $n^\alpha \to +\infty$, so $\sum n^\alpha$ is divergent.

14 If $s_n = \sum_{r=1}^{r=n} a_r$ and $s_n \to s$, then $\sum_{r=k}^{r=n} a_r = s_n - s_{k-1} \to s - s_{k-1}$.
If $\sigma_n = \sum_{r=k}^{r=n} a_r$ and $\sigma_n \to \sigma$, then $\sum_{r=1}^{r=n} a_r = \sigma_n + s_{k-1} \to \sigma + s_{k-1}$.

15 If $\sum_{r=1}^{\infty} a_r = s$, then $(s - s_n) \to 0$ as $n \to \infty$, so $(s - s_{n-1}) \to 0$ and $\left(\sum_{r=n}^{\infty} a_r\right) \to 0$.

16 Given $\varepsilon > 0$, there exists an n_0 such that $\left|\sum_{r=n+1}^{r=n+k} a_r\right| < \varepsilon$, when $n > n_0$, for all positive integers k.

17 From qn 3.44(i).

18 $\sum c \cdot x^n$ is only convergent when $|x| < 1$ or $c = 0$.

19 From qn 3.44(ii).

20 $n \geq 2 \Rightarrow n! \geq 2^{n-1} \Rightarrow 1/n! \leq 1/2^{n-1} \Rightarrow e_n \leq s_n < 3$. Use qn 4.34.

21 $p/q - n/q! = k/q!$ for some integer k.
$1/(q + 1)! + 1/(q + 2)! + \ldots < (1/q!)(1/3 + 1/3^2 + \ldots) = \tfrac{1}{2}/q!$. The proof of this result is set out very fully in Bryant pp. 20–2.

22 $s_{n+1} = s_n + a_{n+1} \geq s_n$. Use qn 4.34.

23 Let $A_n = \sum_{r=1}^{r=n} a_r$ and let $B_n = \sum_{r=1}^{r=n} b_r$.

$a_r \leqslant b_r$ for all $r \Rightarrow A_n \leqslant B_n$ for all n.

$0 \leqslant a_r \leqslant b_r \Rightarrow$ both (A_n) and (B_n) are monotonic increasing.

(B_n) convergent $\Rightarrow (B_n)$ bounded $\Rightarrow (A_n)$ bounded $\Rightarrow (A_n)$ convergent by qn 4.34.

$\sum a_r$ divergent $\Rightarrow (A_n) \to +\infty \Rightarrow (B_n) \to +\infty \Rightarrow \sum b_r$ divergent.

24 $\sum_{n=1}^{\infty} 1/(n+1)^2 = \sum_{n=2}^{\infty} 1/n^2$.

25 $\sqrt{101} - 1$. (i) With the notation of qn 23, $A_n = \sqrt{(n+1)} - 1 \to +\infty$ as $n \to +\infty$.

(iii) $\sqrt{n} < \sqrt{(n+1)} \Rightarrow 2\sqrt{n} < \sqrt{n} + \sqrt{(n+1)} \Rightarrow$
$1/(2\sqrt{n}) > 1/(\sqrt{(n+1)} + \sqrt{n})$.

26 If $\alpha \geqslant 2$, $0 < 1/n^\alpha \leqslant 1/n^2$, so $\sum 1/n^\alpha$ is convergent.

If $0 \leqslant \alpha \leqslant \frac{1}{2}$, $0 < 1/\sqrt{n} \leqslant 1/n^\alpha$, so $\sum 1/n^\alpha$ is divergent.

27 Partial sums are unbounded, so $\sum 1/n$ is divergent.

28 If $\alpha \leqslant 1$, $0 < 1/n \leqslant 1/n^\alpha$, so $\sum 1/n^\alpha$ is divergent.

29 $\alpha > 1 \Rightarrow \alpha - 1 > 0$. $2 > 1 \Rightarrow 2^{\alpha-1} > 1^{\alpha-1}$, though we have only proved this for rational α in qn 2.35 so far. So $1/2^{\alpha-1} < 1$.

All partial sums $< 1/(1 - 1/2^{\alpha-1})$.

30 By qn 10, $\sum (\frac{1}{2})^n$ is convergent so, by qn 23, $\sum (n/(2n+1))^n$ is convergent.

31 $n > 3 \Rightarrow 8n^2 > 72 \Rightarrow (180 - 172)n^2 > 72 \Rightarrow 36(5n^2 - 2) > 43 \cdot 4n^2$. By qn 10, $\sum (36/43)^n$ is convergent so, by qn 23, $\sum (4n^2/(5n^2 - 2))^n$ is convergent for $n > 3$. By qn 14, $\sum (4n^2/(5n^2 - 2))^n$ is convergent.

32 There is an n_0 such that $|\sqrt[n]{a_n} - k| < \frac{1}{2}(1 - k)$, when $n > n_0$. So $0 \leqslant \sqrt[n]{a_n} < \frac{1}{2}(1 + k) = \rho < 1$, for $n > n_0$. Now, by qn 10, $\sum \rho^n$ is convergent. So, by qn 23, $\sum a_n$ is convergent for $n > n_0$. By qn 14, $\sum a_n$ is convergent.

33 $(\sqrt[n]{(n/2^n)}) \to \frac{1}{2}$ and $(\sqrt[n]{(n^2(0.8)^n)}) \to 0.8$, using qn 4.41 in each case.

34 $\sqrt[n]{a_n} \geqslant 1 \Rightarrow a_n \geqslant 1 \Rightarrow (a_n)$ not null. Use qn 11.

35 There is an n_0 such that $|\sqrt[n]{a_n} - k| < \frac{1}{2}(1 - k)$, when $n > n_0$. So $1 < \frac{1}{2}(1 + k) < \sqrt[n]{a_n}$ for $n > n_0$. So $\sum a_n$ is divergent for $n > n_0$ by qn 34. Now $\sum a_n$ is divergent by qn 14.

36 $(\sqrt[n]{(1/n)}) \to 1$ and $\sum 1/n$ is divergent, $(\sqrt[n]{(1/n^2)}) \to 1$ and $\sum 1/n^2$ is convergent.

37 $n \geqslant 3 \Rightarrow 1 + 1/n \leqslant 4/3 \Rightarrow (1 + 1/n)^2 \leqslant 16/9 \Rightarrow a_{n+1}/a_n = \frac{1}{2}(1 + 1/n)^2 \leqslant 8/9$.
$a_4 \leqslant a_3 \cdot (8/9)$. $a_5 \leqslant a_4 \cdot (8/9) \leqslant a_3 \cdot (8/9)^2$. By qn 18, $\sum a_3(8/9)^n$ is

convergent. By qn 23, $\sum a_{n+3}$ is convergent. By qn 14, $\sum a_n$ is convergent.

38 $n \geq 14 \Rightarrow 1 + 1/n \leq 15/14 \Rightarrow (1 + 1/n)^3 \leq 3375/2744$
$\Rightarrow 0.8(1 + 1/n)^3 \leq 674/686 < 0.99$. By qn 18, $\sum a_{14}(0.99)^n$ is convergent.
By qn 23, $\sum a_{n+14}$ is convergent. By qn 14, $\sum a_n$ is convergent.

39 $\sum \rho^n a_1$ is convergent by qn 18, so $\sum a_{n+1}$ is convergent by qn 23 and so $\sum a_n$ is convergent by qn 14.

40 Take $\varepsilon = \frac{1}{2}(1 - k)$.

41 $a_{n+1}/a_n = 2/(n + 1) \to 0$.

42 $a_{n+1}/a_n = 1/(1 + 1/n)^n \to l < \frac{1}{2}$, by qn 4.35.

43 Use qn 11.

44 Choose $\varepsilon = \frac{1}{2}(k - 1)$.

45 $\sqrt[n]{a_n} = 2x(\sqrt[n]{n})^x$. $a_{n+1}/a_n = 2x(1 + 1/n)^x$. Both sequences tend to $2x$, so $\sum a_n$ is convergent when $0 \leq x < \frac{1}{2}$, and divergent when $\frac{1}{2} < x$. If $x = \frac{1}{2}$, $\sum a_n$ is divergent by qn 13.

46 $a_n = 1/n$.

47 $0 < a_n \leq 1/n^2$, so $\sum a_n$ convergent. But $a_{2n}/a_{2n-1} = 2n \to +\infty$.

48 $\dfrac{a_n}{b_n} = \dfrac{1 + 1/\sqrt{n}}{1 + 2/n^2}$.

Take $\varepsilon = \frac{1}{2}$. $\sum b_n$ is convergent, from qn 29. So $\sum \frac{3}{2}b_n$ is convergent, from qn 17. Now $0 < a_n < \frac{3}{2}b_n$, so $\sum a_n$ is convergent, from qn 23.

49 $\dfrac{a_n}{b_n} = \dfrac{1 + 1/\sqrt{n}}{1 + 2/n}$.

Again take $\varepsilon = \frac{1}{2}$. $\sum b_n$ is divergent, from qn 25, so $\sum \frac{1}{2}b_n$ is divergent, from qn 17. Now, $0 < \frac{1}{2}b_n < a_n$, so $\sum a_n$ is divergent, from qn 23.

50 Since $0 < mb_n < a_n$, $\sum a_n$ convergent $\Rightarrow \sum mb_n$ convergent by qn 23 $\Rightarrow \sum b_n$ convergent by qn 17.
Since $0 < a_n < Mb_n$, $\sum b_n$ convergent $\Rightarrow \sum Mb_n$ convergent by qn 17 $\Rightarrow \sum a_n$ convergent by qn 23.

51 $a_n/b_n = 1/n$. Test requires $0 < m < a_n/b_n$.

52 Take $\varepsilon = \frac{1}{2}l$. For some n_0, $\frac{1}{2}l < a_n/b_n < \frac{3}{2}l$, when $n > n_0$. Now apply qn 50 and qn 14.

53 When $n \leq x \leq n + 1$, $1/(n + 1) \leq 1/x \leq 1/n$, and see chapter 10. Sum the inequalities for $n = 1, 2, 3, \ldots, n - 1$.

$s_n - 1 < \log_e n < s_n - 1/n$. $D_1 = 1$, $D_2 = 0.807 \ldots$, $D_3 = 0.735 \ldots$, $D_4 = 0.697 \ldots$.

54 $\big(f(n)\big)$ is monotonic decreasing and bounded below by 0, so the sequence is convergent by qn 4.33. Lower sum $\leqslant$ integral $\leqslant$ upper sum. (s_n) is monotonic increasing and, since $s_n \leqslant f(1) + \int_1^n f(x)dx$, if the integral is bounded above, (s_n) is bounded above and so is convergent by qn 4.34. Conversely, $\int_1^n f(x)dx \leqslant s_{n-1}$ so that, if (s_n) is convergent and so bounded, the integral is bounded above and so convergent by qn 4.34.

55 $\sum_1^n 1/r^2 - \int_1^n dx/x^2 \leqslant 1 \Rightarrow s_n \leqslant 2 - 1/n$.

56 $0 \leqslant \sum_1^n 1/\sqrt{r} - \int_1^n dx/\sqrt{x} \leqslant 1 \Rightarrow 0 \leqslant \sum_1^n 1/\sqrt{r} - [2\sqrt{x}]_1^n \leqslant 1$.

57 $\displaystyle\int_2^n \frac{dx}{x \cdot \log_e x} = [\log(\log x)]_2^n = \log(\log n) - \log(\log 2)$.

$$\int_2^n \frac{dx}{x \cdot \log_e x} + \frac{1}{n \log n} \leqslant \sum_{r=2}^{r=n} \frac{1}{r \log r}.$$

58 The sequence (D_n) is monotonic decreasing, positive, and so bounded below. Thus it is convergent, from qn 4.33.

59 (s_{2n}) is monotonic increasing and bounded above by s_1. (s_{2n+1}) is monotonic decreasing and bounded below by s_2. Let $(s_{2n}) \to L$ and $(s_{2n+1}) \to U$, then $L + 0 = U$ by qn 3.44(ii).

60 Argument reflects qn 59.

61 $s_{2n} = \log 2 + D_{2n} - D_n$, so $s = \log 2 + \gamma - \gamma$.

62 (i) and (ii): use first comparison test with $\sum |a_n|$.
(iii) Qns 17 and 19. (iv) $a_n = u_n - v_n$.

63 $u_n = 0$ when $a_n \leqslant 0$, $u_n = a_n$ when $0 \leqslant a_n$.
$v_n = 0$ when $0 \leqslant a_n$, $v_n = -a_n$ when $a_n \leqslant 0$.
$0 \leqslant u_n$, $v_n \leqslant |a_n|$, so $\sum u_n$, and $\sum v_n$ are convergent by the first comparison test. $\sum (u_n - v_n)$ is convergent by qns 17 and 19.
$u_n - v_n = a_n$.

64 $|a_{n+1}/a_n| \to k < 1 \Rightarrow \sum a_n$ is absolutely convergent.

66 $(\sqrt[n]{|a_n|}) \to k < 1 \Rightarrow \sum a_n$ is absolutely convergent.

67 $\sum a_n$ is convergent. Since $a_n = u_n - v_n$, both $\sum u_n$ and $\sum v_n$ are convergent or both are divergent, for $u_n = a_n + v_n$ and $v_n = u_n - a_n$. If both were convergent, then so would $\sum (u_n + v_n) = \sum |a_n|$ be convergent.

68 $s_{2n} = 0$. $s_{2n+1} = 1/(n+1)$. So $s_n \to 0$. $a_{2n-1} = 1/n$, $a_{2n} = -1/n$.

69 Denote the partial sum of n terms in qn 61 by t_n. $s_{3n} = t_{2n}$, $s_{3n-1} = t_{2n} + 1/n$. $s_{3n-2} = t_{2n} + 1/2n$, so (s_n) and (t_n) have the same limit. $b_{3n} = -1/n$, $b_{3n-1} = 1/2n$, $b_{3n-2} = 1/(2n-1)$.

70 $a_{2n} = b_{3n}$, $a_{4n-1} = b_{3n-1}$, $a_{4n-3} = b_{3n-2}$. $\sum a_n = 0$. $\sum b_n = \log 2$.

71 (i) $1 + 1/\sqrt{3} - 1/\sqrt{2} + 1/\sqrt{5} + 1/\sqrt{7} - 1/\sqrt{4} + 1/\sqrt{9} + 1/\sqrt{11} - 1/\sqrt{6}$.

(iii) Yes, by the alternating series test.

(iv) Compare $\dfrac{2}{\sqrt{n}}$ with $\dfrac{1}{\sqrt{(1 - 3/4n)}} + \dfrac{1}{\sqrt{(1 - 1/4n)}} - \sqrt{2}$.

The first sequence is decreasing and null, the second is decreasing but bounded below by $2 - \sqrt{2}$.

First $< 2 - \sqrt{2}$ when $n = 12$. Multiply inequality by $1/2\sqrt{n}$.

(v) Divergent by comparison with $\sum 1/n$.

73 $\sum u_n$ and $\sum v_n$ are convergent, by comparison with $\sum |a_n|$. $\sum x_n$ is a rearrangement of $\sum u_n$, and $\sum y_n$ is a rearrangement of $\sum v_n$. But $\sum u_n$ and $\sum v_n$ are series of positive terms, so $\sum x_n = \sum u_n$ and $\sum y_n = \sum v_n$.

75 $|a_{n+1}/a_n| = |x|(1 + 1/n) \to |x|$. Series convergent when $|x| < 1$, divergent when $|x| > 1$. Divergent because not null when $|x| = 1$.

76 $\sqrt[n]{|a_n|} = |x|\sqrt[n]{n} \to |x|$.

77 $\sqrt[n]{|a_n|} = |x|(\sqrt[n]{n})^2 \to |x|$. Conclusions as in qn 75.

78 $\sqrt[n]{|a_n|} = |x|/\sqrt[n]{n} \to |x|$. Convergent when $|x| < 1$ and divergent when $|x| > 1$ by Cauchy's nth root test. Convergent when $x = -1$ by the alternating series test. Divergent when $x = 1$: harmonic series.

79 $\sqrt[n]{|a_n|} = |x|/(\sqrt[n]{n})^2 \to |x|$. As before conv./div. as $|x| </> 1$. Absolutely convergent when $|x| = 1$.

80 $\sqrt[n]{|a_n|} = |x|(\sqrt[n]{n})^\alpha \to |x|$. Conv./div. as $|x| </> 1$. When $x = 1$, convergent when $\alpha < -1$ by qn 29. When $x = -1$, convergent when $\alpha < 0$ by the alternating series test.

81 $|a_{n+1}/a_n| = |x|/(n + 1) \to 0$ for all values of x.

82 $|a_{n+1}/a_n| = x^2/(2n + 1)(2n) \to 0$ for all values of x.

83 $|a_{n+1}/a_n| = |x|(n + 1) \to +\infty$ unless $x = 0$.

84 $\sqrt[n]{|a_n|} = |2x|$. Conv./div. as $|x| </> \frac{1}{2}$. Divergent when $|x| = \frac{1}{2}$ since terms not null.

85 $\sqrt[n]{|a_n|} = |2x|/\sqrt[n]{n} \to 2|x|$. Conv./div. as $|x| </> \frac{1}{2}$. Convergent when

$x = -\frac{1}{2}$ by the alternating series test. Divergent when $x = \frac{1}{2}$: harmonic series.

86 $|a_n x^n| = |x/y|^n \cdot |a_n y^n|$ and $|x/y| < 1$.

87 $|a_n x^n| = |x/y|^n \cdot |a_n y^n| < K \cdot |x/y|^n$.

88 If $\sum a_n x^n$ were convergent with $|x| > |y|$, then $\sum a_n y^n$ would be convergent by qn 86. Contradiction.

89 C unbounded $\Rightarrow$ for any given y, there is an x with $|x| > |y|$ and $\sum a_n x^n$ convergent, so $\sum a_n y^n$ must be convergent for all y.
Take $\varepsilon = R - |x|$, and use qn 4.63 to show that such a y exists. Then use qn 86. $|x| \in C \Rightarrow |x| \leqslant R$.

90 d'Alembert's ratio test for absolute convergence implies that $\sum a_n x^n$ is convergent when $|kx| < 1$ and divergent when $|kx| > 1$.

91 $\sqrt[n]{|a_n x^n|} = (\sqrt[n]{|a_n|})|x| \to k \cdot |x|$.

92 Use d'Alembert. $|a_{n+1}/a_n| = (1 + 1/n)^n |x|$, so radius of convergence $= 1/e$.

93 $\left(\dbinom{a}{n+1}\right) \Big/ \left(\dbinom{a}{n}\right) = \dfrac{a - n}{n + 1}$ and $\left|\dfrac{a - n}{n + 1}\right| \to 1$.

94 $\sqrt[n]{|a_n|} > 1/|x| \Rightarrow |a_n x^n| > 1$. So $R = 0$.

95 Take $\varepsilon = \frac{1}{2}/|x|$. $\sqrt[n]{|a_n|} < \varepsilon \Rightarrow |a_n x^n| < (\frac{1}{2})^n$. $R = \infty$.

96 $|x| < 2 \Rightarrow |\frac{1}{2}x| < 1 \Rightarrow \sum (\frac{1}{2}x)^{2n-1}$ absolutely convergent.
Also $|x| < 2 \Rightarrow |\frac{1}{3}x| < 1$.
But $|x| > 2 \Rightarrow |\frac{1}{2}x| > 1 \Rightarrow \sum (\frac{1}{2}x)^{2n-1}$ divergent.

97 Take $\varepsilon = A - 1/|x|$, then a subsequence of $(\sqrt[n]{|a_n|}) \to A$ has infinitely many terms such that $A - \varepsilon < \sqrt[n]{|a_n|}$, and thus infinitely many terms for which $1/|x| < \sqrt[n]{|a_n|}$, giving $1 < \sqrt[n]{|a_n x^n|}$. So the sequence of terms of the power series is not null when $|x| > 1/A$.
Choose $r = \frac{1}{2}(|x| + 1/A)$ for example, and take $\varepsilon = 1/r - A$. Eventually all $\sqrt[n]{|a_n|} < A + \varepsilon = 1/r$.
The radius of convergence is $1/A$.

98 $a_n = n$.

99 $a_n = 1/n$.

100 $a_n = 1/n^2$.

101 The first series is convergent when $-1 < x \leqslant 1$, second and third when $|x| < 1$.

102 $\limsup \sqrt[n]{|na_n|} = \limsup \sqrt[n]{n} \cdot \sqrt[n]{|a_n|} = \limsup |a_n|$.
$\sqrt[n]{(n+1)} \leqslant \sqrt[n]{(2n)} = \sqrt[n]{2} \cdot \sqrt[n]{n} \to 1$.
So $\limsup \sqrt[n]{(a_n/(n+1))} = \limsup \sqrt[n]{|a_n|}/\sqrt[n]{(n+1)} = \limsup \sqrt[n]{|a_n|}$.

103 $\sum |b_n|$ convergent $\Rightarrow (|b_n|)$ null $\Rightarrow |b_n| < 1$ for sufficiently large n. For such n, $0 \leqslant |a_n b_n| \leqslant |a_n|$. Now $\sum |a_n|$ convergent $\Rightarrow \sum |a_n b_n|$ convergent by the first comparison test.

104 $a_n = b_n = (-1)^n/\sqrt{n}$.

105 $a_n = 1/n$.

106 As in qn 104.

107 $\sum 1/n^2$ is convergent by qn 24 so its partial sums are bounded, by qn 22. $\sum (a_n)^2$ is convergent by hypothesis, so its partial sums are bounded, by qn 22. Now, by the Cauchy–Schwarz inequality,

$$\left(\sum_{r=1}^{r=n} \frac{a_r}{r} \right)^2 \leqslant \left(\sum_{r=1}^{r=n} a_r^2 \right) \left(\sum_{r=1}^{r=n} \frac{1}{r^2} \right),$$

so $(\sum a_n/n)^2$ is bounded, and so $\sum a_n/n$ is bounded.
The terms of $\sum a_n/n$ are positive, so the series is convergent by qn 22. If the constraint $a_n \geqslant 0$ is removed then $\sum a_n/n$ is absolutely convergent and so convergent.

108 $0 < c < AB \Rightarrow c/B < A \Rightarrow \frac{1}{2}(A + c/B) < A$, so $\sum a_n x^n$ is absolutely convergent for this value of x. Similarly for the other power series. Clearly

$$AB > \frac{1}{2}(A + c/B) \cdot \frac{1}{2}(B + c/A) = \frac{1}{4}(c^2/AB + 2c + AB)$$

$$= \left(\frac{1}{2}(c/\sqrt{AB} + \sqrt{AB}) \right)^2 > c \text{ by qn 2.39.}$$

Since $\sum a_n(\frac{1}{2}(A + c/B))^n$ and $\sum b_n(\frac{1}{2}(B + c/A))^n$ are absolutely convergent, $\sum a_n b_n (\frac{1}{2}(A + c/B) \cdot \frac{1}{2}(B + c/A))^n$ is absolutely convergent by qn 103, so $\sum a_n b_n c^n$ is absolutely convergent. Since this holds for any positive $c < AB$, the radius of convergence of $\sum a_n b_n x^n$ is at least AB.

109 If $a_n = (\frac{1}{2})^n$ and $b_n = (\frac{1}{3})^n$ then $A = 2$ and $B = 3$. Also $a_n b_n = (1/6)^n$ and radius of convergence of $\sum a_n b_n x^n = AB$.

110 Take $\sum a_n x^n = \sum ((\frac{1}{2}x)^{2n-1} + (\frac{1}{3}x)^{2n})$ and $\sum b_n x^n = \sum ((\frac{1}{3}x)^{2n-1} + (\frac{1}{2}x)^{2n})$, then $A = 2$ and $B = 2$ from qn 96, but the radius of convergence of $\sum a_n b_n x^n = 6$.

111 $c_n = a_0 b_n + a_1 b_{n-1} + \ldots + a_i b_{n-i} + \ldots + a_n b_0$.

112 The two series are convergent by the alternating series test.
$|c_n| \geqslant (n+1)/(\sqrt{(n+1)})(\sqrt{(n+1)}) = 1$. Not convergent by qn 11.

113 Use qn 1.5 for the proof. After completing qn 114 use qn 81 to understand the significance of this result.

114 (i) Use simple algebra. (ii) Use qn 3.44(iv). (iii) Simple algebra.
(iv) Use qn 3.44(iv).
(v) Cauchy product $= d_1 + (d_4 + d_2) + (d_9 + d_3 + d_5) + \ldots$.
(vi) Since $\sum d_n$ is absolutely convergent, any rearrangement is convergent to the same sum, from qn 73.

115 $\sum x^n$ is absolutely convergent when $|x| < 1$, from qn 10. In the Cauchy product, c_n is a sum of $n + 1$ terms.

116 Suppose

$$(1 - x)^{-n}$$

$$= 1 + nx + \frac{n(n + 1)}{1 \cdot 2} x^2 + \ldots + \frac{n(n + 1) \ldots (n + r - 1)}{r!} x^r + \ldots$$

which has radius of convergence 1 by qn 93. The coefficient of x^r in the Cauchy product with $(1 - x)^{-1}$ is

$$1 + n + \frac{n(n + 1)}{1 \cdot 2} + \ldots + \frac{n(n + 1) \ldots (n + r - 1)}{r!}$$

$$= (n + 1) \left(1 + \frac{n}{2} + \ldots + \frac{n(n + 2) \ldots (n + r - 1)}{r!} \right)$$

$$= \frac{(n + 1)(n + 2)}{2} \left(1 + \frac{n}{3} + \ldots + \frac{n(n + 3) \ldots (n + r - 1)}{3 \cdot 4 \cdot \ldots \cdot r} \right)$$

$$\vdots$$

$$= \frac{(n + 1)(n + 2) \ldots (n + r)}{r!},$$

which is in the same form as that assumed for n. When $n = 1$, the series takes the familiar form for a geometric progression. This establishes the expansion for n.

6

Functions and continuity
Neighbourhoods, limits of functions

Preliminary activity: make sure that you have access to graph-drawing facilities on a computer or graphic calculator, and that you can use these facilities with confidence.

Preliminary reading: Leavitt ch. 1.

Concurrent reading: Swann and Johnson, Reade, Smith, Spivak chs 4, 5, 6.

Functions

When you read or hear the phrase 'the function $f(x)$', what comes to your mind? Perhaps a formula, perhaps a graph.

1 Write down what x can stand for, and what is meant by f, in the expression $f(x)$. Compare your answer with the one in the summary at the end of the chapter.

We will introduce some special vocabulary in order to be clear what we mean when talking about functions.

The domain of a function

If $f(x) = x^2$ and the values of x are 0, ±1, ±2, ±3, ..., ±n, ..., then the values of $f(x)$ are 0, 1, 4, 9, ..., n^2, ... The set of possible values of x is called the *domain* of the function. When we say that x is a *variable*, we mean that the symbol x is being used to denote any member of the domain of a function. When the possible values of x are real numbers, the function is called a function of a *real variable*.

The range and co-domain of a function

The set of possible values of $f(x)$ is called the *range* of the function, and any set which contains the range may be declared to be the *co-domain* of the function. We have just given a function with domain $\mathbb{Z}$ and range $\mathbb{N} \cup \{0\}$, which we express symbolically by writing $f: \mathbb{Z} \to \mathbb{N} \cup \{0\}$, with the definition $f(x) = x^2$, (or, $f: x \mapsto x^2$).

The terms *function* and *mapping* are synonymous, and we sometimes say that the function f maps x to $f(x)$. In fact each of the sequences in chapter 3 is a function with domain $\mathbb{N}$. When both the domain and co-domain of f are subsets of $\mathbb{R}$, the function f is called a *real function*.

Arrow diagram of the function f

Domain of $f = \{$values of $x\}$

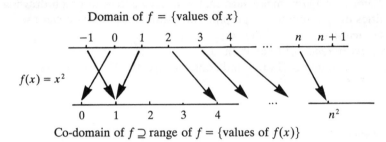

Co-domain of $f \supseteq$ range of $f = \{$values of $f(x)\}$

2 The following functions each have domain and co-domain $\mathbb{R}$. What is the range of each function?

(i) $f(x) = x^2$, (ii) $f(x) = x^3$, (iii) $f(x) = \sin x$,

(iv) $f(x) = 1/(1 + x^2)$, (v) $f(x) = e^x$

[The sine and exponential functions will be defined formally in chapter 11.]

When the co-domain and range of a function are the same, the function is said to be *onto* the co-domain and is called a *surjection*. Check that only one of the functions in question 2 is *onto* $\mathbb{R}$.

The distinctive property of a function or mapping f is that, for a given element x of the domain, $f(x)$ is uniquely determined as a member of the co-domain. Thus $\pm\sqrt{(1 - x^2)}$ does not define a function of x, unless the domain is restricted to $\{\pm 1\}$, for then the range is simply $\{0\}$. But even $f(x) = \sqrt{(1 - x^2)}$ does not define a function with domain and co-domain $\mathbb{R}$, for then $f(x)$ is not defined when $|x| > 1$.

3 Each of the following functions has domain $A \subseteq \mathbb{R}$ and co-domain $\mathbb{R}$. What is the largest possible domain A for the function?

   (i) $f(x) = \sqrt{x}$, (ii) $f(x) = \sqrt{(1 - x^2)}$, (iii) $f(x) = \sqrt{(x^2 - 1)}$,

   (iv) $f(x) = 1/x$, (v) $f(x) = (x^2 - 4)/(x - 2)$,

   (vi) $f(x) = \log x$.

It is sometimes convenient to denote the range of the function $f: A \to \mathbb{R}$ by $f(A)$ and to say that f maps A to $f(A)$.

Although, for every function, f, $x = y \Rightarrow f(x) = f(y)$, it is only sometimes the case that $f(x) = f(y) \Rightarrow x = y$. When this second implication holds, the function f is said to be *one-to-one* or *one-one*, and is called an *injection*.

4 Which of the functions given in questions 2 and 3 are one-one?

Bijections and inverse functions

If a function $f: A \to B$, with domain A and co-domain B, is both one-one *and* onto, not only is each element $a \in A$ matched with a unique element $f(a) \in b$, which is, of course, true for any function f, but also, for each element $b \in B$, there is a unique element $a \in A$ such that $f(a) = b$. A function which is both one-one and onto is called a *bijection*. Such a function has an inverse $g: B \to A$, defined by $g(f(a)) = a$. When this is the case, we will write $g = f^{-1}$.

5 For each of the functions described in qns 2 and 3, identify appropriate subsets A, $B \subseteq \mathbb{R}$, such that $f: A \to B$ is a bijection.

6 The points on the graph of a real function f have the form $(x, f(x))$. If $f: A \to B$ is a bijection, why do the points on the graph of f^{-1} have the form $(f(x), x)$? Sketch the graph of the function f given by $f(x) = x^2$ for positive x and sketch the graph of its inverse function $f^{-1}(x) = \sqrt{x}$. Also sketch the graph of exp (the function E in chapter 11) and its inverse log.

Continuity

Throughout the eighteenth century, functions were given by algebraic formulae, with variables being added, subtracted, multiplied or divided, and combined with exponential, logarithmic or trigonometric functions. Controversy surrounded the differential equation for a plucked string, since in the ordinary starting position two straight-line parts of the string meet at an angle, and thereby require different algebraic formulae for different parts of the string. Cauchy took the example of the function defined by $f(x) = \sqrt{(x^2)}$ to show that even familiar algebraic expressions might lead to a graph with a sharp bend in it.

7 Illustrate on a graph the functions $f: \mathbb{R} \to \mathbb{R}$ given by

(i) $f(x) = 0$, when $x < 0$, and $f(x) = x$ when $x \geqslant 0$;

(ii) $f(x) = 0$, when $x < 0$, and $f(x) = x^2$ when $x \geqslant 0$;

(iii) $f(x) = -x^2$ when $x < 0$, and $f(x) = x^2$ when $x \geqslant 0$.

(iv) $f(x) = x \cdot |x|$.

Allowing a variety of formulae when defining a single function was a significant first step towards a more general notion of function. Further steps were taken when limiting processes were considered.

8 Examine the graphs of $y = x^2$, $y = x^3$, $y = x^4$ and $y = x^{10}$ using a computer or a graphic calculator.
For what values of x does $\lim x^n$ exist as $n \to \infty$?
What is the largest subset $A \subseteq \mathbb{R}$ for which a function $f: A \to \mathbb{R}$ may be defined by $f(x) = \lim x^n$ as $n \to \infty$?
What is the range of this function? Sketch its graph.

9 Examine the graphs of $y = \dfrac{x^n - 1}{x^n + 1}$, for $n = 2, 3, 10, 11$, using a computer or a graphic calculator.
For what values of x does $\lim \dfrac{x^n - 1}{x^n + 1}$ exist as $n \to \infty$?
What is the largest subset $A \subseteq \mathbb{R}$ for which a function $f: A \to \mathbb{R}$ may be defined by $f(x) = \lim \dfrac{x^n - 1}{x^n + 1}$ as $n \to \infty$?
What is the range of this function? Sketch its graph.

10 Examine the graphs of $y = x^2/(x^2 + 1/n)$, for $n = 2, 3, 10, 11$, using a computer or a graphic calculator.
If the function $f: \mathbb{R} \to \mathbb{R}$ is defined by
$f(x) = \lim x^2/(x^2 + 1/n)$ as $n \to \infty$, what is the range of f?
Sketch the graph of f.

11 Identify the function of x defined by $\lim\limits_{n \to \infty} \dfrac{1}{1 + n \sin^2 \pi x}$.

Questions 8, 9, 10 and 11 make it evident that the value of $f(x)$ need not be close to the value of $f(y)$, even when x and y are close to one another. The modern definition of function only requires that for each value of x in the domain there is a unique value of $f(x)$ in the co-domain, and the function is then defined by the pairing $(x, f(x))$, even when something like a fresh formula is needed for each number x.

Definition of continuity by sequences

12 Sketch the graph of the function $f: \mathbb{R} \to \mathbb{Z}$, given by $f(x) = [x]$, the integer part of x, defined before eqn 3.15.

13 Find the limit of the sequence $([1/n])$ as $n \to \infty$.

14 Find the limit of the sequence $([-1/n])$ as $n \to \infty$.

15 Find two sequences $(a_n) \to 1$, and $(b_n) \to 1$, such that $([a_n]) \to 1$ and $([b_n]) \to 0$.

16 For *any* sequence (a_n) which converges to $\frac{1}{2}$, prove that $([a_n]) \to 0$.

The results of questions 13–16 are the basis on which we say that the function f of qn 12 is *continuous* at $x = \frac{1}{2}$ and is *not continuous* at $x = 0$ or $x = 1$.

17 Is the function f of qn 12 continuous at $x = 1\frac{3}{4}$? The critical point to test is whether for *every* sequence $(a_n) \to 1\frac{3}{4}$, the sequence $(f(a_n)) \to 1 = f(1\frac{3}{4})$.

18 Describe the subset of $\mathbb{R}$ of values of x at which the integer function of qn 12 is *not* continuous. Also describe the subset of $\mathbb{R}$ of values of x at which this function *is* continuous.

The ideas in questions 12–18 lead to the following definition.

Definition of continuity
The function $f: A \to \mathbb{R}$ is continuous at $a \in A \subseteq \mathbb{R}$ when, for every sequence (a_n) converging to a, with terms in A, the sequence $(f(a_n))$ converges to $f(a)$.
This is commonly expressed by saying that '$f(x)$ tends to $f(a)$ as x tends to a' which is also expressed symbolically by writing $f(x) \to f(a)$ as $x \to a$.
The function $f: A \to \mathbb{R}$ is said to be *continuous on A*, when it is continuous at every $a \in A$.

Examples of discontinuity

Continuous functions are so familiar that to clarify the meaning of this definition we need some examples of discontinuity, illustrating the absence of continuity. The continuity of a function f at a point a is recognised by the way the function affects *every* sequence (a_n) tending to a. Discontinuity is established by finding just one sequence $(a_n) \to a$ for which $(f(a_n))$ does not tend to $f(a)$.

19 Examine the graph of $y = \sin 1/x$ using a computer or a graphic calculator, paying particular attention to the values of y when x is small.

We define a function $f: \mathbb{R} \to \mathbb{R}$ by
$f(x) = \sin 1/x$ when $x \neq 0$, and $f(0) = 0$.
Let $a_n = 1/(2n\pi)$ and $b_n = 1/((2n + \frac{1}{2})\pi)$.

Which of the four sequences
(i) (a_n), (ii) (b_n), (iii) $(f(a_n))$, (iv) $(f(b_n))$
are null sequences?
Deduce that f is not continuous at $x = 0$.

If $f(0)$ had some value other than 0, might this redefined function f be continuous at $x = 0$?

20 (*Dirichlet*, 1829) Define a function $f: \mathbb{R} \to \mathbb{R}$ by $f(x) = 0$ when x is rational, and $f(x) = 1$ when x is irrational.
If a is a rational number, show that f is not continuous at $x = a$ by considering the sequence (a_n), where $a_n = a + \sqrt{2}/n$.
If a is an irrational number, show that f is not continuous at $x = a$ by considering the sequence $([10^n a]/10^n)$, the infinite decimal for a, of qn 3.43.

Question 18 gives us an example of a function with an infinity of isolated discontinuities, qn 19 an example of a function with a single discontinuity, and qn 20 an example of a function with a dense set of discontinuities. This will be enough to start with, and we now turn to the formal confirmation of continuity in the familiar context of polynomials.

Sums and products of continuous functions

21 If a real function f is defined by $f(x) = c$, that is to say, f is a constant function, prove that f is continuous at each point of its domain.

22 If a real function f is defined by $f(x) = x$, so that f is the identity function, prove that f is continuous at each point of its domain.

23 If real functions f and g have the same domain A, and are both continuous at $x = a \in A$, use qn 3.44(ii) to prove that the real function $f + g$ defined by

$$(f + g)(x) = f(x) + g(x)$$

is also continuous at $x = a$.

24 Use qns 21, 22 and 23 to prove that the function given by $f(x) = x + 1$ is continuous at each point of its domain.

25 If $f_1, f_2, \ldots, f_n$ are all real functions with the same domain A, and are all continuous at $x = a \in A$, prove by induction that the function $f_1 + f_2 + \ldots + f_n$, defined by
$$(f_1 + f_2 + \ldots + f_n)(x) = f_1(x) + f_2(x) + \ldots + f_n(x)$$
is continuous at $x = a$.

26 If real functions f and g have the same domain A, and are both continuous at $x = a \in A$, use qn 3.44(iv) to prove that the real function $f \cdot g$ defined by

$$(f \cdot g)(x) = f(x) \cdot g(x)$$

is also continuous at $x = a$.

27 Which of qns 21–26 are required, and in which order, to prove that the functions $f_i \colon \mathbb{R} \to \mathbb{R}$, given by

  (i) $f_1(x) = 2x$, (ii) $f_2(x) = 2x + 1$, (iii) $f_3(x) = x^2$,

  (iv) $f_4(x) = x^3$, (v) $f_5(x) = 2x^2 + 1$,

are continuous on $\mathbb{R}$?

28 Prove by induction that the function $f \colon \mathbb{R} \to \mathbb{R}$, given by $f(x) = x^n$, where $n \in \mathbb{N}$, is continuous on $\mathbb{R}$.

29 Prove that the function $f \colon \mathbb{R} \to \mathbb{R}$, given by

$$f(x) = a_0 + a_1 x + a_2 x^2 + \ldots + a_n x^n,$$

is continuous on $\mathbb{R}$.
This establishes that polynomial functions are continuous on $\mathbb{R}$.

30 The acid test of understanding the language we have introduced comes when deciding at what points the function given by $f(x) = 1/x$ is continuous. With reference to qn 3, determine the maximum domain of definition of this real function. Use qns 3.54

and 3.55 to prove that this function is continuous at every point where it is defined.

31 Use qn 30 to extend the result of qn 29 to give a family of functions each one of which is continuous on $\mathbb{R}\setminus\{0\}$.

Continuity in less familiar settings

32 Use qn 3.44(v) to prove that the function $f: \mathbb{R} \to \mathbb{R}$ given by $f(x) = |x|$ is continuous on $\mathbb{R}$. Sketch the graph of f near the origin.

33 Sketch the graph of the real function given by $f(x) = \frac{1}{2}(x + |x|)$, and prove that it is continuous on $\mathbb{R}$.

34 Sketch the graph of the real function given by $f(x) = \frac{1}{2}(x - |x|)$, and prove that it is continuous on $\mathbb{R}$.

35 We define a function $f: \mathbb{R} \to \mathbb{R}$ by

$f(x) = 0$ when x is rational,

$f(x) = x$ when x is irrational.

Prove that, for all x

$$\tfrac{1}{2}(x - |x|) \leqslant f(x) \leqslant \tfrac{1}{2}(x + |x|).$$

Locate the graph of the function f in relation to those you have drawn for qns 33 and 34.
Is there a value of x for which $\frac{1}{2}(x - |x|) = \frac{1}{2}(x + |x|)$?

A sandwich theorem

36 If real functions f, g, and h have the same domain A, and

(i) $f(x) \leqslant g(x) \leqslant h(x)$, for all $x \in A$,

(ii) f and h are continuous at $x = a \in A$,

(iii) $f(a) = h(a)$,

use qn 3.44(vi) to prove that g is continuous at $x = a$.

37 Use qn 36 to prove that the function f of qn 35 is continuous at $x = 0$, and use arguments like those of qn 20 to prove that this function is not continuous at any other point in its domain.
This example highlights an important consequence of our definition of continuity since the possibility of continuity at a

single isolated point does not spring from an intuitive view of the concept.

Continuity of composite functions and quotients of continuous functions

38 If $f: \mathbb{R} \to \mathbb{R}$ is defined by $f(x) = \sin x$ and $g: \mathbb{R} \to \mathbb{R}$ is defined by $g(x) = x^2$, what are $f(g(x))$ and $g(f(x))$?
Look at the graphs of these two composite functions on a computer screen.
Generally, if f and g are two real functions and the domain of f contains the range of g, the *composite function* $f \circ g$ is defined on the domain of g by $(f \circ g)(x) = f(g(x))$.
Notice that the composite function $f \circ g$ is generally different from the composite function $g \circ f$.

39 If $\mathrm{ABS}(x) = |x|$ and $f(x) = x + 1$, sketch the graphs of $f \circ \mathrm{ABS}$ and $\mathrm{ABS} \circ f$.

40 If the function $g: A \to \mathbb{R}$ is continuous at $x = a$ and the function $f: g(A) \to \mathbb{R}$ is continuous at $g(a)$, prove that $f \circ g$ is continuous at $x = a$.

41 Use qn 40 to prove that, if a real function f is continuous at $x = a$, then the function $\mathrm{ABS} \circ f$ is also continuous at a, where ABS is defined as in qn 39.

42 For any real function f, we define a function f^+ on the same domain as f by $f^+(x) = \frac{1}{2}(f(x) + |f(x)|)$. If f is continuous at a, prove that f^+ is also continuous at a.
Sketch the graph of f^+ when $f(x) = x^2 - 1$.

43 For any real function f, we define a function f^- on the same domain as f by $f^-(x) = \frac{1}{2}(f(x) - |f(x)|)$. If f is continuous at a, prove that f^- is also continuous at a.
Sketch the graph of f^- when $f(x) = x^2 - 1$.

44 For any two real numbers a and b, we define
$$\max(a, b) = \begin{cases} a & \text{when } b \leqslant a, \\ b & \text{when } a < b. \end{cases}$$
Prove that $\max(a, b) = \frac{1}{2}(a + b) + \frac{1}{2}|a - b|$.
Prove that $f^+(x) = \max(f(x), 0)$.

45 If real functions f and g have the same domain A, and are both continuous at $x = a \in A$, prove that the function $\max(f, g)$, defined by $\max(f, g)(x) = \max(f(x), g(x))$ is also continuous at $x = a$.

46 If $f(x) = x$ and $g(x) = -x$, identify the function $\max(f, g)$.

47 Let $r: \mathbb{R}\backslash\{0\} \to \mathbb{R}$ denote the reciprocal function $r(x) = 1/x$, and let $f(x) = 1 + x^2$. What is the maximum domain of definition of $f \circ r$ and of $r \circ f$? Are both these functions continuous throughout their domains? Use a computer or a graphic calculator to examine the graphs of f, $r \circ f$ and $f \circ r$.

48 Assuming that the sine function is continuous on $\mathbb{R}$ (see chapter 11), determine the largest subset of $\mathbb{R}$ on which the function f of qn 19 is continuous.

49 Use a computer or a graphic calculator to examine the graph of $y = x \sin 1/x$.
A function f is defined on $\mathbb{R}$ by

$f(x) = x \sin 1/x$ when $x \neq 0$,
$f(0) = 0$.

Assuming that the sine function is continuous on $\mathbb{R}$ and that $|\sin x| \leqslant 1$ for all values of x, prove that f is continuous on $\mathbb{R}$. (Hint: use qns 48, 22 and 26 for $x \neq 0$, and qns 32, 32 with 26, and 36 for $x = 0$.)

50 Let $f(x) = (x - 2)(x - 3)$ be defined on $\mathbb{R}$ and let r denote the same function as in qn 47.
What is the maximum domain of definition of $r \circ f$? Is this function continuous throughout its domain? Examine the graphs of f and $r \circ f$ on a computer or a graphics calculator.

51 Let $f(x) = a_0 + a_1 x + a_2 x^2 + \ldots + a_n x^n$, be defined on $\mathbb{R}$. Let r denote the same function as in qn 47. What is the maximum domain of definition of $r \circ f$? Is this function continuous throughout its domain?

52 Let $f: A \to \mathbb{R}$ be a real function which is continuous at $x = a$, with $f(a) \neq 0$. Let r denote the same function as in qn 47. Prove that $r \circ f$ is continuous at $x = a$.
The function $r \circ f$ is also commonly denoted by $1/f$. Give a direct proof that $1/f$ is continuous at a using qn 3.54 and 3.55.

53 Prove that the function given by $f(x) = (x^3 - x)/(x^2 + 1)$ is continuous on $\mathbb{R}$. Examine its graph on a computer or a graphics calculator.

54 Let $f: A \rightarrow \mathbb{R}$ and $g: A \rightarrow \mathbb{R}$ be real functions which are both continuous at a. Let $G = \{x \mid g(x) = 0\}$. If $g(a) \neq 0$, prove that the function $f/g: A \setminus G \rightarrow \mathbb{R}$ defined by $(f/g)(x) = f(x)/g(x)$ is continuous at a. Either consider $f \cdot (r \circ g)$ or use qn 3.56 directly.

55 The function $f: \mathbb{R} \rightarrow \mathbb{R}$ has the property that

$$f(x + y) = f(x) + f(y) \text{ for all } x, y \in \mathbb{R}.$$

  (i) Prove that $f(0) = 0$.
  (ii) Prove that $f(-x) = -f(x)$.
  (iii) If f is continuous at $x = 0$, prove that f is continuous on $\mathbb{R}$.
  (iv) If $f(1) = a$, prove that $f(n) = an$, for $n \in \mathbb{Z}$.
  (v) If $f(1) = a$, prove that $f(p/q) = ap/q$, for $p, q \in \mathbb{Z}$, $q \neq 0$.
  (vi) If $f(1) = a$ and f is continuous on $\mathbb{R}$, prove that $f(x) = ax$.

Neighbourhoods

   We say that a function f is continuous at a point $x = a$ when '$f(x)$ tends to $f(a)$, as x tends to a', which we may think of as meaning 'as x gets near to a, $f(x)$ gets near to $f(a)$'. So far, we have always described *nearness* using sequences and convergence. Another way to describe nearness is to consider *neighbourhoods* of a point a.

56 Illustrate the sets

$$\{x: -1 < x < 3\}, \{x: -2 < x - 1 < 2\}, \{x: |x - 1| < 2\}$$

on a number line. They are, in fact, identical. Such a set is called a 2-neighbourhood of the point 1. The neighbourhood has centre 1 and radius 2. Compare this claim with qn 3.50.

57 Prove that, if $|x - 2| < 0.2$, then $|x^2 - 4| < 1$.
Illustrate this result on a graph of the function $x \mapsto x^2$.

58 Prove that, if $|x - 3| < 0.1$, then $|x^2 - 9| < 1$.

59 Prove that, if $|x - 2| < 1/3$, then $|1/x - 1/2| < 1/10$.

60 Prove that, if $|x - 2| < 0.02$, then $|\sqrt{x} - \sqrt{2}| < 0.01$.

61 By considering $x = -2$ in qn 57, $x = -3$ in qn 58, $x = 2.4$ in qn 59 and $x = 1.975$ in qn 60, show that none of the implications that you have established in qns 57–60 may be reversed.

Each of these implications takes the form:

if $|x - a| < \delta$, then $|f(x) - f(a)| < \varepsilon$.

Expressed rather crudely, each says that if x is near to a then $f(x)$ is near to $f(a)$.

If we want to pinpoint continuity this way, it matters just *how* near is *near*. We can investigate the kind of connection between δ and ε that is needed for continuity by examining a point of discontinuity.

62 Let $f(x) = [x]$, the integer function, and let $a = 2$.

If $|x - 2| < 2$, can you find an ε such that $|f(x) - f(2)| < \varepsilon$?
If $|x - 2| < 1$, can you find an ε such that $|f(x) - f(2)| < \varepsilon$?
If $|x - 2| < \frac{1}{2}$, can you find an ε such that $|f(x) - f(2)| < \varepsilon$?
If $|x - 2| < \frac{1}{4}$, can you find an ε such that $|f(x) - f(2)| < \varepsilon$?

The fact that we can find an ε every time in qn 62 tells us nothing whatever about the continuity of f.

If we turn the question round, it begins to bite.

63 Let $f(x) = [x]$ as before.

Can you find a δ such that, if $|x - 2| < \delta$,
then $|f(x) - f(2)| < 2$?
Can you find a δ such that, if $|x - 2| < \delta$,
then $|f(x) - f(2)| < 1$?
Can you find a δ such that, if $|x - 2| < \delta$,
then $|f(x) - f(2)| < \frac{1}{2}$?
Can you find a δ such that, if $|x - 2| < \delta$,
then $|f(x) - f(2)| < \frac{1}{4}$?

Definition of continuity by neighbourhoods

With our success in finding εs in qn 62 and our failure to find δs in qn 63, it looks as if we can pin down continuity at $x = a$ by saying

> **the function $f: A \to \mathbb{R}$ is continuous at $a \in A \subseteq \mathbb{R}$,**
> **provided that, whatever positive ε we choose,**
> **we can find a δ such that $x \in A$ and**
> **$|x - a| < \delta$ implies $|f(x) - f(a)| < \varepsilon$.**

We first of all see whether this holds in cases that we know are continuous.

64 Find a δ such that $|x - 3| < \delta$ implies $|x^2 - 9| < 0.5$.

65 Find a δ such that $|x - 3| < \delta$ implies $|1/x - 1/3| < 1/6$.

66 Find a δ such that $|x - 3| < \delta$ implies $|\sqrt{x} - \sqrt{3}| < 0.01$.

So we know that the 'neighbourhood definition of continuity' fails in a case (qn 63) where we know the function is not continuous and we have found the neighbourhood definition of continuity holds in three cases (qns 64, 65 and 66) where we know the function is continuous. Can we show that this neighbourhood definition of continuity is equivalent to the sequence definition that we have used in the first half of this chapter?

67 Suppose that the neighbourhood definition of continuity holds for a function f and a point $x = a$ in the domain of the function. Let $(a_n) \to a$ be a sequence in the domain of the function. Can we show that the sequence $(f(a_n)) \to f(a)$?

(i) Interpret the claim that $(f(a_n)) \to f(a)$ in terms of the definition of convergence using qn 3.49.

(ii) Use the neighbourhood definition of convergence to find a neighbourhood of a, such that $|f(a_n) - f(a)| < \varepsilon$, when a_n is in the neighbourhood.

(iii) Use the convergence of (a_n) to show that terms of the sequence (a_n) eventually lie in the neighbourhood you found in part (ii).

Question 67 establishes that neighbourhood continuity at a point implies sequence continuity at that point. Can we show that sequence continuity implies neighbourhood continuity? The sequence definition of continuity considers a collection of sets of points $\{a_n\}$, clustering about a, in the domain of the function f. Although all the sequences converge to a, the definition seems to leave in doubt the question as to whether, if f is sequentially continuous at a, given $\varepsilon > 0$, there must be a δ-neighbourhood of a such that, *whenever* x belongs to that neighbourhood and to the domain of f, $f(x)$ is in an ε-neighbourhood of $f(a)$.

To clarify this, we suppose that sequence continuity holds and neighbourhood continuity fails so that *every* neighbourhood of a contains rogue points x, for which $|f(x) - f(a)| \geq \varepsilon$, and we obtain a contradiction.

68 Suppose that f is a real function which is sequentially continuous at a, and that, for some $\varepsilon > 0$, there is no δ-neighbourhood of a such that, for all x in the neighbourhood, $|f(x) - f(a)| < \varepsilon$.

(i) Why must there be a value of x within the neighbourhood $\{x \mid a - 1/n < x < a + 1/n\}$ such that $|f(x) - f(a)| \geq \varepsilon$?

(ii) If the value of x in (i) is called a_n, how do you know that $(a_n - a)$ is a null sequence?

(iii) With the notation of (i) and (ii), how do you know that the sequence $(f(a_n))$ does *not* tend to $f(a)$? This contradicts our starting point.

The contradiction obtained in qn 68 leads to the conclusion that, if the function f is sequentially continuous at a, then f is neighbourhood continuous at a.

Neighbourhood continuity is in fact used by many authors as the basic definition of continuity. In contrast to the sequence definition, where the infinite process is obvious, the $\varepsilon-\delta$ neighbourhood definition of continuity looks deceptively finite. The infinite process is hidden in the fact that the condition must hold for *any* ε, which must be thought to take an infinity of values, and in particular to get arbitrarily small.

There are two simple properties (that will be needed in the next chapter) which hold for a function continuous at a point, which are easily established with the neighbourhood definition of continuity.

69 If a function f is continuous at a point a, by choosing $\varepsilon = 1$, prove that $f(x)$ is bounded (above *and* below) in some neighbourhood of a.

70 If a function f is continuous at a point a and $f(a) > 0$, by a judicious choice of ε, prove that $f(x) > 0$ in some neighbourhood of a. State and prove an analogous result in case $f(a) < 0$.

The last question in this section (qn 72) is designed to expose one of the most counter-intuitive possibilities that follows from the definition of continuity, namely that the points at which a function is continuous and the points at which a function is not continuous may be densely packed together in the domain of the function.

As a preparation for a new idea to be used in qn 72 we offer qn 71 as an option.

(71) Find a neighbourhood of $\sqrt{2}$ which contains no numbers of the form $n/2$ for any integer n.
Find a neighbourhood of $\sqrt{2}$ which contains no numbers of the form $n/3$ for any integer n.
Find a neighbourhood of $\sqrt{2}$ which contains no numbers of the form $n/4$ for any interger n.
Find a neighbourhood of $\sqrt{2}$ which contains no numbers of the form $n/5$ for any integer n.
Find a neighbourhood of $\sqrt{2}$ which contains no numbers of the form $n/2$, $n/3$, $n/4$ or $n/5$ for any integer n.

72 A function $f: (0, 1] \rightarrow \mathbb{R}$ is defined by

$$f(x) = \begin{cases} 0 & \text{when } x \text{ is irrational,} \\ 1/q & \text{when } x \text{ is rational and } x = p/q \text{ in lowest terms.} \end{cases}$$

This function is sometimes called the *Ruler function* because of the resemblance between parts of its graph and the marks on a pre-decimal ruler.

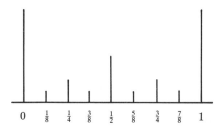

(a) By constructing a sequence of irrational numbers tending to each rational point, prove that *f* is not continuous at any rational point.

No surprises so far. The really astonishing property of this function is that it is *continuous* at every irrational point.

(b) Now let *a* be an irrational number with $0 < a < 1$.
Given $\varepsilon > 0$, we seek a δ such that

$$|x - a| < \delta \Rightarrow |f(x) - f(a)| < \varepsilon.$$

(i) Is there a positive integer *m* such that $1/m < \varepsilon$?
(ii) How many rational numbers p/q can there be between 0 and 1 with $q \leqslant m$? Might any one of them equal *a*?
(iii) Is there a shortest distance $|a - p/q|$ for the rational numbers p/q in part (ii)?
(iv) If we take the shortest distance in (iii) as δ, how big might $|f(x) - f(a)|$ get when $|x - a| < \delta$?
(v) Does this establish continuity by the neighbourhood definition at each irrational point?

One-sided limits

Definition of one-sided limits by sequences

73 When examining the discontinuities of the integer function $f(x) = [x]$, we saw that although the sequence $(1 - 1/n)$ tends to 1, the sequence $(f(1 - 1/n))$ tends to 0 which is not equal to $f(1)$. Construct another sequence (a_n) which tends to 1, but for which the sequence $(f(a_n))$ tends to 0. What property of the sequence (a_n) is necessary for this to be the case?
If $a_n \leqslant 1$ and $(a_n) \to 1$, must $(f(a_n)) \to 0$?
If $a_n < 1$ and $(a_n) \to 1$, must $(f(a_n)) \to 0$?

Some kinds of discontinuity may be identified and discussed by considering the two sides of the point in question separately. If $a_n < a$ and $(a_n) \to a$ implies that $(f(a_n)) \to l$, we write

$$\lim_{x \to a^-} f(x) = l,$$

which is read 'as x tends to a from below, $f(x)$ tends to l', and is also written 'as $x \to a^-$, $f(x) \to l$'.

74 What is $\lim\limits_{x \to a^-} [x]$ for $a = 1, 2, 0$?

75 Prove that if $f: \mathbb{R} \to \mathbb{R}$ is continuous at a, then

$$\lim_{x \to a^-} f(x) = f(a).$$

76 Give a formal definition of what is meant by

$$\lim_{x \to a^+} f(x) = l,$$

which is read 'as x tends to a from above, $f(x)$ tends to l', and is also written 'as $x \to a^+$, $f(x) \to l$', by strict analogy with limits from below.

77 What is $\lim\limits_{x \to a^+} [x]$ for $a = 1, 0, -1$?

78 Prove that if $f: \mathbb{R} \to \mathbb{R}$ is continuous at a, then

$$\lim_{x \to a^+} f(x) = f(a)$$

79 Prove that if a is a point in the domain of the real function f, and

$$\lim_{x \to a^+} f(x) \neq \lim_{x \to a^-} f(x)$$

then f is not continuous at a.

One-sided limits for functions can be defined without knowing the value of the function at the point which seems to be in question, and this gives a way of pinpointing discontinuity at a point. The fact that the value of the function at the point in question does not come into the definition means that a one-sided limit may exist at a point even when the function itself does not.

80 Find $\lim\limits_{x \to 0^-} |x|/x$ and $\lim\limits_{x \to 0^+} |x|/x$.

Notice that although $|x|/x$ is not defined at $x = 0$ the two limits have well-defined values.

If a function $f: \mathbb{R} \to \mathbb{R}$ is defined by $f(x) = |x|/x$ when $x \neq 0$, and $f(0) = k$, use qn 79 to show that f is not continuous at 0 for any choice of k.

There are important limits that cannot be reached which may be interesting in their own right quite apart from questions about continuity.

81 Let f be the function $f: \mathbb{R} \setminus \{0\} \to \mathbb{R}$ given by

$$f(x) = \frac{\sin x}{x}.$$

The function is not defined at $x = 0$, but that does not stop there being a limit from above, or a limit from below as x tends to 0. Explore this function using a calculator. Remember to have x in radians. Investigate the limit of this function as x tends to 0 from above.

Use a calculator to find $f(x)$ when $x = 1, \frac{1}{2}, \frac{1}{4}, \frac{1}{8}$, and continue to find $f(x)$ when $x = (\frac{1}{2})^n$, for $n = 0, 1, 2, \ldots$ until no further change in value takes place, because of the limits of accuracy of your calculator.

Some calculators have a 'Fix' key, by which the number of displayed decimal places is chosen in advance.

(i) Identify the connection between the 'Fix' key and ε-neighbourhoods by deciding what value of ε makes

   $f(x) = 1.00$ *correct to two place of decimals*

   equivalent to

   $|f(x) - 1| < \varepsilon$.

(ii) Also decide what value of ε makes

   $f(x) = 1.000$ *correct to three places of decimals*

   equivalent to

   $|f(x) - 1| < \varepsilon$.

(iii) So far as you can tell using your calculator, is
   $|f(x) - 1| < 0.5$ when $0 < x < \frac{1}{2}$?
   Can you find a $\delta > \frac{1}{2}$ such that $|f(x) - 1| < 0.5$ when $0 < x < \delta$?
   Find a δ such that $|f(x) - 1| < 0.05$ when $0 < x < \delta$.
   Find a δ such that $|f(x) - 1| < 0.005$ when $0 < x < \delta$.

In qn 9.27 we will prove that $\lim\limits_{x \to 0^-} f(x) = 1 = \lim\limits_{x \to 0^+} f(x)$, for the function of qn 81 but, despite qn 79, the function f is not continuous at $x = 0$ because it is not defined there.

82 By choosing a value for $f(0)$ show that it is possible to have a function defined at $x = 0$ for which

$$\lim_{x \to 0^-} f(x) = 1 = \lim_{x \to 0^+} f(x),$$

but which is still not continuous at $x = 0$.

Definition of one-sided limits by neighbourhoods

83 A real function f is defined in a neighbourhood of the point a. If, given any $\varepsilon > 0$, there exists a δ such that, when $a < x < a + \delta$, $|f(x) - l| < \varepsilon$, prove that

$$\lim_{x \to a^+} f(x) = l.$$

The proof may be constructed like that in qn 67.

84 Prove that if $\lim_{x \to a^+} f(x) = l$, it follows that, given $\varepsilon > 0$, there

exists a δ such that, when x is in the domain of f and $a < x < a + \delta$, then $|f(x) - l| < \varepsilon$.
Obtain this proof, by contradiction, by supposing that the sequence definition of limit holds, while the neighbourhood definition fails, and finding an a_n such that $a < a_n < a + 1/n$ and $|f(a_n) - l| \geq \varepsilon$, and examining the sequences (a_n) and $(f(a_n))$. The idea of this proof is used in qn 68.

In qns 83 and 84 we have established that a convergent sequence definition of limit from above and a neighbourhood definition of limit from above are equivalent.

85 State a neighbourhood definition of limit from below equivalent to the convergent sequence definition of limit from below given after qn 73.

Two-sided limits

Definition of continuity by limits

If both $\lim_{x \to a^-} f(x) = l$ and $\lim_{x \to a^+} f(x) = l$

it is customary to write

$$\lim_{x \to a} f(x) = l$$

or 'as $x \to a$, $f(x) \to l$' and to say 'as x tends to a, $f(x)$ tends to l'.

86 If $f: \mathbb{R} \to \mathbb{R}$ is continuous at a, use qns 75 and 78 to show that $\lim_{x \to a} f(x) = f(a)$.

87 Write down the neighbourhood definitions of

$$\lim_{x \to a^-} f(x) = l \quad \text{and} \quad \lim_{x \to a^+} f(x) = l.$$

Use these definitions to show that, if $\lim_{x \to a} f(x) = l$,

then, given $\varepsilon > 0$, there exists a δ such that when x is in the domain of f and $0 < |x - a| < \delta$, then $|f(x) - l| < \varepsilon$.

The important point here is recognising that $0 < |x - a| < \delta$ means

either $a - \delta < x < a$ *or* $a < x < a + \delta$

and the answer to the question consists of showing how to choose the δ.

88 (i) If, given $\varepsilon > 0$, there exists a δ such that when x is in the domain of f and $0 < |x - a| < \delta$, then $|f(x) - l| < \varepsilon$, prove that

$$\lim_{x \to a} f(x) = l.$$

Use neighbourhoods for this proof, not sequences.

(ii) If for every sequence $(a_n) \to a$ in the domain of the function f, with $a_n \neq a$, $(f(a_n)) \to l$, prove that

$$\lim_{x \to a} f(x) = l.$$

89 If, given $\varepsilon > 0$, there exists a δ such that when x belongs to the domain of f and $0 < |x - a| < \delta$, then $|f(x) - f(a)| < \varepsilon$, prove that when x belongs to the domain of f and $|x - a| < \delta$, then $|f(x) - f(a)| < \varepsilon$.

Deduce that if $\lim_{x \to a} f(x) = f(a)$ then f is continuous at a.

Now we have established that a function f is continuous at a point a within a neighbourhood of its domain if and only if

$$\lim_{x \to a} f(x) = f(a).$$

90 Find a value of k such that the function $f: \mathbb{R} \rightarrow \mathbb{R}$ defined by

$$f(x) = \begin{cases} 2x & \text{when } x < 1, \\ k & \text{when } x \geq 1, \end{cases}$$

is continuous at $x = 1$.

91 If the function $f: [a, b] \rightarrow \mathbb{R}$ is continuous at each point of its domain give the value of

$$\lim_{x \to a^+} f(x), \quad \lim_{x \to b^-} f(x) \quad \text{and} \quad \lim_{x \to c} f(x) \quad \text{for} \quad c \in (a, b)$$

92 If

$$\lim_{x \to a^+} f(x) = f(a), \quad \lim_{x \to b^-} f(x) = f(b) \quad \text{and}$$

$$\lim_{x \to c} f(x) = f(c), \quad \text{for every } c \in (a, b),$$

prove that f is continuous at every point of $[a, b]$. Use qn 89, and consider sequences for the one-sided limits.

Theorems on limits

93 Use theorems about sequences to prove that if

$$\lim_{x \to a} f(x) = l \quad \text{and} \quad \lim_{x \to a} g(x) = m,$$

and f and g are real functions on the same domain, then

$$\lim_{x \to a} |f(x)| = |l|,$$

$$\lim_{x \to a} f(x) + g(x) = l + m,$$

$$\lim_{x \to a} f(x) \cdot g(x) = l \cdot m$$

and, provided $l \neq 0$, $\lim_{x \to a} 1/f(x) = 1/l$.

94 Prove that $\lim_{x \to a} c = c$ and $\lim_{x \to a} x = a$,

and use these results to develop theorems about limits of polynomials and rational functions.

95 A function $f: \mathbb{R} \to \mathbb{R}$ is defined by

$f(x) = x^2$, when $x < 2$,

$f(2) = k$,

$f(x) = 3x^2 - 18x + 28$, when $x > 2$.

Is there a value of k which will make f continuous at every point?

96 A function $f: \mathbb{R} \to \mathbb{R}$ is defined by

$f(x) = \dfrac{x^n - a^n}{x - a}$, for $x \neq a$, where n is a positive integer;

$f(a) = k$.

Is there a value of k which would make the function f continuous at every point?

97 If $f(x) \leqslant g(x) \leqslant h(x)$, for all $x \neq a$, prove that

$|g(x) - l| \leqslant \max \{|f(x) - l|, |h(x) - l|\}$, when $x \neq a$.

Deduce that if $\lim\limits_{x \to a} f(x) = l$ and $\lim\limits_{x \to a} h(x) = l$

then $\lim\limits_{x \to a} g(x) = l$.

There is one important theorem about continuous functions which does not carry over into a theorem about limits. If the function g is continuous at the point a and the function f is continuous at the point $g(a)$, then the function $f \circ g$ defined by $(f \circ g)(x) = f(g(x))$ is continuous at the point a.

This is equivalent to the proposition about limits that if

$\lim\limits_{x \to a} g(x) = g(a)$ and $\lim\limits_{y \to g(a)} f(y) = f(g(a))$

then

$\lim\limits_{x \to a} f(g(x)) = f(g(a))$.

However if

$\lim\limits_{x \to a} g(x) = m$ and $\lim\limits_{y \to m} f(y) = l$

it does not necessarily follow that

$\lim\limits_{x \to a} (f \circ g)(x) = l$.

98 Consider the functions f and g defined by $g(x) = 0$ for all x, and, $f(x) = 1$ when $x \neq 0$ and $f(0) = 0$.

Let $a = 0$, then $m = 0$ and $l = 1$ with the notation of the previous paragraph.

Determine the function $f \circ g$ and illustrate the difficulty of applying theorems about limits to composite functions.

Limits as $x \to \infty$ and when $f(x) \to \infty$

99 Use a computer or a graphics calculator to examine the graph of

$$f(x) = \frac{x^2}{x^2 + 1}.$$

Construct a formal definition for

$$\lim_{x \to +\infty} f(x) = l \quad \text{and} \quad \lim_{x \to -\infty} f(x) = l.$$

Give alternative definitions using sequences or neighbourhoods.

100 Use a computer or a graphics calculator to examine the graph of

$$f(x) = \frac{1}{x^2 - 4}.$$

Construct formal definitions for

$$\lim_{x \to a^+} f(x) = +\infty, \quad \lim_{x \to a^+} f(x) = -\infty, \quad \lim_{x \to a^-} f(x) = +\infty,$$

$$\lim_{x \to a^-} f(x) = -\infty.$$

Summary

Definition If a set A and a set B are given, then a *function*
qns 1, 2 $f: A \to B$ is a pairing of each element of A with an
3, 4, 5 element of B. The set A is called the *domain* of the
function. The set B is called the *co-domain* of the
function. The element of B which is paired with
$a \in A$ is denoted by $f(a)$. The set $\{f(a) | a \in A\} \subseteq B$
is called the *range* of the function. When the range
of a function is the whole of its co-domain, the
function is said to be *onto* its codomain and is called
a *surjection*. When $f(x) = f(y) \Rightarrow x = y$, the
function is said to be *one-one* and is called an

injection. A function which is both a surjection and an injection is called a *bijection*.

Definition A function is said to be a *real function* when both its domain and its co-domain are subsets of $\mathbb{R}$.

Theorem If $f: A \to B$ is a bijection, then there is a
qns 5, 6 well-defined inverse function $f^{-1}: B \to A$ such that $f^{-1}(f(a)) = a$, for all $a \in A$.

The sequence definition of continuity

qn 18 A function $f: A \to \mathbb{R}$ is said to be *continuous* at $a \in A$ when, for *every* sequence $(a_n) \to a$ with terms in A, $(f(a_n)) \to f(a)$.

Theorem If the functions $f: A \to \mathbb{R}$ and $g: A \to \mathbb{R}$ are both continuous at $a \in A$, then

qn 41 (i) $|f|$ is continuous at a;

qn 52 (ii) $1/f$ is continuous at a, provided $f(a) \neq 0$;

qn 23 (iii) $f + g$ is continuous at a;

qn 26 (iv) $f \cdot g$ is continuous at a;

qn 54 (v) f/g is continuous at a, provided $g(a) \neq 0$.

Theorem If the function g is continuous at the point a, and
qn 40 the function f is continuous at the point $g(a)$, then the composite function $f \circ g$ is continuous at the point a.

Theorem If for three functions f, g and h
qn 36 (i) $f(x) \leqslant g(x) \leqslant h(x)$ in a neighbourhood of the point a;

(ii) $f(a) = h(a)$;

(iii) f and h are continuous at the point a;
then the function g is continuous at a.

Theorem A constant function is continuous at each point of its
qn 21 domain.

Theorem The identity function is continuous at each point of
qn 22 its domain.

Theorem A polynomial function is continuous at each point of
qn 29 its domain.

Theorem *The neighbourhood definition of continuity*
qns 67 A function $f: A \to \mathbb{R}$ is continuous at $a \in A$ if and
68 only if, given $\varepsilon > 0$, there exists a δ such that when $x \in A$ and $|x - a| < \delta$, then $|f(x) - f(a)| > \varepsilon$.

Theorem If a function f is continuous at a point a, then it is
qn 69 bounded in some neighbourhood of a.

Theorem If a function is continuous at a point a and $f(a) > 0$,
qn 70 then f is positive in some neighbourhood of a.

Definition One-sided limits
qns 73   If $(f(a_n)) \to l$ whenever $(a_n) \to a$ with $a_n < a$, then
76   we write $\lim_{x \to a^-} f(x) = l$, and say that $f(x)$ tends to
         l as x tends to a from below.
         If $(f(a_n)) \to l$ whenever $(a_n) \to a$ with $a_n > a$, then
         we write $\lim_{x \to a^+} f(x) = l$, and say that $f(x)$ tends to
         l as x tends to a from above.

Theorem $\lim_{x \to a^-} f(x) = l$ if and only if
qns 83, given $\varepsilon > 0$, there exists a δ such that when x is in
84, 85   the domain of f and $a - \delta < x < a$, then $|f(x) - l|$
         $< \varepsilon$. $\lim_{x \to a^+} f(x) = l$ if and only if, given $\varepsilon > 0$,
         there exists a δ such that when x is in the domain of
         f and $a < x < a + \delta$, then $|f(x) - l| < \varepsilon$.

Definition Two-sided limits
         $\lim_{x \to a} f(x) = l$
         when both $\lim_{x \to a^-} f(x) = l$ and $\lim_{x \to a^+} f(x) = l$.

Theorem $\lim_{x \to a} f(x) = l$ if and only if, given $\varepsilon > 0$, there
qn 88   exists a δ such that when x is in the domain of f and
         $0 < |x - a| < \delta$, then $|f(x) - l| < \varepsilon$.

Theorem *The definition of continuity by limit*
qns 75, The function f is continuous at a if and only if
78, 89 $\lim_{x \to a} f(x) = f(a)$.

Theorem If $\lim_{x \to a} f(x) = l$ and $\lim_{x \to a} g(x) = m$, then
qn 93

   (i) $\lim_{x \to a} |f(x)| = |l|$;

   (ii) $\lim_{x \to a} f(x) + g(x) = l + m$,

   (iii) $\lim_{x \to a} f(x) \cdot g(x) = l \cdot m$,

   (iv) $\lim_{x \to a} 1/f(x) = 1/l$, provided $l \neq 0$.

Theorem If for three functions f, g and h
qn 97

   (i) $f(x) \leqslant g(x) \leqslant h(x)$ except possibly when $x = a$,

   (ii) $\lim_{x \to a} f(x) = l$ and $\lim_{x \to a} h(x) = l$,

   then $\lim_{x \to a} g(x) = l$.

The Axiom of Completeness has not been used in this chapter, so all the definitions and theorems here are valid for any Archimedean ordered field.

Historical note

Throughout the seventeenth and eighteenth centuries, and for many mathematicians well on into the nineteenth century, the only functions which were under consideration were polynomials, rational functions, trigonometric and exponential functions, and combinations of these. Although the verbal definitions of a function used during the eighteenth century sound like modern definitions, only functions defined by a formula were under consideration. Such functions are continuous and infinitely differentiable except possibly at isolated points (e.g. at $x = 0$ for the function $x \mapsto 1/x$). Even continuous functions given by different formulae on different segments of their domain, such as the equation of a plucked string, were only admitted with discomfort.

It was the consideration of limiting processes by Fourier and Cauchy, among others, which began to widen the field of functions worthy of consideration. Following his study of series of trigonometric functions, Fourier insisted (1822) that the values of a function need not be smoothly connected. In 1829 Dirichlet proposed a function like that of qn 20 in his discussion of integration and it was in 1837 that Dirichlet gave what amounts to the modern definition of function in which each value of x gives a unique value of $f(x)$, which need not depend in any way on the value given for a different x, and the function is defined by the pairing $(x, f(x))$.

The notation $f(x)$ or $\varphi(x)$ dates from Euler (1734) though separating the variable in brackets only became standard practice after Cauchy (1821).

Until the early years of the present century a real function might take the value ∞, so that, for example, if $f(x) = 1/x$, $f(0) = \infty$. In 1820, Herschel used the symbol f^{-1} for the inverse function, but this did not become standard practice until the present century. Describing functions in terms of their domains and co-domains followed the set-theoretic formulation of mathematics in the early years of the present century. The terms injection, surjection and bijection were devised by the corporate French mathematician Bourbaki.

It has always been accepted that an equation like $x^2 + y^2 = 1$ determines a functional relationship between y and x. But a given value of x between ± 1 gives two possible values of y, since $y = \pm\sqrt{(1 - x^2)}$. Such *many-valued* functions appear in the literature right through the nineteenth century, and functions were taken to be *single-valued* only when they were said to be so. However for the purposes of defining continuity, limits, and inverse functions, single-valued functions are essential and so during the present century *functions* have come to mean *single-valued* functions, though this convention is still not universal.

In order to formulate the Intermediate Value Theorem both Bolzano (1817) and Cauchy (1821) needed a definition of continuity, and said that f was continuous when $f(x + h) - f(x)$ became small with h. Their definitions are recognisable as our neighbourhood definition of continuity though they only considered functions which were continuous at least on an interval, not just at an isolated point. For his proof of the Intermediate Value Theorem, Bolzano needed the result that a function which is continuous and takes a non-zero value at a point must be non-zero in a neighbourhood of that point. The first $\varepsilon-\delta$ description of a limit appears in Cauchy's description of derivative (1823) without symbolic use of absolute values or inequalities. The $\varepsilon-\delta$ definition of limit was formulated by Weierstrass using the modern notation for absolute value, which he introduced, in his first lectures in Berlin (1859). The equivalence of the neighbourhood definition and the sequential definition was established by Cantor in 1871 and the sequential definition of continuity was proposed by Cantor and Heine in 1872. It was Weierstrass who really paved the way for the modern study of continuous functions by his construction of an everywhere continuous and nowhere differentiable function, which appeared in print in 1874. Until Weierstrass, considerations of continuity had not been clearly distinguished from those of differentiability. In fact, Bolzano had found an everywhere continuous but nowhere differentiable function in 1834, but was not able to publish. In 1851 Riemann proved that a continuous function was bounded in a neighbourhood of one of its points.

The use of 'lim' when referring to limits dates back to L'Huilier in 1786. Cauchy, in speaking of limits as x tends to 0, said that $\lim \sin x$ had one value, $\lim 1/x$ had two values, and $\lim \sin 1/x$ had an infinity of values. The study of limits from above and from below had been stimulated by considerations of Fourier series and in 1837 Dirichlet proposed the notation $f(a + 0)$ and $f(a - 0)$ for $\lim_{x \to a^+} f(x)$ and $\lim_{x \to a^-} f(x)$, and his notation was used by many authors through the nineteenth century. Weierstrass wrote $\lim_{x=a} f(x)$, and the replacing of the equals sign by an arrow, which is now universal, comes from a suggestion of the British mathematician J. G. Leathem in 1905.

Answers

2 (i) $\mathbb{R}^+ \cup \{0\}$, (ii) $\mathbb{R}$, (iii) $[-1, 1]$, (iv) $(0, 1]$, (v) $\mathbb{R}^+$.

3 (i) $\mathbb{R}^+ \cup \{0\}$, (ii) $[-1, 1]$, (iii) $|x| \geq 1$, (iv) $\mathbb{R} \setminus \{0\}$, (v) $\mathbb{R} \setminus \{2\}$, (vi) $\mathbb{R}^+$.

4 2. (ii) and (v). 3. (i), (iv), (v) and (vi).

5 2. (i) $\mathbb{R}^+ \cup \{0\} \to \mathbb{R}^+ \cup \{0\}$, (ii) $\mathbb{R} \to \mathbb{R}$, (iii) $[-\frac{1}{2}\pi, \frac{1}{2}\pi] \to [-1, 1]$,
   (iv) $\mathbb{R}^+ \cup \{0\} \to (0, 1]$, (v) $\mathbb{R} \to \mathbb{R}^+$.
   3. (i) $\mathbb{R}^+ \cup \{0\} \to \mathbb{R}^+ \cup \{0\}$, (ii) $[0, 1] \to [0, 1]$,
   (iii) $[1, \infty) \to [0, \infty)$. (iv) $\mathbb{R} \setminus \{0\} \to \mathbb{R} \setminus \{0\}$, (v) $\mathbb{R} \setminus \{2\} \to \mathbb{R} \setminus \{4\}$,
   (vi) $\mathbb{R}^+ \to \mathbb{R}$.

6 Points on the graph of f^{-1} have the form $(b, f^{-1}(b))$, which if $b = f(a)$ have the form $(f(a), a)$. See Fig. 6.1.

7 (i) See Fig 6.7. (iii) and (iv) are identical.
   See Fig. 6.2.

8 $A = (-1, 1]$, $f(A) = \{0, 1\}$. See Fig. 6.3

9 $A = \mathbb{R} \setminus \{-1\}$, $f(A) = \{-1, 0, 1\}$. See Fig. 6.4.

10 $f(\mathbb{R}) = \{0, 1\}$. See Fig. 6.5.

11 $f(\mathbb{Z}) = \{1\}$, $f(\mathbb{R} \setminus \mathbb{Z}) = \{0\}$.

13 0.

14 -1.

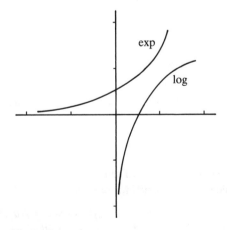

Figure 6.1

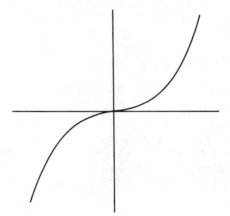

Figure 6.2

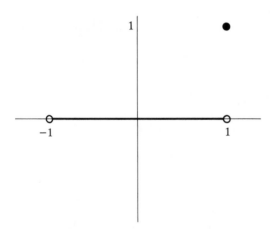

Figure 6.3

15 $a_n = 1 + 1/n$, $b_n = 1 - 1/n$.

16 If $(a_n) \to \frac{1}{2}$, taking $\varepsilon = \frac{1}{2}$, $0 < a_n < 1$ eventually, so $[a_n] = 0$ eventually.

17 If $(a_n) \to 1\frac{3}{4}$, taking $\varepsilon = \frac{1}{4}$, $1\frac{1}{2} < a_n < 2$ eventually, so $[a_n] = 1$ eventually.

18 Not continuous on $\mathbb{Z}$. Continuous on $\mathbb{R} \setminus \mathbb{Z}$.

19 (i), (ii) and (iii) are null. (iv) $\to 1$. For all null sequences (a_n) there is no single value of $f(0)$ for which $(f(a_n)) \to f(0)$ for continuity.

20 Let a be rational: $(a_n) \to a$, but $f(a_n) = 1$, so $(f(a_n)) \to 1 \neq 0 = f(a)$. Let a be irrational: $([10^n a]/10^n) \to a$, but $f([10^n a]/10^n) = 0$ for all n, so $(f([10^n a]/10^n)) \to 0 \neq 1 = f(a)$.

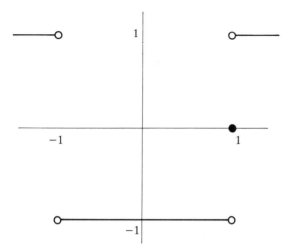

Figure 6.4

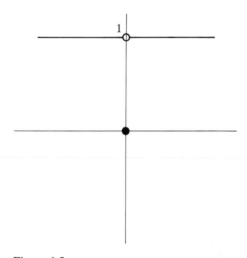

Figure 6.5

21 Let $(a_n) \to a$. $f(a_n) = c$, so $(f(a_n)) \to c = f(a)$.

22 Let $(a_n) \to a$. $f(a_n) = a_n$, so $(f(a_n)) = (a_n) \to a = f(a)$.

24 By qn 21, $g: x \mapsto 1$ is continuous at every point.
By qn 22, $f: x \mapsto x$ is continuous at every point.
So, by qn 23, $f + g: x \mapsto x + 1$ is continuous at every point.

27 (i) 21, 22, 26. (ii) 21, (i), 23. (iii) 22, 26. (iv) 22, 26, 26.
(v) 21, 22, 26, 26, 21, 23.

28 Qn 22 gives basis for induction. Qn 26 gives the inductive step.

29 Repeated use of qns 21, 28, 26 and 25.

30 $\mathbb{R} \setminus \{0\}$. Let $(a_n) \to a \neq 0$, then, by qns 3.54 and 3.55,
$(f(a_n)) = (1/a_n) \to 1/a = f(a)$.

31 $f(x) = a_0 + a_1 + a_2 x^2 + \ldots + a_n x^n + b_1/x + b_2/x^2 + \ldots + b_m/x^m$.

32 See Fig. 6.6.

33 See Fig. 6.7. Use qns 22, 32, 23, and 26.

34 See Fig. 6.8. Proof as qn 33.

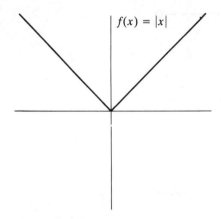

Figure 6.6

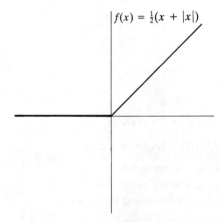

Figure 6.7

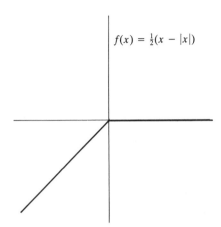

$f(x) = \frac{1}{2}(x - |x|)$

Figure 6.8

35 $-|x| \leqslant x \leqslant |x| \Rightarrow -\frac{1}{2}|x| \leqslant \frac{1}{2}x \leqslant \frac{1}{2}|x| \Rightarrow \frac{1}{2}x - \frac{1}{2}|x| \leqslant x \leqslant \frac{1}{2}x + \frac{1}{2}|x|$.
Also $-\frac{1}{2}|x| \leqslant \frac{1}{2}x \leqslant \frac{1}{2}|x| \Rightarrow \frac{1}{2}x - \frac{1}{2}|x| \leqslant 0 \leqslant \frac{1}{2}x + \frac{1}{2}|x|$.
$\frac{1}{2}(x - |x|) = \frac{1}{2}(x + |x|) \Rightarrow x = 0$.

36 Let $(a_n) \to a$, then $(f(a_n)) \to f(a)$ and $(h(a_n)) \to h(a) = f(a)$.
So $(g(a_n)) \to f(a) = g(a) = h(a)$ from qn 3.44(vi).

37 In qn 36, take $f(x) = \frac{1}{2}(x - |x|)$, $g = f$ in qn 35, and $h(x) = \frac{1}{2}(x + |x|)$.
By qns 33 and 34, f and h are continuous. By qns 35 and 36, g is
continuous at 0.
Let $a \neq 0$ be a rational number. If $a_n = a + \sqrt{2}/n$, $g(a_n) = a_n$, so
$(g(a_n)) = (a_n) \to a \neq 0 = g(a)$, so g is not continuous at a.
Let a be an irrational number then $([10^n a]/10^n) \to a$. But
$g([10^n a]/10^n) = 0$, so $(g([10^n a]/10^n)) \to 0 \neq a = g(a)$. So g is not
continuous at a.

38 $x \mapsto \sin(x^2)$. $x \mapsto (\sin x)^2$.

39 See Fig. 6.9.

40 Let $(a_n) \to a$, then by the continuity of g at a, $(g(a_n)) \to g(a)$, so by the
continuity of f at $g(a)$, $(f(g(a_n))) \to f(g(a))$.

41 ABS is continuous from qn 32.

42 f^+ is continuous by qns 41, 23, 26. See Fig. 6.10.

43 f^- is continuous by qns 41, 26, 23, 26. See Fig. 6.11.

44 Consider the cases $b \leqslant a$ and $a < b$ separately.

45 $\max(f, g)(x) = \frac{1}{2}(f(x) + g(x)) + \frac{1}{2}|f(x) - g(x)|$. Use qns 22, 23, 26, 41.

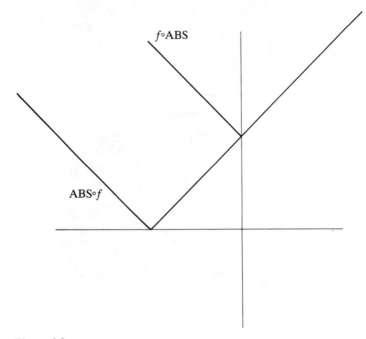

$f \circ \text{ABS}$

$\text{ABS} \circ f$

Figure 6.9

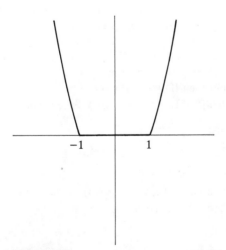

-1 1

Figure 6.10

46 ABS.

47 Maximum domain for $f \circ r$ is $\mathbb{R} \backslash \{0\}$; for $r \circ f$ is $\mathbb{R}$. Both composite functions are continuous at every point by qns 29, 30 and 40.

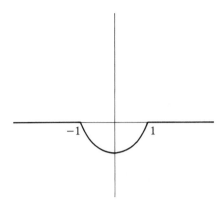

Figure 6.11

48 Qn 19 shows discontinuity at 0. When $x \neq 0$, the function is sine $\circ\, r$
which is continuous on $\mathbb{R} \setminus \{0\}$.

50 Maximum domain for $r \circ f$ is $\mathbb{R} \setminus \{2, 3\}$. The composite function is
continuous at every point, by qns 29, 30, 40.

51 Maximum domain for $r \circ f$ is $\mathbb{R} \setminus A$, where $A = \{x \,|\, f(x) = 0\}$.
Continuous, by qns 29, 30, 40.

52 Since r is continuous at $f(a)$, $r \circ f$ is continuous at a, by qn 40.
If $(a_n) \to a$ then $(f(a_n)) \to f(a)$ since f is continuous at a. Now, by
qns 3.54 and 3.55, $(1/f(a_n)) \to 1/f(a)$ since $f(a) \neq 0$. Thus $1/f$ is
continuous at a.

53 If $g(x) = x^3 - x$ and $h(x) = x^2 + 1$, the function is $g \cdot (r \circ h)$.

54 If $(a_n) \to a$ is a sequence in $A \setminus G$ then $(a_n) \to a$ is a sequence in A so
$(f(a_n)) \to f(a)$ since f is continuous at a, and
$(g(a_n)) \to g(a)$ since g is continuous at a.
Also $(f(a_n)/g(a_n)) \to f(a)/g(a)$
since $g(a) \neq 0$ by qn 3.56. So f/g is continuous at a.

55   (i) $f(x) = f(x + 0) = f(x) + f(0) \Rightarrow f(0) = 0$.
   (ii) $0 = f(0) = f(x - x) = f(x) + f(-x) \Rightarrow f(-x) = -f(x)$.
   (iii) Let $(a_n) \to a$. Then $(a_n - a)$ is a null sequence, so
      $(f(a_n - a)) \to f(0)$, since f is continuous at 0.
      So $(f(a_n) - f(a)) \to 0$, and so $(f(a_n)) \to f(a)$, which establishes
      continuity on $\mathbb{R}$.
   (iv) $f(1) = a \Rightarrow f(1 + 1) = a + a$, so by induction $f(n) = na$ for $n \in \mathbb{N}$.
      Now use (i) and (ii).
   (v) $q \cdot f(p/q) = f(q(p/q)) = f(p) = pa \Rightarrow f(p/q) = (p/q)a$.

(vi) Already proved that $f(x) = ax$ for rational x. Suppose x is irrational. $([10^n x]/10^n) \to x$. But f is continuous so $(f([10^n x]/10^n)) \to f(x)$. Now $f([10^n x]/10^n) = a[10^n x]/10^n$ and $(a[10^n x]/10^n) \to ax$. So $f(x) = ax$.

57 $1.8 < x < 2.2 \Rightarrow 3.24 < x^2 < 4.84 \Rightarrow 3 < x^2 < 5$.
Neither of these implications may be reversed.

58 $2.9 < x < 3.1 \Rightarrow 8.41 < x^2 < 9.61 \Rightarrow 8 < x^2 < 10$.

59 $5/3 < x < 7/3 \Rightarrow 3/7 < 1/x < 3/5 \Rightarrow 4/10 < 1/x < 6/10$.

60 $1.98 < x < 2.02 \Rightarrow 1.407 < \sqrt{x} < 1.422 \Rightarrow -0.008 < \sqrt{x} - \sqrt{2} < 0.008$.

62 Any $\varepsilon > 2$ for the first question.
Any $\varepsilon > 1$ for the second, third and fourth.

63 $|x - 2| < 1 \Rightarrow |[x] - 2| < 2$.
There is no δ such that $|x - 2| < \delta \Rightarrow |[x] - 2| < 1$ or $\frac{1}{2}$ or $\frac{1}{4}$.

64 Any $\delta \le 0.08$.

65 Any $\delta \le 1$.

66 Any $\delta \le 0.03$.

67　(i) Given $\varepsilon > 0$, there is an n_0 such that

$$|f(a_n) - f(a)| < \varepsilon \text{ when } n > n_0.$$

(ii) Given $\varepsilon > 0$, there is a δ such that

$$|f(a_n) - f(a)| < \varepsilon \text{ when } |a_n - a| < \delta.$$

(iii) Given $\delta > 0$, there is an n_0 such that $|a_n - a| < \delta$ when $n > n_0$.

68　(i) $\delta = 1/n$ does not ensure that $|f(x) - f(a)| < \varepsilon$ for all x satisfying $|x - a| < \delta$. So for at least one x in this δ-neighbourhood of a $|f(x) - f(a)| \ge \varepsilon$.

(ii) $|a_n - a| < 1/n$ ensures that $(a_n - a)$ is a null sequence by qns 3.28 and 3.27.

(iii) $(a_n) \to a$. But $|f(a_n) - f(a)| \ge \varepsilon$ means $(f(a_n))$ does not tend to $f(a)$.

69 There is a δ such that $|x - a| < \delta \Rightarrow |f(x) - f(a)| < 1$

$$\Rightarrow f(a) - 1 < f(x) < f(a) + 1.$$

70 Choose $\varepsilon = \frac{1}{2} f(a)$.
There is a δ such that $|x - a| < \delta \Rightarrow |f(x) - f(a)| < \frac{1}{2} f(a)$

$$\Rightarrow \tfrac{1}{2} f(a) < f(x) < 1\tfrac{1}{2} f(a).$$

When $f(a) < 0$, choose $\varepsilon = -\frac{1}{2} f(a)$.

71 $|x - \sqrt{2}| < \delta$, where $\delta_2 = |3/2 - \sqrt{2}|$, $\delta_3 = |4/3 - \sqrt{2}|$, $\delta_4 = |6/4 - \sqrt{2}|$, $\delta_5 = |7/5 - \sqrt{2}|$, $\delta = \min(\delta_2, \delta_3, \delta_4, \delta_5)$.

72 (a) If $a \in (0, 1]$ is rational and $a_n = a - \sqrt{2}/n$, then (a_n) is eventually in the domain of the function and $f(a_n) = 0$. So $(a_n) \to a$ and $(f(a_n)) \to 0 \neq f(a)$.

    (b) (i) By Archimedean order there is an integer $m > 1/\varepsilon$.

        (ii) For each q, there are q possible values for p and so not more than $1 + 2 + \ldots + m = \frac{1}{2}m(m + 1)$ rational numbers between 0 and 1 with denominator $\leqslant m$.

        (iii) Of the $\frac{1}{2}m(m + 1)$ or less numbers, $|a - p/q|$, none of which are zero, there is a least.

        (iv) If x is irrational, then $|f(x) - f(a)| = 0$.
If x is rational and $|x - a| < \delta$, then $x = p/q$ with $q > m$.
So $|f(x) - f(a)| = f(x) = 1/q < 1/m < \varepsilon$.

73 For example $a_n = 1 - (\frac{1}{2})^n$. $a_n < 1$.
If $a_n = 1$, then $(a_n) = 1$, but $(f(a_n)) \to 1$.

74 $0, 1, -1$.

75 If for *every* sequence $(a_n) \to a$, $(f(a_n)) \to f(a)$, then for every sequence $(a_n) \to a$ with $a_n < a$, $(f(a_n)) \to f(a)$.

76 Answer in summary.

77 $1, 0, -1$.

78 See answer to qn 75.

79 If f were continuous at a then both limits would equal $f(a)$ from qns 75 and 78.

80 $-1, 1$.

81 (i) $\varepsilon = 0.005$.
    (ii) $\varepsilon = 0.0005$.
    (iii) Yes. $\delta \leqslant 1.895$, $\delta \leqslant 0.551$, $\delta \leqslant 0.173$.

82 Let $f(x) = 1$ when $x \neq 0$, and let $f(0) = 2$.

83 Let $(a_n) \to a$ where $a < a_n$. Given $\varepsilon > 0$, eventually $a < a_n < a + \delta$, so $|f(a_n) - l| < \varepsilon$ and this proves that $(f(a_n)) \to l$.
So the neighbourhood definition of limit implies the sequence definition of limit.

84 From $a < a_n < a + 1/n$, $(a_n) \to a$ by qn 3.44(vi). But $|f(a_n) - l| \geqslant \varepsilon$ implies that $(f(a_n))$ does not tend to l, which contradicts our hypothesis.
So the sequence definition of limit implies the neighbourhood definition of limit.

85 Answer in summary.

87 Given $\varepsilon > 0$, the limit from below says that, for some δ_1, x in the domain of f and $a - \delta_1 < x < a \Rightarrow |f(x) - l| < \varepsilon$,
and the limit from above says that, for some δ_2, x in the domain of f and $a < x < a + \delta_2 \Rightarrow |f(x) - l| < \varepsilon$.
So if $0 < |x - a| < \min(\delta_1, \delta_2)$, it follows that $|f(x) - l| < \varepsilon$.

88 (i) $a - \delta < x < a \Rightarrow 0 < |x - a| < \delta \Rightarrow |f(x) - l| < \varepsilon$.
$a < x < a + \delta \Rightarrow 0 < |x - a| < \delta \Rightarrow |f(x) - l| < \varepsilon$.
(ii) The two conditions for one-sided limits are obviously satisfied.

89 $0 < |x - a| < \delta$ or $x = a \Leftrightarrow |x - a| < \delta$.
$|f(x) - f(a)| < \varepsilon$ is trivial when $x = a$.
First proposition is equivalent to the limit (qn 88).
Second proposition implies continuity (qn 67).

90 $k = 2$. Calculate limits from above and below $x = 1$ and use qns 79 and 89.

91 $f(a)$ from qn 78, $f(b)$ from qn 75 and $f(c)$ from qn 86.

92 Continuous at c from qn 89.
For every sequence $(a_n) \to a$ in the domain $[a, b]$, $a \leqslant a_n$. If $a < a_n$ for only a finite number of terms then $(f(a_n))$ is eventually constant with terms equal to $f(a)$. If $a < a_n$ for an infinite number of terms, $n = n_i$, then $(f(a_{n_i})) \to f(a)$ for this subsequence because of the limit.
So $(f(a_n) - f(a))$ is null.
Similarly for b.

93 First result from qn 3.44(v), second from qn 3.44(ii), third from qn 3.44(iv) and the fourth from qns 3.54 and 3.55.

94 The development is like that for continuous functions in qns 21, 22, 23, 25, 26, 28, 29, and then uses qn 93 parts three and four for rational functions. Composite functions must not be used; see qn 98.

95 Use limits from above and below $x = 2$. Take $k = 4$.

96 $x^n - a^n = (x - a)(x^{n-1} + ax^{n-2} + \ldots + a^{n-2}x + a^{n-1})$.
So $\lim_{x \to a} f(x) = na^{n-1} = k$.

97 $f(x) - l \leqslant g(x) - l \leqslant h(x) - l$.
If $0 \leqslant g(x) - l$, then $|g(x) - l| \leqslant |h(x) - l|$.
If $g(x) - l \leqslant 0$, then $|g(x) - l| \leqslant |f(x) - l|$.
Given $\varepsilon > 0$, there is a δ_1 such that $0 < |x - a| < \delta_1 \Rightarrow |f(x) - l| < \varepsilon$, and there is a δ_2 such that $0 < |x - a| < \delta_2 \Rightarrow |h(x) - l| < \varepsilon$.

So, if $0 < |x - a| < \min(\delta_1, \delta_2)$,
$|g(x) - l| \leq \max(|f(x) - l|, |h(x) - l|) < \varepsilon$.

The result follows from qn 87.

98 $f \circ g(x) = f(g(x)) = f(0) = 1$, so $\lim_{x \to 0} f \circ g(x) = 1$.
However, $\lim_{x \to 0} g(x) = 0$ and $\lim_{y \to 0} f(y) = 0$.

99 $\lim_{x \to +\infty} f(x) = l$ means, for any sequence $(a_n) \to +\infty$, $(f(a_n)) \to l$:
or, given $\varepsilon > 0$, there exists a number M such that
$|f(x) - l| < \varepsilon$ when $x > M$.
$\lim_{x \to -\infty} f(x) = l$ means, for any sequence $(a_n) \to -\infty$, $(f(a_n)) \to l$:
or, given $\varepsilon > 0$, there exists a number M such that
$|f(x) - l| < \varepsilon$ when $x < M$.

100 $\lim_{x \to a^+} f(x) = +\infty$ means, for all sequences $(a_n) \to a$ with $a < a_n$,
$(f(a_n)) \to +\infty$; or, given any number M, there exists a δ such that
$M < f(x)$ when $a < x < a + \delta$.
$\lim_{x \to a^+} f(x) = -\infty$ means, for all sequences $(a_n) \to a$ with $a < a_n$,
$(f(a_n)) \to -\infty$; or, given any number M, there exists a δ such that
$f(x) < M$ when $a < x < a + \delta$.
$\lim_{x \to a^-} f(x) = +\infty$ means for all $(a_n) \to a$ with $a_n < a$, $(f(a_n)) \to +\infty$;
or, given any number M, there exists a δ such that
$M < f(x)$ when $a - \delta < x < a$.
$\lim_{x \to a^-} f(x) = -\infty$ means for all sequences $(a_n) \to a$ with $a_n < a$,
$(f(a_n)) \to -\infty$; or, given any number M, there exists a δ such that
$f(x) < M$ when $a - \delta < x < a$.

7

Continuity and completeness
Functions on intervals

Concurrent reading: Spivak ch. 7.
Further reading: Rudin ch. 4.

Monotonic functions: one-sided limits

Increasing and decreasing functions, like increasing and decreasing sequences, have interesting properties and it is useful to be able to name such functions.

If $f(x) \leqslant f(y)$, when $x < y$, the function f is said to be *increasing*, or *monotonic increasing*.

If $f(x) < f(y)$, when $x < y$, the function f is said to be *strictly increasing*, or *strictly monotonic increasing*.

1 Which of the following functions with domain $\mathbb{R}$ are monotonic increasing:

(i) $x \mapsto x^2$; (ii) $x \mapsto x^3$; (iii) $x \mapsto [x]$; (iv) $x \mapsto x - [x]$;
(v) $x \mapsto \tan x$, $(x \neq (n + \frac{1}{2})\pi)$; (vi) $x \mapsto e^x$?

For each one that is *not* monotonic increasing, name a subset of the domain on which it *is* monotonic increasing.

2 Define what is meant by saying that a real function is *decreasing* or *monotonic decreasing*.

Also define what is meant by saying that a real function is *strictly decreasing* or *strictly monotonic decreasing*.

3 Which of the following functions with domain $\mathbb{R}$ are monotonic decreasing:

(i) $x \mapsto -x^2$; (ii) $x \mapsto -x^3$; (iii) $x \mapsto 1/x$ *and* $0 \mapsto 0$; (iv) $x \mapsto e^{-x}$?

For each one that is not monotonic decreasing, name a subset of the domain on which it is monotonic decreasing.

4 If $f: \mathbb{R} \to \mathbb{R}$ is monotonic increasing throughout its domain and the set $V = \{f(x) \mid x \in \mathbb{R}\}$ is bounded, prove that $\lim_{x \to \infty} f(x) = \sup V$. In this case the line $y = \sup V$ is called an *asymptote* to the graph of the function.

5 Let $f: \mathbb{R} \to \mathbb{R}$ be a function which is monotonic increasing throughout its domain, and let a be a point of the domain. Let $L = \{f(x) \mid x < a\}$ and let $U = \{f(x) \mid a < x\}$.
Find an upper bound for L and a lower bound for U.
Deduce that $\sup L$ and $\inf U$ exist.
Prove that $\sup L = \lim_{x \to a^-} f(x)$ and $\inf U = \lim_{x \to a^+} f(x)$.
Thus one-sided limits exist for monotonic increasing functions whether they are continuous or not. What are the values of these limits for the function $x \mapsto [x]$ at integer points of the domain?

6 Formulate an analogue of qn 5 for monotonic decreasing functions, and give an example of a discontinuous decreasing function and of its one-sided limits at points of discontinuity.

7 If, in qn 5, f is continuous at a, show that $\sup L = \inf U$. If conversely, $\sup L = \inf U$, show that f is continuous at a. If $\sup L \neq \inf U$, explain why $\sup L < \inf U$, and deduce that a rational number must lie between these two numbers. Deduce that a monotonic function can have at most a countable number of discontinuities.

Intervals

8 Find the range of each of the functions defined below. The domain in each case is $\mathbb{R}$.

(i) $f_1(x) = 1$.

(ii) $f_2(x) = \sin x$.

(iii) $f_3(x) = \arctan x$.

(iv) $f_4(x) = \dfrac{1}{1 + x^2}$.

(v) $f_5(x) = e^x$.

(vi) $f_6(x) = x^2$.

(vii) $f_7(x) = x$.

The ranges of these seven functions exhibit the seven kinds of *interval* on the real line.

Intervals, or connected sets, on the real line are subsets of the real line which contain all the real numbers lying between any two points of the subset. The set I is an *interval* if when r and $s \in I$, and $r < s$, every x such that $r < x < s$ also belongs to I.

The seven types of interval are distinguished by their boundedness and their boundaries.

Bounded above and below

   (i) singleton point $\{a\}$

   (ii) closed interval $[a, b] = \{x \mid a \leqslant x \leqslant b\}$

   (iii) open interval $(a, b) = \{x \mid a < x < b\}$

   (iv) half-open interval $[a, b) = \{x \mid a \leqslant x < b\}$

                                    $(a, b] = \{x \mid a < x \leqslant b\}$

*Bounded above **or** below, but not both*

   (v) open half-ray $(a, +\infty) = \{x \mid a < x\}$

                                    $(-\infty, a) = \{x \mid x < a\}$

   (vi) closed half-ray $[a, +\infty) = \{x \mid a \leqslant x\}$

                                    $(-\infty, a] = \{x \mid x \leqslant a\}$

Unbounded

(vii) the whole real line $\mathbb{R}$

The word *open* as we have used it here indicates that there is space *within* each open set around each point of the set. For example, if $c \in (a, b)$, then $a < \frac{1}{2}(a + c) < c < \frac{1}{2}(c + b) < b$. This is also described by saying that an open set contains a neighbourhood of each of its points.

The word *closed* as we have used it here indicates that every convergent sequence within each closed set converges to a point of the set. (See qn 3.66.)

The whole real line is both open and closed!

9 Sketch graphs of continuous functions $f: (a, b) \to \mathbb{R}$ for which the range is

 (i) a singleton point,
 (ii) a closed interval
 (iii) an open interval,
 (iv) a half-open interval,
 (v) an open half-ray,
 (vi) a closed half-ray,
 (vii) $\mathbb{R}$.

10 Sketch the graph of a continuous function $f: A \to \mathbb{R}$ for which the range of the function *is not* an interval.

What would you conjecture about the domain A of the function in such a case?

If we wish to try to prove that continuous functions always map intervals onto intervals, it is worth noticing first that this result fails if we are working with only the rational numbers.

11 Let the function $f: \mathbb{Q} \to \mathbb{Q}$ be defined by $f(x) = x^2 - 2$. This is certainly a continuous function. Now restrict the domain to the set $[1, 2] \cap \mathbb{Q}$. The numbers -1 and 2 are in the range, but what rational number between them is not?

This result suggests that it will be necessary to use the completeness of the real numbers in a proof that continuous functions map intervals onto intervals.

Intermediate Value Theorem

A simple form of the problem which we are facing is this: suppose that A is an interval and that $f: A \to \mathbb{R}$ is a continuous function. Suppose further that a and b are points in A and that $f(a) < 0$ and $f(b) > 0$. Can we be sure that there is a point c in the interval (a, b) such that $f(c) = 0$?

12 Let $f: [a, b] \to \mathbb{R}$ be a continuous function with $f(a) < 0$ and $f(b) > 0$. Sketch the graph of such a function. Try different sketches which illustrate the same conditions. Notice that neither of the sets $\{x \mid f(x) < 0\}$ or $\{x \mid f(x) > 0\}$ need be an interval. We construct sequences in the domain of the function which either reach a point x where $f(x) = 0$ or converge to such a point by locating smaller and smaller intervals on each of which the function f changes sign.

Let $a_1 = a$, $b_1 = b$ and $d = \frac{1}{2}(a_1 + b_1)$. If $f(d) = 0$ we have found the point we want; if $f(d) \neq 0$, explain how to choose a_2 and b_2 so that $f(a_2) < 0$, $f(b_2) > 0$ and $b_2 - a_2 = \frac{1}{2}(b_1 - a_1)$. Extend this process to an inductive definition of a_n, b_n where $f(a_n) < 0 < f(b_n)$, and $a_1 \leqslant a_n < b_n \leqslant b_1$, with $b_{n+1} - a_{n+1} = \frac{1}{2}(b_n - a_n)$.

13 Let $f: [a, b] \to \mathbb{R}$ be a continuous function with $f(a) < 0$ and $f(b) > 0$. With the notation of qn 12, suppose that the process of repeated bisection of the interval $[a, b]$ has not located a point at which the function is 0.
   (i) How do you know the sequences (a_n) and (b_n) are both convergent?
   (ii) If $(a_n) \to A$ and $(b_n) \to B$, why must $f(A) \leqslant 0 \leqslant f(B)$?
   (iii) Why must $A = B$ and $f(A) = 0$?
       We summarise the result of qn 13 by saying that, if
       $f: [a, b] \to \mathbb{R}$ is a continuous function, $f(a) < 0$, $f(b) > 0$,
       there exists a $c \in (a, b)$ such that $f(c) = 0$.

14 By applying the result of qn 13 to a suitably chosen function, prove that, if $f: [a, b] \to \mathbb{R}$ is a continuous function and $f(a) < k < f(b)$, then there exists a $c \in (a, b)$ such that $f(c) = k$.

15 By applying the result of qn 14 to a suitably chosen function, prove that if $f: [a, b] \to \mathbb{R}$ is a continuous function and $f(a) > k > f(b)$, then there exists a $c \in (a, b)$ such that $f(c) = k$.

16 Give an example of a discontinuous function $f: [a, b] \to \mathbb{R}$ which takes every value between $f(a)$ and $f(b)$. The existence of such a function shows that the converse of the Intermediate Value Theorem (qn 13) is false.

17 Prove that a continuous function $f: \mathbb{R} \to \mathbb{Z}$ is necessarily constant.

18 Prove that a continuous function $f: \mathbb{R} \to \mathbb{Q}$ is necessarily constant.

19 For any real numbers a and b show that the function defined on $\mathbb{R}$ by $f(x) = x^3 + ax + b$ has a real root. To what class of polynomial functions may this result be extended?

20 If $f: [0, 1] \to [0, 1]$ is a continuous function, by applying the result of qn 13 to an appropriately chosen function, prove that there is at least one $c \in [0, 1]$ such that $f(c) = c$. Give an example to show that this result might fail if the domain of f were $(0, 1)$. Generalise this result for a continuous function $f: [a, b] \to [a, b]$.

21 Can the Intermediate Value Theorem be used to prove that the image of an interval under a continuous function is necessarily an interval? Or, in other words, under a continuous function, is a connected set always mapped onto a connected set?

Inverses of continuous functions

22 If $f: [a, b] \to \mathbb{R}$ is a continuous and one–one function with $f(a) < f(b)$, prove that f is strictly monotonic increasing. Deduce that the range of f is $[f(a), f(b)]$.
What is the analogous result if $f(a) > f(b)$?

23 If $f: [a, b] \to \mathbb{R}$ is strictly monotonic, must f be
 (i) one–one;
 (ii) invertible;
 (iii) continuous?

24 If $f: [a, b] \to \mathbb{R}$ is continuous and invertible (see qn 6.5) must f be strictly monotonic?

25 Suppose that $f: [a, b] \to \mathbb{R}$ is a continuous and strictly monotonic function,
 (i) How do you know that the range of f is a closed interval $[c, d]$, say?
 (ii) How do you know that f has an inverse function $g: [c, d] \to [a, b]$?
 (iii) How do you know that g must be strictly monotonic?
 (iv) Let x be any point in the domain of f and let $y = f(x)$, so that $a \leqslant x \leqslant b$ and $c \leqslant y \leqslant d$.
 How do you know that $\lim_{t \to y^-} g(t)$ and $\lim_{t \to y^+} g(t)$ exist?
 (v) Let $(x_n) \to x$ be an increasing sequence in $[a, b]$ and let $(s_n) \to x$ be a decreasing sequence in $[a, b]$. Why must $(f(x_n)) \to f(x)$ and $(f(s_n)) \to f(x)$, and of these two sequences why must one be increasing and one decreasing?
 (vi) Deduce that there is one sequence $(t_n) \to y$ from below and one tending to y from above for which $(g(t_n)) \to g(y)$.
 (vii) Why does this establish that g is continuous at y?

We have now proved that a function which is continuous and strictly monotonic on an interval has a continuous inverse. The use of completeness in qn 25 is unavoidable. Consider $f: \mathbb{Q} \to \mathbb{Q}$ given by $f(x) = x$ when $x < \sqrt{2}$ and $f(x) = x + 1$ when $\sqrt{2} < x$. An example of a one–one continuous function $\mathbb{Q} \to \mathbb{Q}$, whose inverse is discontinuous everywhere, is given in the book by Dieudonné cited in the bibliography.

26 Does the function $f: \mathbb{R} \to \mathbb{R}$ defined by $f(x) = x^2$ have an inverse? Find a maximal interval $A \subseteq \mathbb{R}$ for which the function $f: A \to \mathbb{R}$ defined by $f(x) = x^2$ is monotonic. With this A, is the function $f: A \to A$ a bijection? Is the function $f^{-1}: A \to A$ continuous? How is it usually denoted?

27 Why does the function $f: \mathbb{R}^+ \cup \{0\} \to \mathbb{R}$ defined by $f(x) = x^n$ have an inverse whether the positive integer n is odd or even? Say why there is a continuous inverse $f^{-1}: \mathbb{R}^+ \cup \{0\} \to \mathbb{R}^+ \cup \{0\}$. This inverse function is normally denoted by $x \mapsto \sqrt[n]{x}$ or $x^{1/n}$.

We established the existence of nth roots for positive real numbers in qn 4.39. The argument here establishes that the function $x \mapsto x^{1/n}$ is continuous on $\mathbb{R}^+ \cup \{0\}$.

28 Find the limit

$$\lim_{x \to 1} \frac{1 - x}{1 - x^{m/n}},$$

when m and n are positive integers with $m \neq n$.

Continuous functions on a closed interval

29 What does the Intermediate Value Theorem allow us to claim about the possible ranges of a continuous function $f: [a, b] \to \mathbb{R}$?

30 Attempt to draw graphs of continuous functions $f: [a, b] \to \mathbb{R}$ with the seven different ranges as in qn 9. Use pencil and paper, not a computer.

We now investigate whether the fact that the ranges found in qn 30 were all bounded above and bounded below, is a necessary consequence of the conditions, or an indication of our lack of imagination. We suppose that there exists a continuous function $f: [a, b] \to \mathbb{R}$ whose range is unbounded above and see whether any contradiction arises.

31 We suppose that the range of the continuous function

$f: [a, b] \to \mathbb{R}$ is unbounded above. The unboundedness of the range of f means that, whatever integer n we choose, we can find an x in $[a, b]$ such that $f(x) > n$. If such an x can be found we call it x_n. This gives us a sequence (x_n).

(i) Is there any reason why the sequence (x_n) should be convergent?

(ii) Is the sequence (x_n) bounded?

(iii) Must the sequence (x_n) contain a convergent subsequence?

(iv) If the subsequence (x_{n_i}) converges to c, why must $c \in [a, b]$?

(v) What does the continuity of f allow us to say about the sequence $(f(x_{n_i}))$?

(vi) For sufficiently large n_i can we be sure that $|f(x_{n_i}) - f(c)| < 1$?

(vii) How does this lead to a contradiction, which therefore undermines our hypothesis of the unboundedness of the sequence $(f(x_n))$?

32 If the range of a continuous function $f: [a, b] \to \mathbb{R}$ were presumed to be unbounded below, outline how you would establish a contradiction.

33 If a $f: \mathbb{Q} \to \mathbb{Q}$ is defined by $f(x) = 1/(x^2 - 2)$,

(i) show that f is continuous throughout its domain,

(ii) explain why the behaviour of f on the interval $[1, 2]$ does not contradict the result established in qn 31.

Questions 31 and 32 eliminate $\mathbb{R}$, $(a, +\infty)$, $[a, +\infty)$, $(-\infty, a)$ and $(-\infty, a]$ as possible ranges for a continuous function $f: [a, b] \to \mathbb{R}$. This leaves open intervals, closed intervals, half-open intervals and singleton points as the only possibilities. We know that closed intervals and singleton points are real possibilities. Can we now eliminate the open intervals and the half-open intervals as our investigation in qn 30 would suggest we might?

34 Let $f: [a, b] \to \mathbb{R}$ be a continuous function and let

$$V = \{f(x) \mid a \leqslant x \leqslant b\}.$$

(i) How do we know that V is an interval?

(ii) How do we know that V is bounded above and below?

(iii) Can you be sure that sup V and inf V exist? If we let sup $V = M$, then $M \geqslant f(x)$ for all $x \in [a, b]$. As a result of our graph drawing, we conjecture that there has to be a value of x at which $M = f(x)$.

(iv) Why, given $\varepsilon > 0$ must there be an $x \in [a, b]$ such that
$M - \varepsilon < f(x) \leqslant M$?

(v) Define $x_n \in [a, b]$ so that $M - 1/n < f(x_n) \leqslant M$.
Why must the sequence (x_n) contain a convergent
subsequence?

(vi) Suppose $(x_{n_i}) \to c$. Why must $c \in [a, b]$?

(vii) Why must $(f(x_{n_i}))$ be convergent?

(viii) By considering the limit of the sequence $(f(x_{n_i}))$ prove that
$M = f(c)$.

(ix) Having proved that the function f attains its supremum,
indicate how to show that the function must also attain its
infimum.

35 Explain why function $f: \mathbb{Q} \to \mathbb{R}$ defined by $f(x) = \sin x$ does not
attain its bounds when examined on the domain $[0, 6]$. Does this
contradict the result of the previous question?

The result of questions 31 and 32 is usually described by saying
that a continuous function on a closed interval is *bounded*, and the
result of qn 34 by saying that a continuous function on a closed
interval *attains its bounds*. These two results together are sometimes
called the *Minimax Theorem*. Although both results depend on
completeness it is not necessary for the Intermediate Value Theorem
to precede their proof. Taken on their own, they imply for
continuous real functions that a closed and bounded set is mapped to
a closed and bounded set. From the standpoint of topological spaces
this generalises to the theorem that, under continuous maps, a
compact set is mapped to a compact set.

36 Give examples to show that continuous functions on unbounded
intervals or open intervals or half-open intervals need not be
bounded and, even if bounded, need not attain their bounds.

Uniform continuity

Both the sequence definition of continuity and the neighbourhood
definition of continuity define the continuity of a function at a *point*,
and we only claim continuity on an *interval* when the function is
continuous at every point of that interval. Moreover, when
determining continuity at a point with the neighbourhood definition
of continuity, given an ε, the necessary δ may be different for
continuity at a point a from what is needed to establish continuity at
a point b. We examine now, when, with a given ε, a choice of δ may
be made which will establish continuity throughout an interval. The
rôle of closed intervals again turns out to be critical.

37 If $f: [-10, 10] \to \mathbb{R}$ is defined by $f(x) = x^2$, prove that if $|x - y| < 1/20$ then $|f(x) - f(y)| < 1$. Extend this result to show that, for any positive ε, if $|x - y| < \varepsilon/20$ then $|f(x) - f(y)| < \varepsilon$.

38 If $f: (0, 1) \to \mathbb{R}$ is defined by $f(x) = 1/x$, $f(x)$ is unbounded. Find a δ_1 such that if $|x - \frac{1}{2}| < \delta_1$, then $|f(x) - f(\frac{1}{2})| < 1$. Find a δ_2 such that if $|x - \frac{1}{4}| < \delta_2$, then $|f(x) - f(\frac{1}{4})| < 1$. Show that, if $|x - a| < \delta$ implies $|f(x) - f(a)| < 1$, then $\delta \leqslant a^2/(1 + a)$.

Deduce that there is no constant δ such that $|x - a| < \delta$ implies $|f(x) - f(a)| < 1$ for all x and a in the domain of the function f.

A function $f: A \to \mathbb{R}$ is said to be *uniformly continuous on A* when, given $\varepsilon > 0$, there exists a δ such that

$$|f(x) - f(y)| < \varepsilon \text{ whenever } |x - y| < \delta.$$

39 Can you be sure that a function which is uniformly continuous on A is necessarily continuous at each point of A?

In qn 37, we have proved that the function $x \mapsto x^2$ *is* uniformly continuous on $[-10, 10]$. In qn 38, we have proved that the function $x \mapsto 1/x$ *is not* uniformly continuous on $(0, 1)$.

40 By considering $x = n + 1/n$ and $y = n$, and taking $\varepsilon = 1$, prove that the function defined by $f(x) = x^2$ *is not* uniformly continuous on $\mathbb{R}$. Notice the effect of changing the domain from question 37.

41 By considering $x = 1/2n\pi$ and $y = 1/(2n + \frac{1}{2})\pi$, and taking $\varepsilon = \frac{1}{2}$, prove that the function defined by $f(x) = \sin(1/x)$ *is not* uniformly continuous on the interval $(0, 1)$. This shows that continuous functions (see qn 6.48) may fail to be uniformly continuous even when they are bounded.

42 If a function $f: A \to \mathbb{R}$ satsifies a *Lipschitz condition,* namely that there is a constant real number L such that

$$|f(x) - f(y)| \leqslant L \cdot |x - y|$$

for any $x, y \in A$, show that f is uniformly continuous on A, by finding an appropriate δ for a given ε. Verify that the function of qn 37, satisfies a Lipschitz condition with $L = 20$.

After studying the Mean Value Theorem in chapter 9 it will be clear that any function with bounded derivatives necessarily satisfies a Lipschitz condition, with an upper bound on the absolute value of the derivatives as the Lipschitz constant.

43 We seek to prove that a continuous function $f: [a, b] \to \mathbb{R}$ is necessarily uniformly continuous, and we do so by supposing that it is not and establishing a contradiction.

When a function *is not* uniformly continuous, it means that, for some $\varepsilon > 0$, no δ can be found for which $|x - y| < \delta$ implies that $|f(x) - f(y)| < \varepsilon$. That is, for each δ there are points x and y in the domain such that $|x - y| < \delta$ but $|f(x) - f(y)| \geq \varepsilon$.

Suppose that f *is not* uniformly continuous and construct two sequences (x_n) and (y_n) for which the difference $|x_n - y_n| < 1/n$ but for which $|f(x_n) - f(y_n)| \geq \varepsilon$.

(i) Is there any reason why the sequence (x_n) should be convergent?

(ii) Is the sequence (x_n) bounded?

(iii) Must the sequence (x_n) contain a convergent subsequence?

(iv) If the subsequence (x_{n_i}) converges to c, must the subsequence $(y_{n_i}) \to c$?

(v) Why must $c \in [a, b]$?

(vi) What can be said about the sequences $(f(x_{n_i}))$ and $(f(y_{n_i}))$, because of the continuity of f?

(vii) Use the inequality

$$|f(x) - f(y)| \leq |f(x) - f(c)| + |f(c) - f(y)|$$

to obtain the contradiction we seek, by showing that the right-hand side may be made less than the left-hand side.

44 In qns 38, 40 and 41 we found continuous functions which were not uniformly continuous. Each of these functions was, however, either unbounded or else had unbounded slope. The function $f: [0, \infty) \to \mathbb{R}$ defined by $f(x) = \sqrt{x}$ is unbounded and has unbounded slope near the origin. Prove that this function is uniformly continuous by appealing to qns 26 and 43 on $[0, 1]$ and by noting that, if either x or $y > 1$, then $|\sqrt{x} - \sqrt{y}| < |x - y|$.

45 If $f: (a, b) \to \mathbb{R}$ is continuous and $\lim_{x \to a^+} f(x)$ and $\lim_{x \to b^-} f(x)$ both exist and are finite, prove that f is uniformly continuous by constructing a function g which is continuous on $[a, b]$ and for which $g(x) = f(x)$ on (a, b).

Extension of functions on $\mathbb{Q}$ to functions on $\mathbb{R}$

46 Is the function $f: [0, 1] \cap \mathbb{Q} \to \mathbb{R}$ defined by $f(x) = 1/(2x^2 - 1)$ continuous? uniformly continuous? If you attempted to extend f to a function $g: [0, 1] \to \mathbb{R}$ such that $f(x) = g(x)$ for rational x, is there any value that you could take for $g(1/\sqrt{2})$ which would make the function g continuous at the point $1/\sqrt{2}$?

47 The function $f: [0, 1] \cap \mathbb{Q} \to \mathbb{R}$ is uniformly continuous.
   (i) If (a_n) is a Cauchy sequence of rational numbers in $[0, 1]$, show that $(f(a_n))$ is a Cauchy sequence.
   (ii) If (a_n) and (b_n) are rational Cauchy sequences in $[0, 1]$ with the same real limit, prove that $(f(a_n))$ and $(f(b_n))$ have the same limit.
   (iii) Is there a well-defined function $g: [0, 1] \to \mathbb{R}$ such that $g(x) = f(x)$ for rational x and $(f(a_n)) \to g(a)$, when $(a_n) \to a$ is a Cauchy sequence of rational numbers with an irrational limit a?
   (iv) Finally, show that g is uniformly continuous on $[0, 1]$. Given $\varepsilon > 0$, choose δ such that $|x - y| < \delta$ implies $|f(x) - f(y)| < \frac{1}{3}\varepsilon$, for rational x and y. Let a and b be real numbers in $[0, 1]$ such that $|a - b| < \delta$, and let $(a_n) \to a$ and $(b_n) \to b$ be Cauchy sequences of rational numbers wholly within the interval $[a, b]$. Use the inequality

$$|g(a) - g(b)|$$
$$\leq |g(a) - f(a_n)| + |f(a_n) - f(b_n)| + |f(b_n) - g(b)|$$

to complete the proof that the function f may be extended to a continuous function on the whole interval $[0, 1]$.

The Intermediate Value Theorem and the Minimax Theorem, like the theorem that a continuous function on a closed interval is uniformly continuous, can only be proved using the completeness of the real numbers. However, uniform continuity, like continuity, can be defined whether the domain of the function is complete or not. What we have shown in qns 46 and 47 is that uniform continuity is the necessary and sufficient condition that a continuous function defined on a dense subset of a closed interval may be extended to a continuous function on the whole interval. As a final optional exercise we indicate how this theorem can be used to define exponents.

(48) For positive $a \neq 1$ and positive rational x we can construct a^x by qn 4.39, and we can define $a^0 = 1$ and $a^{-x} = 1/a^x$.
   (i) Prove that $f: \mathbb{Q} \to \mathbb{R}^+$ defined by $f(x) = a^x$ is strictly monotonic. (In fact it is monotonic increasing when $1 < a$, and monotonic decreasing when $0 < a < 1$.)

(ii) For $x \neq 0$, and $a > 1$, let $g(x) = (a^x - 1)/x$. Use qn 2.48(iii) to show that $g(n) < g(m)$ for two positive integers $n < m$. By putting $b = a^{qs}$ prove that $g(p/q) < g(r/s)$ for two rational numbers $0 < p/q < r/s$, so that $g: \mathbb{Q}^+ \to \mathbb{R}^+$ is monotonic increasing. Use qn 4.40 to extend g to a monotonic increasing function on $\mathbb{Q}^+ \cup \{0\}$.

(iii) Prove that f satisfies a Lipschitz condition on any closed interval $[0, B] \cap \mathbb{Q}$, where B is a positive rational and so f is uniformly continuous on that interval by qn 43.

(iv) Deduce from qn 47 that f may be extended to a continuous function on $[0, B]$ in a unique way.

(v) Use the defintion of a^{-x} to extend f to a continuous function on $[-B, B]$.

Summary

Definition qns 1, 2	*Monotonic functions* A real function f is said to be monotonic increasing when $x < y \Rightarrow f(x) \leqslant f(y)$. A real function f is said to be strictly monotonic increasing when $x < y \Rightarrow f(x) < f(y)$. A real function f is said to be monotonic decreasing when $x < y \Rightarrow f(x) \geqslant f(y)$. A real function f is said to be strictly monotonic decreasing when $x < y \Rightarrow f(x) > f(y)$.
Theorem qns 5, 6, 7	A monotonic function has one-sided limits at each point of its domain. It is continuous where the two one-sided limits are equal.
Definition qns 8, 9	*Intervals* A subset I of $\mathbb{R}$ is said to be connected if when $a, b \in \mathbb{R}$, and $a < b$, then $x \in I$ for every x satisfying $a < x < b$. A connected set is called an interval. Bounded intervals are classified as closed intervals $[a, b]$, open intervals (a, b), half-open intervals $(a, b]$ and singletons $\{a\}$. Unbounded intervals are classified as closed half-rays $[a, +\infty)$ or $(-\infty, a]$, open half-rays $(a, +\infty)$ or $(-\infty, a)$, or the whole line $\mathbb{R}$.

The Intermediate Value Theorem
qns 12, 13, 14, 15	If a continuous function, f, defined on an interval takes positive and negative values then, for some c in the interval, $f(c) = 0$.

Theorem *qn 21*	The range of a continuous function defined on an interval is always an interval.
Theorem *qns 22, 23, 24, 25*	If a continuous function, f, is defined on an interval, f has an inverse function if and only if f is strictly monotonic. In this case the inverse function is continuous and strictly monotonic.
The Minimax	*Theorem*
qns 31, 32, 34	A continuous real function defined on a closed interval is bounded, and attains is bounds.
Definition *qns 34–42*	*Uniform continuity* A function $f: A \to \mathbb{R}$ is said to be uniformly continuous on A when, given $\varepsilon > 0$, there exists a δ such that $|f(x) - f(y)| < \varepsilon$ whenever $|x - y| < \delta$.
Theorem *qn 43*	If a real function is continuous on a closed interval then it is uniformly continuous on that interval.
Theorem *qn 47*	If a function is uniformly continuous on a dense subset of a closed interval, including the end points, then it may be extended to a continuous function on the whole interval.

Historical note

The Intermediate Value Theorem had been assumed from geometrical perceptions during the eighteenth century as the basis of work on approximations to roots of equations. There were mathematicians who regarded it as the essential characterisation of continuity. (We saw how wrong that was in qn 16.) In 1817 Bolzano insisted that this was a theorem which required analytical proof. He provided the first formal definition of a continuous function, showed that if a continuous function was positive at a point it must be positive in a neighbourhood of that point and likewise if it is negative at a point it must be negative in a neighbourhood of that point. Taking the function to be positive at the upper end of the domain and negative at the lower end he claimed that the set of points of the domain at which the function was negative was bounded above and so must have a least upper bound. At this least upper bound only a zero value for the function is not contradictory. Bolzano proved the least upper bound property using a Cauchy sequence, though not by that name (!), derived from repeated bisections. His proof that Cauchy sequences converge was incomplete because he thought it followed from the impossibility of the convergence of such a sequence to two distinct limits. Probably quite independently of Bolzano, Cauchy also produced a formal definition of continuity in 1821 and the proof of the Intermediate Value Theorem that we constructed in this chapter was essentially his. He repeatedly divided his domain into m equal parts and then selected one part on which a change of sign had taken place. We took $m = 2$. Cauchy assumed without proof that monotonic bounded sequences were convergent.

The theorem that continuous functions on a closed interval were bounded and attained their bounds had been familiar from the seventeenth century but under much stronger differentiability conditions. It was Weierstrass, insisting on the distinction between continuous and differentiable functions, who first proved that a continuous function on a closed interval was bounded and attained its bounds in his lectures in Berlin about 1869. Weierstrass affirmed the importance of considering continuous but not necessarily differentiable functions by his approximation theorem that every continuous function is the uniform limit of a sequence of polynomials.

The distinction between continuity at a point and continuity on an interval was not clear in Cauchy's work. In his proof of the integrability of continuous functions (1823), Cauchy assumed continuity but used uniform continuity. Dirichlet proved that a function which was continuous on a closed interval was uniformly continuous on that interval in his lectures in Berlin in 1854. The first proof of this result in a Weierstrassian context was published by E. Heine in 1872.

Answers

1 (i) $\mathbb{R}^+ \cup \{0\}$. (ii) Yes. (iii) Yes. (iv) $[0, 1)$ for example. (v) $(-\frac{1}{2}\pi, \frac{1}{2}\pi)$ for example. (vi) Yes.

2 If $f(x) \geq f(y)$ when $x < y$, the function f is said to be decreasing or monotonic decreasing.
If $f(x) > f(y)$ when $x < y$, the function is said to be strictly decreasing or strictly monotonic decreasing.

3 (i) $\mathbb{R}^+ \cup \{0\}$. (ii) Yes. (iii) $\mathbb{R}^+$. (iv) Yes.

4 $\sup V$ exists, by qn 4.63. By 4.63, given $\varepsilon > 0$, for some x_1,
$\sup V - \varepsilon < f(x_1) \leq \sup V$, but f is monotonic increasing so, for $x \geq x_1$,
$\sup V - \varepsilon < f(x_1) \leq f(x) \leq \sup V$, so $|f(x) - \sup V| < \varepsilon$
and $\lim\limits_{x \to +\infty} f(x) = \sup V$.

5 $f(a)$ is an upper bound for L and a lower bound for U, so $\sup L$ and $\inf U$ exist by qns 4.62 and 63. Given $\varepsilon > 0$, for some $x_1 < a$,
$\sup L - \varepsilon < f(x_1) \leq \sup L$. But f is monotonic increasing so, if
$x_1 < x < a$,
$\sup L - \varepsilon < f(x_1) \leq f(x) \leq \sup L$, and $|f(x) - \sup L| < \varepsilon$.
Similarly, for some $x_2 > a$, $\inf U \leq f(x_2) < \inf U + \varepsilon$. But f is monotonic increasing so, if $a < x < x_2$, $\inf U \leq f(x) \leq f(x_2) < \inf U + \varepsilon$, and $|f(x) - \inf U| < \varepsilon$.
$\lim\limits_{x \to n^-} [x] = n - 1$ and $\lim\limits_{x \to n^+} [x] = n$.

6 If f is monotonic decreasing then, with the notation of qn 5, $f(a)$ is a lower bound for L and an upper bound for U. Proceed as in qn 5 with appropriate inequalities reversed.

7 If f is continuous at a, then limits from above and below at a are equal.
$\sup L \leq f(a) \leq \inf U$ so, if $\sup L = \inf U$, then
$\lim\limits_{x \to a^-} f(x) = f(a) = \lim\limits_{x \to a^+} f(x)$, and so f is continuous by qn 6.89.
If $\sup L < \inf U$, then there is a rational number between these two by qn 4.6. Because the function is monotonic this locates a distinct rational number in each discontinuity. A set of rationals is countable, so the set of discontinuities of a monotonic function is countable.

8 (i) $\{1\}$, (ii) $[-1, 1]$, (iii) $(-\frac{1}{2}\pi, \frac{1}{2}\pi)$, (iv) $(0, 1]$, (v) $\mathbb{R}^+$, (vi) $\mathbb{R}^+ \cup \{0\}$, (vii) $\mathbb{R}$.

9 Make your own sketches without considering possible formulae.

10 Take A as the union of (two) disjoint and separated intervals.

11 The function is continuous by qn 6.29. The number 0 is not in the range.

12 If $f(d) \neq 0$, then either $f(d) > 0$ or $f(d) < 0$. If $f(d) > 0$, then take $a_2 = a_1$ and $b_2 = d$. If $f(d) < 0$, take $a_2 = d$ and $b_2 = b_1$. Now suppose $f(a_n) < 0$ and $f(b_n) > 0$ and let $d = \frac{1}{2}(a_n + b_n)$. If $f(d) = 0$ we have finished. If $f(d) > 0$, take $a_{n+1} = a_n$ and $b_{n+1} = d$. If $f(d) < 0$, take $a_{n+1} = d$ and $b_{n+1} = b_n$.

13 (i) (a_n) is monotonic increasing and bounded above by b, and thus is convergent, by qn 4.34. (b_n) is monotonic decreasing and bounded below by a, and thus is convergent, by qn 4.33.

(ii) $(f(a_n)) \to f(A)$ by the continuity of f and $f(A) \leqslant 0$ by qn 3.66.

(iii) From qns 12 and 3.44(iii), $B - A = \frac{1}{2}(B - A)$, so $B = A$.
$f(A) \leqslant 0 \leqslant f(A) \Rightarrow f(A) = 0$.

14 Define $g(x) = f(x) - k$, then $g(a) < 0$ and $g(b) > 0$ and g is continuous by qn 6.23. So $g(c) = 0$ for some $c \in (a, b)$, and $f(c) - k = 0$, so $f(c) = k$.

15 Define $g(x) = -f(x)$, then $g(a) < -k < g(b)$ and g is continuous, by qn 6.26. So $g(c) = -k$ for some $c \in (a, b)$, and $f(c) = k$.

16 The following function reaches every value betwen 0 and 1, but is discontinuous everywhere. $f: [0, 1] \to [0, 1]$.
$f(0) = \frac{1}{2}$; $f(\frac{1}{2}) = 0$; $f(x) = x$ when x is rational and $\neq 0, \frac{1}{2}$;
$f(x) = 1 - x$ when x is irrational.
Also, the function of qn 6.19 reaches every value between $f(0)$ and $f(x)$, but is discontinuous at 0.

17 If $f(a) < f(b)$ but both are integers then, by qn 14, for some $c \in (a, b)$, $f(c) = f(a) + \frac{1}{2}$ which is not an integer.

18 If $f(a) < f(b)$ but both are rational then, by qn 14, for some $c \in (a, b)$,
$$f(c) = \frac{f(a) + \sqrt{2}f(b)}{1 + \sqrt{2}}$$
which is irrational. See qn 4.20.

19 If $x = |a| + |b| + 1$, then $f(x) > 0$.
If $x = -|a| - |b| - 1$, then $f(x) < 0$. So there is a root of the equation $x^3 + ax + b = 0$ by the Intermediate Value Theorem. This result may be extended to any polynomial of odd degree.

20 Choose $g(x) = x - f(x)$ which is continuous, by qn 6.23. Then $g(0) = -f(0) \leqslant 0$ and $g(1) = 1 - f(1) \geqslant 0$. So $g(c) = 0$ for some $c \in [0, 1]$, and $f(c) = c$.

21 If a and b are any two points in an interval domain of f, then $f(a)$ and $f(b)$ are two points in the range of f. The Intermediate Value Theorem says that, for any k, $f(a) \leqslant k \leqslant f(b)$ or $f(a) \geqslant k \geqslant f(b)$, there is a $c \in [a, b]$ such that $f(c) = k$. So the range is connected and is therefore an interval.

22 If $a < x < b$, then $f(a) < f(x) < f(b)$: because if $f(a) \geqslant f(x)$ then, for some $c \in [x, b)$, $f(c) = f(a)$ and f is not one–one; and if $f(x) \geqslant f(b)$ then, for some $c \in (a, x]$, $f(c) = f(b)$ and, again, f is not one–one. Applying this result to the interval $[x, b]$ we have $a < x < y < b$ implies $f(a) < f(x) < f(y) < f(b)$, and f is strictly monotonic increasing. If $f(a) > f(b)$ and f is continuous and one–one, then f is strictly monotonic decreasing.

23 (i) Yes, immediate consequence of being strictly monotone.
  (ii) Yes, consequence of (i) from qns 6.5 and 6.6. The domain of f^{-1} must be the range of f.
  (iii) No. For example, consider $f: [0, 2] \to [0, 3]$ given by $f(x) = x$ on $[0, 1]$ and $f(x) = x + 1$ on $(1, 2]$.

24 Yes, from qn 22.

25 (i) The range is an interval from qn 21. But, for $x \in [a, b]$, $f(a) \leqslant f(x) \leqslant f(b)$, taking f as increasing. So $c = f(a)$ and $d = f(b)$.
    If f is decreasing then $c = f(b)$ and $d = f(a)$.
  (ii) Strictly monotonic functions are one–one and so invertible.
  (iii) $x < y \Leftrightarrow f(x) < f(y) \Rightarrow g(f(x)) < g(f(y)) \Leftrightarrow f(x) < f(y)$.
  (iv) From qn 5, using completeness.
  (v) The sequences are convergent from the continuity of f. They are increasing and decreasing because f is strictly monotone.
  (vi) Let $t_n = f(x_n)$ from below and $t_n = f(s_n)$ from above.
  (vii) $\lim\limits_{t \to y^-} g(t)$ exists from (iv) and equals $g(y)$ from (vi).
    Similarly for the limit from above, so $\lim\limits_{t \to y} g(t) = g(y)$ and g is continuous at y by qn 6.89.

26 f only has an inverse on $\mathbb{R}^+ \cup \{0\}$ or on $\mathbb{R} \backslash \mathbb{R}^+$. On $\mathbb{R}^+ \cup \{0\}$, $x \mapsto x^2$ is continuous by qn 6.29, and strictly monotonic by qn 2.22. The range is an interval by the Intermediate Value Theorem and is unbounded above, so f is a bijection.
$f^{-1}: x \mapsto \sqrt{x}$ is continuous and monotonic on $\mathbb{R}^+ \cup \{0\}$, by qn 25.

27 f is continuous by qn 6.29, strictly monotonic by qn 2.35, and unbounded above, so f is a bijection. Then f has a continuous inverse by qn 25.

28 Let $a = \sqrt[n]{x}$, then

$$f(x) = \frac{1-x}{1-x^{m/n}} = \frac{1-a^n}{1-a^m} = \frac{1 + a + a^2 + \ldots + a^{n-1}}{1 + a + a^2 + \ldots + a^{m-1}}.$$

As $x \to 1$, $a \to 1$, by the continuity of $x \mapsto \sqrt[n]{x}$.
Now by qns 6.29, 6.54 and 6.86 $\lim\limits_{x \to 1} f(x) = n/m$.

29 The range must be an interval, by qn 21.

30 The only possible ranges are a singleton and a closed interval.

31 (i) None whatsoever. (ii) $a \leqslant x_n \leqslant b$. (iii) Yes, by qn 4.46.
(iv) By qn 3.66. (v) Its limit is $f(c)$. (vi) Take $\varepsilon = 1$.
(vii) Choose $n_i > f(c) + 1$, then, from the definition of x_n,
$$f(x_{n_i}) > n_i > f(c) + 1,$$ and this contradicts (vi), so our original
hypothesis was wrong.

32 As in qn 31, but starting from a sequence (x_n) in the domain of the
function such that $f(x_n) < -n$.

33 (i) Denominator not zero on $\mathbb{Q}$, so the function is continuous by
qn 6.51.
(ii) Because $\mathbb{Q}$ is not complete, a bounded sequence in $\mathbb{Q}$ need not
contain a subsequence which converges to a point of $\mathbb{Q}$.

34 (i) From qn 21. (ii) From qn 31.
(iii) Every non-empty set of real numbers which is bounded above has a
least upper bound: qn 4.63. (iv) qn 4.60 interpreted in qn 4.63.
(v) Since $a \leqslant x_n \leqslant b$, (x_n) is bounded, and so must contain a
convergent subsequence by qn 4.46. (vi) By qn 3.66.
(vii) Because f is continuous.
(viii) Since $|f(x_n) - M| < 1/n$, the limit of the subsequence $(f(x_{n_i}))$ is
M, so $M = f(c)$.
(ix) If $\inf V = m$, construct a sequence (y_n) such that
$m \leqslant f(y_n) < m + 1/n$. Argue as in (iv)–(viii).

35 $\sin x = \pm 1$ only when $x = (n + \frac{1}{2})\pi$ which is irrational. But the domain of
the function is dense in $\mathbb{R}$ so it attains values arbitrarily close to ± 1.
Since $\mathbb{Q}$ is not complete, the argument of qn 34(v) fails on $\mathbb{Q}$.

36 For unbounded domains use the functions of qn 8.
For bounded domains consider $x \mapsto 1/x$ on $(0, 1)$ or $(0, 1]$, and for
bounded ranges consider $x \mapsto x$ on $(0, 1)$ or $(0, 1]$.

37 $|(x^2 - y^2)/(x - y)| = |x + y| \leqslant |x| + |y| \leqslant 20$, so $|x^2 - y^2| \leqslant 20|x - y|$.

38 $\delta_1 \leqslant 1/6.$ $\delta_2 \leqslant 1/20.$
$|1/x - 1/a| < 1 \Leftrightarrow |x - a| < |xa| \Leftrightarrow \delta \leqslant (a - \delta)a \Leftrightarrow \delta \leqslant a^2/(1 + a).$ As a gets near to 0, an ever smaller δ is required, so there is no one δ for all a.

39 Taking $y = a$ gives the neighbourhood definition of continuity.

40 For the suggested x and y, $|x - y| < 1/n$, but $|x^2 - y^2| > 2$, so there is no universal δ for $\varepsilon = 2$.

41 For the suggested x and y, $|x - y| < 1/n^2$, but $f(x) = 0$ and $f(y) = 1$, so $|f(x) - f(y)| = 1 > \frac{1}{2}$, so there is no universal δ for $\varepsilon = \frac{1}{2}$.

42 Take $\delta = \varepsilon/L.$

43 (i) None at all. (ii) $a \leqslant x_n \leqslant b$. (iii) Yes, by qn 4.46.
(iv) Yes, use qn 3.44(iii) and the fact that $(x_n - y_n)$ is a null sequence.
(v) By qn 3.66. (vi) Both tend to $f(c)$. (vii) For sufficiently large n_i both $|f(x_{n_i}) - f(c)| < \frac{1}{2}\varepsilon$ and $|f(y_{n_i}) - f(c)| < \frac{1}{2}\varepsilon.$

44 f is continuous by qn 26, and therefore uniformly continuous on $[0, 1]$ by qn 43. $|x - y| = |\sqrt{x} - \sqrt{y}| \cdot |\sqrt{x} + \sqrt{y}| > |\sqrt{x} - \sqrt{y}|$ provided either x or $y > 1$. To establish uniform continuity on $\mathbb{R}^+ \cup \{0\}$, for a given ε, choose the lesser of the δ from $[0, 1]$ and $\delta = \varepsilon$.

45 Define $g(a) = \lim_{x \to a^+} f(x)$ and $g(b) = \lim_{x \to b^-} f(x)$ and then g is continuous on $[a, b]$ by qn 6.92. So g is uniformly continuous on $[a, b]$ by qn 43 and therefore uniformly continuous on the subset (a, b) where g coincides with f.

46 Because the denominator is not zero at any rational point, the function is continuous by qn 6.51.
f is not uniformly continuous because, choosing rational x and y such that $\frac{1}{2} < y^2 < \frac{1}{2} + 1/6n < \frac{1}{2} + 1/4n < x^2 < \frac{1}{2} + 1/2n,$ $f(y) - f(x) > n$ and $x^2 - y^2 < 1/2n,$ but $\frac{1}{2} < x^2 \Rightarrow \frac{1}{2} < x \Rightarrow 1 < x + y$
$\Rightarrow x - y < x^2 - y^2 < 1/2n.$
If $(a_n) \to 1/\sqrt{2}$ then a continuous g would make $(g(a_n)) \to g(1/\sqrt{2}).$
But, if the sequence (a_n) only has rational terms, $g(a_n) = f(a_n).$
However $(f(a_n)) \to +\infty.$

47 (i) If, given $\varepsilon > 0$, $|f(x) - f(y)| < \varepsilon$ when $|x - y| < \delta$; for sufficiently large n, $|a_{n+m} - a_n| < \delta$, so $|f(a_{n+m}) - f(a_n)| < \varepsilon.$
(ii) From qn 3.44(iii) $(a_n - b_n)$ is a null sequence, so, for sufficiently large n, $|a_n - b_n| < \delta$, so $|f(a_n) - f(b_n)| < \varepsilon.$ Thus $(f(a_n) - f(b_n))$ is a null sequence.
(iii) By (ii), every rational sequence $(a_n) \to a$ gives a sequence $(f(a_n))$ converging to the same limit. So $g(a)$ is well defined.

(iv) $a \leqslant a_n \leqslant b_n \leqslant b$. $|a - b| < \delta \Rightarrow |a_n - b_n| < \delta \Rightarrow$
$|f(a_n) - f(b_n)| < \frac{1}{3}\varepsilon$. For sufficiently large n, $|g(a) - f(a_n)| < \frac{1}{3}\varepsilon$
and $|f(b_n) - g(b)| < \frac{1}{3}\varepsilon$ So $|a - b| < \delta \Rightarrow |g(a) - g(b)| < \varepsilon$.

48 Take $a > 1$, and work with positive x. Let $p, q, r, s \in \mathbb{Z}^+$.

(i) $p/q < r/s \Leftrightarrow ps < qr \Leftrightarrow a^{ps} < a^{qr} \Leftrightarrow a^{p/q} < a^{r/s}$, using qn 2.35.

(ii) $ps < qr \Leftrightarrow g(ps) < g(qr) \Leftrightarrow g(p/q) < g(r/s)$ changing from a to b.
From qn 4.40, $g(1/n) \to L$ (say) as $n \to \infty$, so, since g is strictly
increasing $\lim_{x \to 0^+} g(x) = L$. Define $g(0) = L$.

(iii) $\dfrac{a^x - a^y}{x - y} = a^y \left(\dfrac{a^{x-y} - 1}{x - y} \right) < a^B \left(\dfrac{a^B - 1}{B} \right)$, supposing $y < x$.

(v) Define $f(-x) = 1/f(x)$.

8

Derivatives
Tangents

Preliminary reading: Bryant ch. 4, Courant and John ch. 2.
Concurrent reading: Spivak ch. 9.
Further reading: Tall (1982).

It is easy enough to say when a straight line is a tangent to a circle or
an ellipse. For these curves, a straight line – infinitely extended –
meets the curve in 0, 1 or 2 points. Each line with a unique point of
intersection is a tangent. However if we try to use such a test to
identify tangents to other curves we are in for a disappointment, and
on several counts.

1 At how many points does the line $x = 1$ intersect the parabola
   $y = x^2$? Draw a sketch. Is this line a tangent to the curve?

2 At how many points does the line $y = -2$ intersect the cubic
   curve $y = x^3 - 3x$? Draw a sketch. Is this line a tangent to the
   curve?

From qn 2 we learn that whether a line is a tangent to a curve or
not is a local question, which must be asked relative to a particular
point of intersection, that is, inside a sufficiently small neighbourhood
of that point.

3 At how many points does the line $y = mx$ intersect the cubic
   curve $y = x^3$? For how many values of m might the line be a
   tangent to the curve?
   Check that the line $y = h^2x$ is the chord joining the two points
   $(0, 0)$ and (h, h^3) on the curve, provided $h \neq 0$.
   If $h \to 0$, to what does the slope (or gradient) of the chord tend?

4 Write down the equation of the line with slope m through the point $(a, f(a))$.
Write down the equation of the chord joining the points $(a, f(a))$ and $(a + h, f(a + h))$.
If

$$\lim_{h \to 0} \frac{f(a + h) - f(a)}{h} = m,$$

how would you describe the line $y - f(a) = m(x - a)$?

On the strength of the ideas in qn 4, we will define the *derivative* of a function at a point a of its domain.

Definition of derivative

If a is a cluster point of the domain of the real function f and

$$\lim_{h \to 0} \frac{f(a + h) - f(a)}{h} = m, \quad \text{for some real number } m,$$

then m is called the *derivative* of f at a, usually denoted by $f'(a)$, and f is said to be *differentiable* at a.

The definition is designed to give a formal definition of the slope of the tangent to $y = f(x)$ at $x = a$, and hence to make it possible to define analytically what is meant by a tangent to a curve. On a distance–time graph the slope of a chord gives the average velocity between two points, while the slope of a tangent gives the velocity at a point.

5 Although the motivation for our study of derivatives has been the geometric notion of tangent, there is still one circumstance when a tangent to the graph of a function with domain $\mathbb{R}$ may exist without a derivative of the function at the point in question. Examine the definition of derivative carefully in order to identify the circumstance in question.

6 If f is a constant function with domain $\mathbb{R}$, what is $f'(a)$?

The attempt to establish a converse to qn 6 exposes some unexpected subtleties, and will be examined in qn 9.17 using the Mean Value Theorem.

7 If $f(x) = mx + c$, what is $f'(a)$?

You will have noticed in your calculations for qns 6 and 7, how critical it is that in finding the limit of a function as $h \to 0$ we pay no regard to the value of that function when $h = 0$. In fact the 'slope of the chord' function is not defined when $h = 0$. An equation like $h/h = 1$ is only valid when $h \neq 0$.

8 Show that the two limits

$$\lim_{h \to 0} \frac{f(a + h) - f(a)}{h}, \quad \text{and} \quad \lim_{x \to a} \frac{f(x) - f(a)}{x - a},$$

are equivalent.

9 If $f(x) = x^2$, what is $f'(a)$?

Sums of functions

10 If two real functions f and g, on the same domain, are both differentiable at a, prove using qn 6.93 that the function $f + g$ defined by $f + g: x \mapsto f(x) + g(x)$ is differentiable at a, and that

$$(f + g)'(a) = f'(a) + g'(a).$$

Explain how this result may be generalised to show that the sum of n real functions, each differentiable at a, is also differentiable at a.

11 If f is differentiable at a, and $g(x) = k \cdot f(x)$, prove that

$$g'(a) = k \cdot f'(a).$$

Questions 10 and 11 give linearity in determining derivatives, in the sense that if the functions f and g are differentiable at a, then the function given by $x \mapsto l \cdot f(x) + m \cdot g(x)$ has derivative

$l \cdot f'(a) + m \cdot g'(a)$ at a.

The product rule

12 By considering the product

$$f(x) - f(a) = \frac{f(x) - f(a)}{x - a} \cdot (x - a), \quad \text{when} \quad x \neq a,$$

and, using qn 6.93, show that if the real function f is differentiable at a, then f is continuous at a.

13 If two real functions f and g, on the same domain, are both differentiable at a, prove using qns 6.93 and 12 that the function $f \cdot g$ defined by $f \cdot g: x \mapsto f(x) \cdot g(x)$ is differentiable at a, and that

$$(f \cdot g)'(a) = f(a) \cdot g'(a) + f'(a) \cdot g(a).$$

14 Prove by induction that, if $f(x) = x^n$, then $f'(a) = n \cdot a^{n-1}$, for any positive integer n.
Obtain the same result by considering the equation
$$x^n - a^n = (x - a)(x^{n-1} + x^{n-2}a + x^{n-3}a^2 + \ldots + a^{n-1}).$$

15 By using qns 10, 11, 13 and 14 find $f'(a)$ when
$f(x) = b_0 + b_1 x + b_2 x^2 + \ldots + b_n x^n$.

16 If $f(x) = 1/x^n$, prove that $f'(a) = -n/a^{n+1}$, for any positive integer n, provided $a \neq 0$.

The quotient rule

17 If two real functions f and g, on the same domain, are both differentiable at a, prove using qns 6.93 and 12 that the function f/g defined by $f/g: x \mapsto f(x)/g(x)$ is differentiable at a, provided $g(a) \neq 0$, and that
$$\left(\frac{f}{g}\right)'(a) = \frac{g(a) \cdot f'(a) - f(a) \cdot g'(a)}{\left(g(a)\right)^2}$$

The chain rule

18 If the real function g is differentiable at a, and the real function f is differentiable at $g(a)$, we seek to prove that the composite function $f \circ g$ defined by $f \circ g: x \mapsto f(g(x))$ is differentiable at a, and that
$$(f \circ g)'(a) = f'(g(a)) \cdot g'(a).$$

  (i) First suppose that there is a neighbourhood of a in which $g(x) \neq g(a)$, unless $x = a$, and use the continuity of g at a and the equation
$$\frac{f(g(x)) - f(g(a))}{x - a} = \frac{f(g(x)) - f(g(a))}{g(x) - g(a)} \cdot \frac{g(x) - g(a)}{x - a},$$
  for $x \neq a$.

  (ii) Secondly, suppose that in every neighbourhood of a there are points $x \neq a$ for which $g(x) = g(a)$.
  Prove that $g'(a) = 0$.

Let
$$F(x) = \frac{f(g(x)) - f(g(a))}{x - a}, \quad \text{for } x \neq a.$$
Let (a_n) be any sequence which tends to a, in the domain of $f \circ g$, but which does not contain a among its terms. We seek to show
$$(F(a_n)) \to 0 = f'(g(a)) \cdot 0 = f'(g(a)) \cdot g'(a), \quad \text{as } n \to \infty.$$
If $g(a_n) = g(a)$, show that $F(a_n) = 0$.
If (a_{n_i}) is the subsequence of (a_n) of terms for which $g(a_n) \neq g(a)$, use the equation in (i) and qn 3.44(iv) to show that
$$(F(a_{n_i})) \to f'(g(a)) \cdot g'(a) = 0.$$
Why must $(F(a_n))$ be a null sequence?
Why does this establish the required limit?

Differentiability and continuity

19 Sketch the graph of the real function f given by $f(x) = |x|$. For this function find
$$\lim_{h \to 0^+} \frac{f(h) - f(0)}{h} \quad \text{and} \quad \lim_{h \to 0^-} \frac{f(h) - f(0)}{h}.$$
Deduce that f is not differentiable at $x = 0$.

20 Define the function $f: \mathbb{R} \to \mathbb{R}$ by
$$f(x) = \sin(1/x) \text{ when } x \neq 0,$$
$$f(0) = k.$$
By considering the two null sequences $(1/2n\pi)$ and $(1/(2n + \frac{1}{2})\pi)$ prove that the function f is not continuous at $x = 0$ whatever value k may have.

21 Define the function $f: \mathbb{R} \to \mathbb{R}$ by
$$f(x) = x \sin(1/x) \text{ when } x \neq 0,$$
$$f(0) = 0.$$
Using the fact that $-|x| \leq f(x) \leq |x|$ for all x, prove that f is continuous at $x = 0$.
Prove from the definition of derivative that f is not differentiable at $x = 0$.

22 Define the function $f: \mathbb{R} \to \mathbb{R}$ by

$f(x) = x^2 \sin(1/x)$ when $x \neq 0$,

$f(0) = 0$.

By methods similar to those of qn 21, prove that f is continuous at $x = 0$.
Prove also that f is differentiable at $x = 0$.

In qn 12 we proved that a function that is differentiable at a point is necessarily continuous at that point. We now have two examples (in qns 19 and 21) which illustrate that the converse of this theorem is false. In fact the independence of differentiability and continuity is remarkable, for a function may be continuous at every point and differentiable at no points, and a computer program exhibiting such a function has been made by David Tall. Such a function is given in qn 12.46.

23 If a function f is differentiable at the point a, prove that

$$\lim_{h \to 0} \frac{f(a + h) - f(a - h)}{2h} = f'(a).$$

Give an example to show that this limit may exist when f is not differentiable at a.

When we have a formal definition, however well it seems to match our intuition, it is important to press the meaning of that definition as far as possible. We therefore ask a question which focuses on differentiability at a single point, by considering functions which are, for the most part, not continuous, and therefore not differentiable.

24 The real functions f, g and h are defined on $\mathbb{R}$, such that

$$f(x) = \begin{cases} 0 & \text{when } x \text{ is rational,} \\ 1 & \text{when } x \text{ is irrational;} \end{cases}$$

$$g(x) = \begin{cases} 0 & \text{when } x \text{ is rational,} \\ x & \text{when } x \text{ is irrational;} \end{cases}$$

$$h(x) = \begin{cases} 0 & \text{when } x \text{ is rational,} \\ x^2 & \text{when } x \text{ is irrational.} \end{cases}$$

Clearly none of these functions is continuous when $x \neq 0$, and therefore none is differentiable when $x \neq 0$.
Determine whether either f, g or h is
(i) continuous at $x = 0$; (ii) differentiable at $x = 0$.

Another conflict with intuition comes with the discovery that a function with a positive derivative at a point need not be increasing in any neighbourhood of that point.

25 Draw the graphs of $y = x + 2x^2$ and $y = x - 2x^2$ near the origin. Define the function f by

$$f(x) = x + 2x^2 \cos(1/x) \text{ when } x \neq 0$$

$$f(0) = 0.$$

Use the method of qn 22 to show that $f'(0) = 1$.
Prove that f is not increasing in any neighbourhood of 0 by showing that $f(1/2n\pi) > f(1/(2n - 1)\pi)$ for any $n \in \mathbb{Z}^+$. Show also that $f'(1/(2n - \frac{1}{2})\pi) = -1$.

Derived functions

26 For the real function f given by $f(x) = |x|$, which is defined for all $x \in \mathbb{R}$, find the subset of the domain which consists of those points of $\mathbb{R}$ at which f is differentiable.

If a real function $f: A \to \mathbb{R}$ is differentiable at each point of a subset $B \subseteq A$, with B maximal, we define the *derived function* of f, $f': B \to \mathbb{R}$ by $f': x \mapsto f'(x)$, and we say that f is differentiable on B.

27 Sketch the graph of the derived function of f where f is defined by $f(x) = |x|$.

28 Sketch the graph of the function given by $f(x) = x^2 - 2x + 2$ and of its derived function. How does the point where the graph of the derived function cuts the x-axis relate to the shape of the graph of f?

29 Define the terms *local maximum* and *local minimum* in such a way that they describe respectively the points $(-1, 5)$ and $(1, 1)$ on the graph of the real function defined by $f(x) = x^3 - 3x + 3$. What is $f'(-1)$ and $f'(1)$?

30 If a real function f has a local maximum or a local minimum at a point a of its domain, and f is differentiable at a, prove that $f'(a) = 0$ by considering limits from above and below.

31 Give an example of a function f which shows that it is possible to have $f'(a) = 0$ without the function f having a local maximum or a local minimum at a.

32 Give an example of a function f which shows that it is possible for the function to have a local maximum at $x = a$ without being differentiable at a.

Questions 15 and 17 enable us to find derived functions for polynomials and rational functions. In chapter 11 we will give a formal definition of logarithmic and exponential functions and obtain $\log'(x) = 1/x$, for positive x, and $\exp'(x) = \exp(x)$, for all real x, and we will give a formal definition of the circular (or trigonometric) functions sine and cosine and obtain $\sin'(x) = \cos(x)$ and $\cos'(x) = -\sin(x)$ for all real x.

According to the Leibnizian description of derived functions, when, for example, $y = x^3$,

$$\frac{dy}{dx} = 3x^2.$$

For the product rule the Leibnizian expression is

$$\frac{d(uv)}{dx} = u \cdot \frac{dv}{dx} + v \cdot \frac{du}{dx}.$$

For the quotient rule the Leibnizian expression is

$$\frac{d\left(\dfrac{u}{v}\right)}{dx} = \frac{v \cdot \dfrac{du}{dx} - u \cdot \dfrac{dv}{dx}}{v^2}.$$

The notation of Leibniz is particularly suggestive when describing the chain rule:

$$\frac{dy}{dx} = \frac{dy}{du} \cdot \frac{du}{dx}.$$

33 Apply the chain rule to find the derived function of f where $f(x) = \sin x^2$ by considering $y = \sin u$ and $u = x^2$.

34 Give algebraic expressions for the derived function f' of qn 22.
Is the function f' continuous at $x = 0$?

Second derivatives

35 A real function f is defined by

$$f(x) = \begin{cases} 0 & \text{when } x < 0, \\ x & \text{when } 0 \leq x < 1, \\ \frac{1}{2}x^2 + \frac{1}{2} & \text{when } 1 \leq x. \end{cases}$$

Sketch the graphs of (i) f, (ii) f', (iii) the derived function of f'.

When the real function $f: A \to \mathbb{R}$ has a derived function $f': B \to \mathbb{R}$, and the derived function $f': B \to \mathbb{R}$ has a derived function $(f')': C \to \mathbb{R}$, with $C \subseteq B \subseteq A$, we write $(f')' = f''$ and call f'' the *second derivative* of f.
When $f''(a)$ exists, we say that f is twice differentiable at a.

36 How do you know that, when $f''(a)$ exists, f' is continuous at a?

37 A function $f: \mathbb{R} \to \mathbb{R}$ is defined by

$$f(x) = \begin{cases} -x^2 & \text{when } x < 0 \text{ and} \\ x^2 & \text{when } x \geq 0. \end{cases}$$

Determine whether
 (i) f is continuous at 0,
 (ii) $f'(0)$ exists,
 (iii) f' is continuous at 0,
 (iv) $f''(0)$ exists.

38 Define the third derivative of f, and inductively, the nth derivative of f, which is written $f^{(n)}$.

Inverse functions

39 Find the derivative of the function f given by $f(x) = \sqrt{x}$ at some point $a > 0$, from first principles. Assume that f is continuous at a.
If $g(x) = x^2$, attempt to relate $f'(a)$ to $g'(f(a))$.

40 If the real function $f: A \to B$ is a bijection, and $g: B \to A$ is its inverse, so that $f(g(b)) = b$ for any $b \in B$, and $g(f(a)) = a$ for any $a \in A$, and if further we suppose that f is differentiable at a, and g is differentiable at $f(a)$, use the chain rule to prove that

$$g'(f(a)) = \frac{1}{f'(a)}.$$

The Leibnizian expression is again particularly suggestive here:

$$\frac{dx}{dy} = 1 \bigg/ \left(\frac{dy}{dx}\right).$$

41 If f and g are inverse functions as in qn 40, and $f'(a) = 0$, prove that g cannot be differentiable at $f(a)$.
Indicate by a sketch why this should be expected and relate your answer to qn 5.

42 Let f be a continuous bijection $f: A \to B$ where A and B are open intervals, and let $g: B \to A$ be the inverse of f. We saw in chapter 7 that g is also a continuous bijection. Now suppose that at some point $a \in A$, f is differentiable and $f'(a) \neq 0$.
  (i) Prove that

$$\lim_{x \to a} \frac{g(f(x)) - g(f(a))}{f(x) - f(a)} = \frac{1}{f'(a)}.$$

  (ii) Now let $f(a) = b$, and let (b_n) be a sequence in B which tends to b, but which does not contain any term equal to b. Why must there exist a unique $a_n \in A$, $a_n \neq a$, such that $f(a_n) = b_n$?
  (iii) Use the continuity of g to prove that $(a_n) \to a$.
  (iv) Use the limit above to prove that

$$\left(\frac{g(b_n) - g(b)}{b_n - b}\right) \to \frac{1}{f'(a)}, \text{ as } n \to \infty.$$

  (v) Deduce that g is differentiable at $f(a)$.

The only definition or theorem in this chapter which depends on the Axiom of Completeness is qn 42. Completeness is needed for part (iii). This was necessary because, on rational domains, differentiable bijections do not have to have continuous inverses.

43 If the function f is defined by $f(x) = x^n$ for $x \geqslant 0$ and a positive integer n, identify the inverse function f^{-1}, and find its derived function.
If, for positive x, and positive integers p and q, we define $x^{p/q} = (x^p)^{1/q}$,
use the chain rule to evaluate the derived function.
Extend this result to show that for any rational number $r \neq 0$, the

derived function of g, where $g(x) = x^r$, for positive x, is given by $g'(x) = rx^{r-1}$.

For what rational numbers r is g differentiable at $x = 0$?

44 Find a maximal interval domain containing zero for which the sine function has a well-defined inverse. Find the derived function of this inverse function (known as arcsine) and state the maximal interval domain for this derived function. You may use any of the properties of the circular functions which are established in chapter 11.

45 Defining $\tan x = (\sin x)/(\cos x)$, provided $\cos x \neq 0$, identify a maximal interval domain containing zero for which the tangent function has a well-defined inverse. Find the derived function of this inverse function (known as arctan) and state the maximal domain for this derived function. You may use the properties of the circular functions established in chapter 11.

Derivatives at end points

We say that a function f is differentiable on a set A when f is differentiable at each point of A. If A is an open interval, the meaning is clear enough, since an open interval contains a neighbourhood of each of its points. But functions such as $x \mapsto \sqrt{x}$ may be defined on $[0, \infty)$ and in such a case it may be important to be able to affirm or deny the differentiability of the function at 0.

46 Look back to the definition of derivative and propose a criterion to determine whether a function $f: [a, b] \to \mathbb{R}$ is differentiable at a. Modify your criterion to determine differentiability (or otherwise) at b.

Summary

Definition If a is a cluster point of the domain of a real
qn 4 function f, then, when the limit

$$\lim_{x \to a} \frac{f(x) - f(a)}{x - a}$$

exists, f is said to be *differentiable* at a and the value of the limit is denoted by $f'(a)$.

Theorem If a function f is differentiable at a, then f is
qn 12 continuous at a.

Theorem   If functions f and g are both differentiable at a,
              then
qns 10, $(f + g)'(a) = f'(a) + g'(a)$,
    13, $(f \cdot g)'(a) = f(a) \cdot g'(a) + f'(a) \cdot g(a)$,
    17   and $(f/g)'(a) = \big(f'(a) \cdot g(a) - f(a) \cdot g'(a)\big)/\big(g(a)\big)^2$,
              provided $g(a) \neq 0$.
Theorem   If $f(x) = b_0 + b_1 x + b_2 x^2 + \ldots + b_n x^n$,
qn 15   then $f'(a) = b_1 + 2b_2 a + 3b_3 a^2 + \ldots + nb_n a^{n-1}$.
Theorem   If the function g is differentiable at a and the
qn 18   function f is differentiable at $g(a)$,
              then $(f \circ g)'(a) = f'\big(g(a)\big) \cdot g'(a)$.
Definition A function f is said to have a *local maximum* at a if,
qn 29   for all x in some neighbourhood of a, $f(x) \leqslant f(a)$.
              A function f is said to have a *local minimum* at a if
              for all x in some neighbourhood of a, $f(x) \geqslant f(a)$.
Theorem   If the function f has a local maximum or a local
qn 30   minimum at a, and f is differentiable at a, then
              $f'(a) = 0$.
Definition If $f: A \to \mathbb{R}$ is differentiable at each point of $B \subseteq A$,
qns 36,   then $f': B \to \mathbb{R}$ is called the *derived function* or
    38   *derivative* of f.
              By induction we define the nth *derivative* of f,
              denoted by $f^{(n)}$, as the derivative of $f^{(n-1)}$.
Theorem   If f is a bijection and g is its inverse, and if f is
qn 40   differentiable at a and g is differentiable at $f(a)$,
              then $g'\big(f(a)\big) = 1/f'(a)$.

Historical note

From one point of view, most of the material of this chapter was
known by mathematicians before Newton's creative burst in 1664–6.
From another point of view, none of the results of this chapter were
proved in the sense in which we take them today until the time of
Weierstrass, 1860–70.

The essential backcloth to the calculus is Descartes' method of
converting geometric problems into algebraic ones (coordinate
geometry) and his notation for exponents (1637), which we use today.
Fermat (about 1640) knew how to find the slope of the tangent to the
graph of a polynomial using the limit of the slope of a chord, and
found maxima and minima by examining the slope of the chord near
the point. There were in fact a host of special methods found for
dealing with special curves.

Both Newton (when he was away from Cambridge for fear of the plague) in 1665 and 1666, and Leibniz in Paris in 1675, working quite independently, systematically established the product, quotient and chain rules and discovered the inverse relationship between slopes of tangents and areas under curves. Having discovered the Binomial Theorem for rational index, Newton was strongly motivated by what he could do by differentiating and integrating power series term by term. Newton considered his variables as real entities which were changing with time, and he expressed the derivative of the 'fluent' x, as the 'fluxion' $\dot{x}$. Leibniz had originally been motivated by the use of differences in summing series (as in qns 5.3 and 5.4). In this notation, dx originally meant the difference between x and its next value, and dy/dx was originally a ratio of infinitesimals. Leibniz' great discovery was of the results he could generate using the characteristic triangle with sides dx, dy and ds. Leibniz' dy/dx was expressed by Newton in the form $\dot{y}/\dot{x}$.

The first calculus book was published in 1696. This was *Analyse des infiniments petits* by the Marquis de l'Hôpital. It was repeatedly reprinted during the eighteenth century. The book consists of lectures on the differential calculus which Johann Bernoulli gave to de l'Hôpital in 1691, on Leibniz' work. Some of Newton's calculus appeared in his *Principia* (1687), more of his calculus was in the appendix to his *Opticks* (1704), but his original exposition *De Analysi* (written in 1669 and circulated to friends) was not published until 1736, after his death. Both Newton and Leibniz originally thought of limits in terms of infinitesimals, but later in life tried to avoid them. Newton spoke of prime and ultimate ratios (the slope of the chord and tangent respectively). Leibniz eventually came to speak of infinitesimals as convenient fictions.

In Bishop Berkley's tract of 1734, the apparently self-contradictory nature of limits was highlighted. In affirming that, as x tends to a, $(x^2 - a^2)/(x - a)$ tends to $2a$, x is not equal to a at the beginning of the computation, and this allows the division to be performed, and then x is put equal to a at the end of the computation to obtain the answer. This paradox was not resolved during the eighteenth century, but it did not discourage mathematicians in their work. Indeed, an immediate chronological successor to Bishop Berkley was L. Euler, the most ingenious manipulator of infinitesimals there has ever been. Berkley's criticisms were not forgotten and in 1764 d'Alembert offered the useful notion of the value of the function differing from the limit 'by as little as one wishes', a phrase which was taken up by Cauchy.

The origin of both Newton's and Leibniz' calculus was geometric and it was only late in the eighteenth century that J. L. Lagrange

(1797) tried to make algebraic definitions of its fundamental notions. Having, as he thought, proved that every function has a Taylor series he showed that $f(a + h) = f(a) + hf'(a) + hr(h)$ where $r(h)$ tends to 0 as h tends to 0, seemingly without reference to infinitesimals, and used this to prove (the need for proof being itself a remarkable insight) that a differentiable function with positive derivatives was increasing. It was a small step from this to Cauchy's definition. It is to Lagrange that we owe the notion of the derived function $f'(x)$ as against the ratio of infinitesimals dy/dx, and also the word *derivative*.

In 1823, Cauchy offered an ε–δ neighbourhood definition of derivatives, and this largely resolved the problem of the definition of limits and therefore made it possible to construct rigorous proofs using limits. Cauchy used infinitesimals, but he defined them as variables which were tending to zero, not as indivisibles as they had been considered in the seventeenth century. However, continuous functions were the context of Cauchy's work on differentiation, so that the proposition that a differentiable function was necessarily continuous was for Cauchy a matter of definition, not a theorem. Indeed some followers of Cauchy believed that they had successfully proved that continuous functions must be differentiable! Also Cauchy's functions were continuous, or differentiable, on intervals, so that he did not conceive of a function that was defined everywhere, but continuous at only one point. Defining continuity and differentiability in terms of the behaviour of functions at a single point followed the work of Riemann (1854) and Weierstrass (1860). Likewise until the time of Weierstrass, with the awareness of the completeness of the real numbers and of the possibility of continuous but non-differentiable functions, it was not realised that the result of our qn 42 was distinct from that of qn 40 on the invertibility of derivatives.

The advent of rigour did not come to the calculus all at once. Distinguished mathematicians in the present century have failed to provide a rigorous treatment of the chain rule. (See Hardy, 9th edition, ch. 6, §114.)

Answers

1 One. No.

2 Two. Yes.

3 Three if $m > 0$. One if $m \leq 0$. Only $m = 0$. Slope $\rightarrow 0$.

4 $y - f(a) = m(x - a)$.

$y - f(a) = \dfrac{f(a + h) - f(a)}{h}(x - a)$. With the given m this is a tangent.

5 There may still be a tangent when the slope of the chord $\rightarrow \infty$.

6 $f'(a) = 0$.

7 $f'(a) = m$.

8 If, given $\varepsilon > 0$, $0 < |h| < \delta \Rightarrow \left| \dfrac{f(a + h) - f(a)}{h} - l \right| < \varepsilon$

then $0 < |x - a| < \delta \Rightarrow \left| \dfrac{f(x) - f(a)}{x - a} - l \right| < \varepsilon$; and conversely.

9 $2a$.

10 Generalise by induction as in qn 6.25.

11 $\dfrac{g(x) - g(a)}{x - a} = k \cdot \dfrac{f(x) - f(a)}{x - a}$ and use qn 6.93.

12 By qn 6.93 $\lim_{x \to a} f(x) - f(a) = f'(a) \cdot 0$. So $\lim_{x \to a} f(x) = f(a)$ and f
is continuous at a by qn 6.89.

13 $\dfrac{f(x) \cdot g(x) - f(a) \cdot g(a)}{x - a} = f(x) \cdot \dfrac{g(x) - g(a)}{x - a} + g(a) \cdot \dfrac{f(x) - f(a)}{x - a}$.

This tends to $f(a) \cdot g'(a) + g(a) \cdot f'(a)$ by qn 6.93 and the continuity of f.

14 Trivial for $n = 1$. If $f(x) = x^n \Rightarrow f'(a) = na^{n-1}$, then for $g(x) = x^n x$,
$g'(a) = a^n \cdot 1 + na^{n-1} \cdot a = (n + 1)a^n$ by qn 13.
$\lim_{x \to a} x^{n-1} + x^{n-2}a + \ldots + a^{n-1} = a^{n-1} + a^{n-1} + \ldots + a^{n-1} = na^{n-1}$ by
qn 6.93.

15 $f'(a) = b_1 + 2b_2a + 3b_3a^2 + \ldots + nb_na^{n-1}$.

16 $\dfrac{f(x) - f(a)}{x - a} = \dfrac{1/x^n - 1/a^n}{x - a} = \dfrac{a^n - x^n}{a^nx^n(x - a)} \to \dfrac{-na^{n-1}}{a^na^n}$, using qn 14.

17 $\dfrac{f(x)/g(x) - f(a)/g(a)}{x - a} = \dfrac{f(x)\cdot g(a) - f(a)\cdot g(x)}{g(x)\cdot g(a)\cdot(x - a)}$

$$= \dfrac{g(a)\big(f(x) - f(a)\big) - f(a)\big(g(x) - g(a)\big)}{g(x)\cdot g(a)\cdot(x - a)}$$

which gives the required result by qn 6.93 and the continuity of $1/g$ at a.

18 (ii) $\lim\limits_{x\to a}\dfrac{g(x) - g(a)}{x - a} = g'(a)$.

If, in every neighbourhood of a, there are points $x \ne a$ such that $g(x) = g(a)$ then $g'(a) = 0$: because, if $g'(a) = l \ne 0$, then

$$\left|\dfrac{g(x) - g(a)}{x - a} - l\right| < \varepsilon = \tfrac{1}{2}|l|$$

would not be implied by $0 < |x - a| < \delta$, for any δ.
If $|F(a_{n_i})| < \varepsilon$ for $n_i > n_0$, then $|F(a_n)| < \varepsilon$ for $n > n_0$.
Since $\big(F(a_n)\big) \to 0$ if $\big(a_n\big) \to a$ from above and from below, $\lim\limits_{x\to a} F(x) = 0$.

From the definition of derivative, $(f \circ g)'(a) = 0$.

19 First limit $= 1$, second limit $= -1$, so two-sided limit does not exist.

20 If $a_n = 1/(2n\pi)$ then $f(a_n) = 0$. If $b_n = 1/\big((2n + \tfrac{1}{2})\pi\big)$ then $f(b_n) = 1$.
So $\big(f(a_n)\big)$ and $\big(f(b_n)\big)$ have different limits and therefore f is not continuous at 0 whatever the value of $f(0)$. Compare with qn 6.19.

21 $x \mapsto |x|$ is continuous by qn 6.32.
$x \mapsto -|x|$ is continuous, by qns 6.32 and 6.26.
So f is continuous, by qn 6.36. But

$$\dfrac{f(x) - f(0)}{x - 0} = \sin\dfrac{1}{x},$$

and this has no limit as $x \to 0$.

22 $x \mapsto x^2$ and $x \mapsto -x^2$ are both continuous, by qn 6.29.
$-x^2 \le f(x) \le x^2$ so f is continuous at $x = 0$, by qn 6.36.

$$\dfrac{f(x) - f(0)}{x - 0} = x\sin\dfrac{1}{x} \text{ when } x \ne 0 \text{ and from qn 21 this} \to 0 \text{ as } x \to 0.$$

23 If $\lim\limits_{h\to 0}\dfrac{f(a + h) - f(a)}{h} = f'(a)$, then $\lim\limits_{h \to 0} \dfrac{f(a-h)-f(a)}{-h} = f'(a)$.

The result follows from qn 6.93.
The limit exists when $a = 0$ for $f(x) = |x|$, but this function is not differentiable at 0 by qn 19.

24 None of these functions is continuous when $x \neq 0$, using arguments like that of qn 6.20.
Because f is not continuous at 0 by qn 6.20, f is not differentiable at 0 by qn 12.
The function g is continuous at 0 by qn 6.35. But g is not differentiable at $x = 0$ because

$$\frac{g(x) - g(0)}{x - 0} = f(x) \quad \text{when } x \neq 0,$$

and f has no limit as $x \to 0$.
Since $0 \leqslant h(x) \leqslant x^2$ for all x, h is continuous at 0 by qn 6.36.

$$\frac{h(x) - h(0)}{x - 0} = g(x) \text{ when } x \neq 0, \quad \text{and} \quad \lim_{x \to 0} g(x) = g(0)$$

so h is differentiable at 0.

25 $\dfrac{f(x) - f(0)}{x - 0} = 1 + 2x \cos \dfrac{1}{x} \to 1$ as $x \to 0$, as in qn 21.

$$f\left(\frac{1}{2n\pi}\right) > f\left(\frac{1}{(2n-1)\pi}\right) \Leftrightarrow \frac{1}{2n} + \frac{2}{4n^2\pi} > \frac{1}{2n-1} - \frac{2}{(2n-1)^2\pi},$$

$$\Leftrightarrow \frac{2}{\pi}\left(2 + \frac{1}{4n^2 - 2n}\right) > 1.$$

When f' is positive throughout an interval, this counter-intuitive possibility cannot occur, as we shall see in the next chapter.

26 $\mathbb{R}\backslash\{0\}$.

27 When $x > 0$, $f'(x) = 1$. When $x < 0$, $f'(x) = -1$.

28 $f'(1) = 0$. $f(x) \geqslant f(1)$ for all x. So $f(1)$ is the minimum value of the function.

29 If for all x in some neighbourhood of a, $f(x) \leqslant f(a)$ then f has a *local maximum* at a. If for all x in some neighbourhood of a, $f(x) \geqslant f(a)$ then f has a *local minimum* at a. $f'(-1) = f'(1) = 0$.

30 If f has a local maximum at a then in some neighborhood of a

$$\frac{f(x) - f(a)}{x - a} \geqslant 0 \text{ when } x < a, \quad \text{and} \quad \frac{f(x) - f(a)}{x - a} \leqslant 0 \text{ when } x > a.$$

Therefore $\lim\limits_{x \to a^-} \dfrac{f(x) - f(a)}{x - a} \geqslant 0$ and $\lim\limits_{x \to a^+} \dfrac{f(x) - f(a)}{x - a} \leqslant 0$.

Since the two-sided limit exists it equals 0.

31 $f(x) = x^3$ at $x = 0$.

32 For example, $x \mapsto -|x|$ at $x = 0$.

33 $f'(x) = (\cos x^2)(2x)$.

34 $f'(x) = 2x \sin(1/x) - \cos(1/x)$, when $x \neq 0$,
$f'(0) = 0$.
f' is not continuous at 0, since $x \mapsto \cos(1/x)$ is like the function in qn 20.
So f' is not differentiable at 0.

35 f has domain $\mathbb{R}$. f' has domain $\mathbb{R}\backslash\{0\}$. f'' has domain $\mathbb{R}\backslash\{0, 1\}$.

36 Apply qn 12 to f'.

37 Consider limits from below and limits from above at 0.
(i) Yes. (ii) Yes. (iii) Yes. (iv) No.

38 f''' is the derived function of f''.
$f^{(n)}$ is the derived function of $f^{(n-1)}$.

39 $\dfrac{f(x) - f(a)}{x - a} = \dfrac{1}{\sqrt{x} + \sqrt{a}} \to \dfrac{1}{2\sqrt{a}}$ as $x \to a$.

So $f'(a) = 1/(2\sqrt{a})$ and $g'(f(a)) = 2\sqrt{a}$.

41 $g'(f(a)) \cdot f'(a) = 1$. But if $f'(a) = 0$, there is no possible value for
$g'(f(a))$. Consider the graphs of the functions f and g in qn 39 at 0.

42 (i) $\dfrac{g(f(x)) - g(f(a))}{f(x) - f(a)} = \dfrac{x - a}{f(x) - f(a)} = 1 \bigg/ \left(\dfrac{f(x) - f(a)}{x - a} \right)$.

Use the fact that $f'(a)$ exists and is non-zero, with qn 6.93.
(ii) Because f is a bijection.
(iii) $(b_n) \to b$ by definition. This may be rewritten $(f(a_n)) \to f(a)$.
By the continuity of g at $b = f(a)$, $(g(f(a_n))) \to g(f(a))$. So
$(a_n) \to a$.
(iv) From the limit in (i),
$$\left(\dfrac{g(f(a_n)) - g(f(a))}{f(a_n) - f(a)} \right) \to \dfrac{1}{f'(a)},$$
so $\left(\dfrac{g(b_n) - g(b)}{b_n - b} \right) \to \dfrac{1}{f'(a)}$.

(v) From the definition of (b_n), limits from above and below b exist and are equal, so the two-sided limit

$$\lim_{y \to b} \frac{g(y) - g(b)}{y - b} = \frac{1}{f'(a)}.$$

Thus $g'(f(a))$ exists and equals $1/f'(a)$.

43 If $f(x) = x^n$, then $f^{-1}(x) = \sqrt[n]{x}$. From qn 42, f^{-1} is differentiable except at 0.
$f^{-1}{}'(a) = f^{-1}{}'(f(\sqrt[n]{a})) = 1/f'(\sqrt[n]{a}) = 1/(n(\sqrt[n]{a})^{n-1}) = (1/n)a^{(1/n)-1}$.
Now let $P(x) = x^p$ and $Q(x) = x^{1/q}$, then $Q(P(x)) = x^{p/q}$.
$(Q \circ P)'(x) = Q'(P(x)) \cdot P'(x) = (1/q)(x^p)^{(1/q)-1} px^{p-1} = (p/q)x^{(p/q)-1}$.
Use qn 17 to extend to negative rational indices.
g is differentiable at $x = 0$ provided $r \geq 1$.

44 $[-\frac{1}{2}\pi, \frac{1}{2}\pi] \to [-1, 1]$.
$\sin'(\arcsin x) \cdot \arcsin' x = 1$, so $\arcsin' x = 1/\cos(\arcsin x) = 1/\sqrt{(1 - x^2)}$.

45 $(-\frac{1}{2}\pi, \frac{1}{2}\pi)$. $\tan' x = (\cos^2 x + \sin^2 x)/(\cos^2 x) = 1 + \tan^2 x$.
$\tan'(\arctan x) \cdot \arctan' x = 1$, so $\arctan' x = 1/(1 + x^2)$.

46 $f: [a, b] \to \mathbb{R}$ is differentiable at a if $\lim_{x \to a^+} \dfrac{f(x) - f(a)}{x - a}$ exists
and is finite.
Use the limit from below for b.

9

Differentiation and completeness
Mean Value Theorems, Taylor's Theorem

Concurrent reading: Burkill ch. 4 and §5.8, Courant and John ch. 5.
Further reading: Spivak chs 11, 19.

Both Rolle's Theorem and the Mean Value Theorem are
geometrically transparent. Each claims, with slightly more generality
in the case of the Mean Value Theorem, that for the graph of a
differentiable function, there is always a tangent parallel to a chord.
It is something of a surprise to find that such intuitive results can lead
to such powerful conclusions: namely, de l'Hôpital's rule and the
existence of power series convergent to a wide family of functions.

Rolle's Theorem

1 Sketch the graphs of some differentiable functions

$$f: [a, b] \to \mathbb{R},$$

for which $f(a) = f(b)$.
Can you find a point c, with $a < c < b$, such that $f'(c) = 0$, for
each function f which you have sketched?
What words could you use to describe the point c in relation to
the function f?

2 Must *any* differentiable function

$$f: [a, b] \to \mathbb{R}$$

have a maximum and a minimum value? Why?
Must *any* such function, for which $f(a) = f(b)$, have a maximum
and a minimum value in the *open* interval (a, b)?

Must *any* such function, for which $f(a) = f(b)$, have a maximum *or* a minimum value in the *open* interval (a, b)?

3   If $f: [a, b] \to \mathbb{R}$ is differentiable and, for some c such that $a < c < b$, $f(x) \leqslant f(c)$ for all $x \in [a, b]$, must $f'(c) = 0$? Why?

4   If $f: [a, b] \to \mathbb{R}$ is differentiable,

$$f(a) = f(b),$$

$$M = \sup \{f(x) | \ a \leqslant x \leqslant b\},$$

and $m = \inf \{f(x) | \ a \leqslant x \leqslant b\}$,

use qn 2 to show that there is a c with $a < c < b$ for which $f(c) = M$, or a c for which $f(c) = m$, or possibly distinct cs satisfying each of these conditions.
Deduce from qn 3 that $f'(c) = 0$.

5   Sketch the graph of a continuous function

$f: [a, b] \to \mathbb{R}$

which is differentiable on the open interval (a, b), but not differentiable at the points a or b.

6   Identify the two places in the proof of the result of qn 4 at which the differentiability of the function f has been invoked.

7   If the function $f: [a, b] \to \mathbb{R}$ satisfies the conditions
   (i) f is continuous on $[a, b]$,
   (ii) f is differentiable on (a, b),
   (iii) $f(a) = f(b)$,
show that there is a point c in the open interval (a, b) such that $f'(c) = 0$. This result is known as *Rolle's Theorem*.

8   Sketch the graphs of functions for which one or more of the conditions (i), (ii) and (iii) in qn 7 fail, illustrating how the conditions are necessary for Rolle's Theorem, and also how, even in their absence, there may still exist a point $c \in (a, b)$ for which $f'(c) = 0$.

(9)   If $a_0 + a_1/2 + a_2/3 + \ldots + a_n/(n + 1) = 0$, prove that the polynomial function

$x \mapsto a_0 + a_1 x + a_2 x^2 + \ldots + a_n x^n$

has a zero in the interval $[0, 1]$.

10 If the function $f: \mathbb{R} \to \mathbb{R}$ is twice differentiable and $f(a) = f(b) = f(c) = 0$, with $a < b < c$, prove that for some $d \in (a, c)$, $f''(d) = 0$. Generalise.

11 (i) The function $f: \mathbb{Q} \to \mathbb{Q}$ is defined by $f(x) = 1/(x^2 - 2)^2$. Show that the function f satisfies the conditions (i), (ii) and (iii) of qn 7 on $[0, 2] \cap \mathbb{Q}$. Why does the conclusion of Rolle's Theorem fail in this case?

(ii) The function $g: \mathbb{Q} \to \mathbb{Q}$ is defined by $g(x) = x^3 - 21x + 20$. Show that the function g satisfies the conditions (i), (ii) and (iii) of qn 7 on $[1, 4] \cap \mathbb{Q}$. Why does the conclusion of Rolle's Theorem fail in this case?

An intermediate value theorem for derivatives

Although we have seen that differentiable functions do not necessarily have continuous derivatives (in qn 8.34), derivatives have an intermediate value property which we might only expect from a continuous function.

12 *Darboux's Theorem*
If $[a, b]$ is contained in the domain of a real differentiable function f and $f'(a) < k < f'(b)$, determine whether the function g defined by

$$g(x) = f(x) - kx$$

is

(i) a real function which is continuous on $[a, b]$,
(ii) a function which is bounded and attains its bounds on $[a, b]$,
(iii) a function which is differentiable on $[a, b]$,
(iv) a function which is not constant on $[a, b]$.

Determine the derivative of g at a point where it attains a bound, and deduce that, for some c in the open interval (a, b), $f'(c) = k$.

Check the compatability of this result with the differentiable function in qn 8.22 having discontinuous derivative, using graph drawing facilities on a computer.

The Mean Value Theorem

The next question looks at a slanted version of Rolle's Theorem, and identifies conditions under which an arc has a tangent parallel to a chord joining two points of the arc.

13 A function $f: [a, b] \to \mathbb{R}$ is continuous on the closed interval $[a, b]$ and differentiable on the open interval (a, b).

(i) Construct a function D which measures the distance (in a direction parallel to the y-axis) between the graph of the function f and the chord joining $(a, f(a))$ to $(b, f(b))$. Does D satisfy the conditions of Rolle's Theorem? What is the result of applying Rolle's Theorem to D, when expressed in terms of the function f?

(ii) Let the slope of the chord from $(a, f(a))$ to $(b, f(b))$ be

$$\frac{f(b) - f(a)}{b - a} = K.$$

Define the function $F: [a, b] \to \mathbb{R}$ by

$$F(x) = f(b) - f(x) - K(b - x).$$

Check that F satisfies the conditions of Rolle's Theorem on $[a, b]$.

Apply Rolle's Theorem to the function F, to obtain a value for K in terms of the function f'.

You should have obtained the same result from qn 13(i) and 13(ii). This result is called the *Mean Value Theorem*. One application is that, on a continuous and smooth journey, the average speed of the journey must actually be reached at least once during the journey. While this result looks obvious enough, it has powerful and significant applications.

14 Show that the conclusion of the Mean Value Theorem applied to a function f on an interval $[a, a + h]$ may be expressed by saying that, under the appropriate conditions, there is a θ, with $0 < \theta < 1$, such that $f(a + h) - f(a) = hf'(a + \theta h)$.

15 A function $f: [a, b] \to \mathbb{R}$ is continuous on the closed interval $[a, b]$ and differentiable on the open interval (a, b).

Prove that if $f'(x) > 0$ for all values of x in (a, b), then the function f is strictly monotonic increasing.

Show also that if $f'(x) < 0$ for all values of x, then the function f is strictly monotonic decreasing.

(16) Functions f_1, f_2, f_3 and f_4 are defined on $[0, 2\pi]$ by

$$f_1(x) = x - \sin x, \quad f_2(x) = \cos x - 1 + \frac{x^2}{2},$$

$$f_3(x) = \sin x - x + \frac{x^3}{6}, \quad f_4(x) = 1 - \frac{x^2}{2} + \frac{x^4}{24} - \cos x.$$

Prove that each of these four functions is strictly increasing on the given domain.

Deduce that

$$x > \sin x > x - \frac{x^3}{6},$$

and

$$1 - \frac{x^2}{2} + \frac{x^4}{24} > \cos x > 1 - \frac{x^2}{2}, \quad \text{on } (0, 2\pi).$$

Use graph-drawing facilities on a computer to compare the graphs of sine and cosine with those of the polynomials we have found.

17 A function $f \colon [a, b] \to \mathbb{R}$ is continuous on the closed interval $[a, b]$ and differentiable on the open interval (a, b).

Prove that if $f'(x) = 0$ for all values of x, then the function f is constant.

Construct a real function f which is not constant, but for which $f'(x) = 0$ for every point x of its domain.

18 A function $f \colon [a, b] \to [a, b]$ is continuous on the closed interval $[a, b]$ and differentiable on the open interval (a, b).

If $|f'(x)| \le L$ for all values of x, prove that

$$|f(x) - f(y)| \le L \cdot |x - y| \quad \text{for all } x \text{ and } y \text{ in the domain,}$$

so that f satisfies a Lipschitz condition, and is uniformly continuous as in qn 7.42.

If a sequence (a_n) is defined in $[a, b]$ by $a_{n+1} = f(a_n)$, prove that $|a_{n+m} - a_{n+1}| \le L^n |a_m - a_1| \le L^n |b - a|$.

If, further, $L < 1$, deduce that (a_n) is a Cauchy sequence, and hence convergent.

If (b_n) is a sequence in $[a, b]$ defined by $b_{n+1} = f(b_n)$ and $(a_n) \to A$, and $(b_n) \to B$, explain why A and B lie in $[a, b]$, why $f(A) = A$ and $f(B) = B$, and prove that $A = B$.

The condition $|f'(x)| < 1$, makes the function f an example of a *contraction mapping* of the domain, which, as we have shown, has a unique fixed point, satisfying $f(x) = x$. This result is particularly useful for calculating better approximations to the roots of given equations and provides a reason for the convergence of sequences defined by an iterative formula.

19 Prove that the equation $x = \cos x$ has exactly one root in the interval $[0, \tfrac{1}{2}\pi]$.

20 Prove that the function defined by $f(x) = \sqrt{(x + 2)}$ is a contraction mapping on the interval $[1, 2]$. Deduce that the sequence defined in qn 3.67 is convergent when the first term is taken in that interval.

21 For a continuous function $f: [a, b] \to [a, b]$ the existence of a root for $f(x) = x$ is guaranteed, as we found in qn 7.20. Illustrate how, in the absence of the condition $|f'(x)| < 1$, either (a) multiple roots for $f(x) = x$ may exist, or (b) the sequence (a_n) may not converge.

22 A function $f: [a, b] \to \mathbb{R}$ is continuous on the closed interval $[a, b]$ and differentiable on the open interval (a, b). If, for some c, with $a < c < b$, $f'(c) = 0$, and $f''(c)$ exists and > 0, prove that there is a neighbourhood of c within which $f'(x) > 0$ when $x > c$, and $f'(x) < 0$ when $x < c$. Deduce that f has a local minimum at c.

The result of qn 22 may seem deceptively straightforward. However, the two conditions, $f'(c) = 0$ and $f''(c) > 0$, do not, by themselves, imply the continuity of derivatives other than at $x = c$.

(23) Let real functions $f_n: \mathbb{R} \to \mathbb{R}$ be defined by

$$f_n(x) = 1 - x + \frac{x^2}{2!} - \frac{x^3}{3!} + \ldots + (-1)^n \frac{x^n}{n!}.$$

Check that $f'_{n+1} = -f_n$.
If n is even and $f_n(x) > 0$ for all x, prove that f_{n+1} is monotonic decreasing and has one and only one root. The existence of at least one root comes from the argument of qn 7.19.
Deduce that there is exactly one value of x at which $f'_{n+2}(x) = 0$, that f_{n+2} is minimal at this point, and that $f_{n+2}(x) > 0$ at this point so that f_{n+2} is positive everywhere.
Now use induction to prove that f_n is positive when n is even and has a unique root when n is odd.

24 A function $f: [a, b] \to \mathbb{R}$ is continuous on the closed interval $[a, b]$ and twice differentiable on the open interval (a, b). Prove that if $f''(x) > 0$, for all values of x, then for any θ such that $0 < \theta < 1$,

$$f(\theta x + (1 - \theta)y) < \theta f(x) + (1 - \theta)f(y).$$

Sketch this result on a graph.

Cauchy's Mean Value Theorem

25 Functions $f: [a, b] \to \mathbb{R}$ and $g: [a, b] \to \mathbb{R}$ are continuous on the closed interval $[a, b]$ and differentiable on the open interval (a, b) and $g'(x) \neq 0$ for any $x \in [a, b]$.
Let

$$\frac{f(b) - f(a)}{g(b) - g(a)} = K.$$

Prove that the conditions of Rolle's Theorem hold for the function $F: [a, b] \to \mathbb{R}$ defined by

$$F(x) = f(b) - f(x) - K(g(b) - g(x)).$$

Apply Rolle's Theorem to F to show that, for some c such that $a < c < b$,

$$K = \frac{f'(c)}{g'(c)}.$$

de l'Hôpital's rule

26 Functions $f: [a, b] \to \mathbb{R}$ and $g: [a, b] \to \mathbb{R}$ are continuous on the closed interval $[a, b]$ and differentiable on the open interval (a, b).
We further assume that $f(a) = g(a) = 0$ and $g'(x) \neq 0$ for any x in $[a, b]$.
What does Cauchy's Mean Value Theorem give in this case?
Now suppose that

$$\lim_{x \to a^+} \frac{f'(x)}{g'(x)} = l.$$

Prove that, given $h > 0$,

$$\left| \frac{f(a + h)}{g(a + h)} - l \right| = \left| \frac{f'(a + \theta h)}{g'(a + \theta h)} - l \right|$$

for some θ, with $0 < \theta < 1$.
Deduce that if $0 < k < \delta$ implies

$$\left| \frac{f'(a + k)}{g'(a + k)} - l \right| < \varepsilon,$$

then $0 < h < \delta$ implies

$$\left| \frac{f(a + h)}{g(a + h)} - l \right| < \varepsilon,$$

so that

$$\lim_{x \to a^+} \frac{f(x)}{g(x)} = l.$$

It is easy to see that a result analogous to that of qn 26 can be obtained using limits from the left if appropriate conditions apply. When the functions f and g are differentiable at a, either result may be used.

27 Use de l'Hôpital's rule, qn 26, to find

(i) $\displaystyle\lim_{x \to 0} \frac{\sqrt{(x + 2)} - \sqrt{2}}{\sqrt{(x + 1)} - 1}$,

(ii) $\displaystyle\lim_{x \to 0} \frac{e^x - 1}{x}$

(iii) $\displaystyle\lim_{x \to 0} \frac{\sin x}{x}$, hence $\displaystyle\lim_{x \to 0} \frac{1 - \cos x}{x^2}$,

and hence

$$\lim_{x \to 0} \frac{x - \sin x}{x^3}.$$

28 If, in qn 26, we had had

$$\lim_{x \to a^+} \frac{f'(x)}{g'(x)} = +\infty,$$

instead of

$$\lim_{x \to a^+} \frac{f'(x)}{g'(x)} = l.$$

would we have been able to deduce that

$$\lim_{x \to a^+} \frac{f(x)}{g(x)} = +\infty?$$

29 A function $f \colon [a, b] \to \mathbb{R}$ is continuous on the closed interval $[a, b]$ and differentiable on the open interval (a, b).
By applying de l'Hôpital's rule to

$$\frac{f(c + h) + f(c - h) - 2f(c)}{h^2},$$

prove that if f'' exists at $c \in (a, b)$ then

$$f''(c) = \lim_{h \to 0} \frac{f(c + h) + f(c - h) - 2f(c)}{h^2}.$$

By considering the function f for which $f(x) = x^2$ when $x \geq 0$, and $f(x) = -x^2$ when $x < 0$, show that this limit may exist when f is not twice differentiable.

It may happen that $\lim_{x \to a^+} f'(x)/g'(x)$ does not exist,

when $\lim_{x \to a^+} f(x)/g(x)$ does.

30 Let $f(x) = x^2 \sin(1/x)$, when $x \neq 0$, $f(0) = 0$ and $g(x) = x$.
Show that $\lim_{x \to 0^+} f(x)/g(x) = 0$.

Show also that $\lim_{x \to 0^+} f'(x)/g'(x)$ does not exist.
Check that the conditions for Cauchy's Mean Value Theorem hold here for any interval $[0, x]$, with $x \neq 0$. Why may the argument in the proof of de l'Hôpital's rule *not* be used to establish the existence of $\lim f'(x)/g'(x)$ when $\lim f(x)/g(x)$ is known?

The Second and Third Mean Value Theorems

31 A function $f: [a, b] \to \mathbb{R}$ has a continuous derivative on the closed interval $[a, b]$ and is twice differentiable on the open interval (a, b).
Let

$$\frac{f(b) - f(a) - (b - a)f'(a)}{(b - a)^2} = K.$$

Apply Rolle's Theorem to the function $F: [a, b] \to \mathbb{R}$ given by

$$F(x) = f(b) - f(x) - (b - x)f'(x) - K(b - x)^2$$

to show that, for some c with $a < c < b$, $K = \frac{1}{2}f''(c)$.
Deduce that

$$f(b) = f(a) + (b - a)f'(a) + \frac{(b - a)^2}{2}f''(c).$$

This is called the *Second Mean Value Theorem*.

32 Rewrite the Second Mean Value Theorem replacing b by $a + h$. Then, by a judicious choice of a and h, show that for any twice differentiable function $f: \mathbb{R} \to \mathbb{R}$

$$f(x) = f(0) + xf'(0) + \frac{x^2}{2}f''(c) \quad \text{for some } c \in (0, x).$$

33 The Third Mean Value Theorem states that, if a function $f: [a, b] \to \mathbb{R}$ has a continuous second derivative on the closed interval $[a, b]$ and is three times differentiable on the open interval (a, b), then there exists a c, with $a < c < b$, such that

$$f(b) = f(a) + (b - a)f'(a) + \frac{(b - a)^2}{2!}f''(a) + \frac{(b - a)^3}{3!}f'''(c).$$

Indicate how to prove the Third Mean Value Theorem by defining a constant K and a function F analogous to those in qn 31, and applying Rolle's theorem.

Reformulate the theorem, replacing b by $a + h$.

34 If $f(x) = a_0 + a_1x + a_2x^2 + \ldots + a_nx^n$, prove that

$$a_0 = f(0), \ a_1 = f'(0), \ a_2 = \tfrac{1}{2}f''(0), \ldots, a_n = \frac{f^{(n)}(0)}{n!}.$$

We now extend the Second and Third Mean Value Theorems to an nth Mean Value Theorem to compare functions, which may be differentiated any number of times, with polynomials.

Taylor's Theorem

Taylor's Theorem or nth Mean Value Theorem

35 The function $f: [a, b] \to \mathbb{R}$ has a continuous $(n - 1)$th derivative on the closed interval $[a, b]$ and is differentiable n times on the open interval (a, b).

Let

$$\frac{f(b) - f(a) - (b - a)f'(a) - \ldots - \dfrac{(b - a)^{n-1}}{(n - 1)!}f^{(n-1)}(a)}{(b - a)^n} = K.$$

Apply Rolle's Theorem to the function $F: [a, b] \to \mathbb{R}$ given by

$$F(x) = f(b) - f(x) - (b - x)f'(x) - \frac{(b - x)^2}{2!}f''(x) - $$
$$\ldots - \frac{(b - x)^{n-1}}{(n - 1)!}f^{(n-1)}(x) - K(b - x)^n,$$

to show that, for some c with $a < c < b$, $K = f^{(n)}(c)/n!$.
So

$$f(b) = f(a) + (b - a)f'(a) + \frac{(b - a)^2}{2!}f''(a) +$$

$$\ldots + \frac{(b - a)^{n-1}}{(n - 1)!}f^{(n-1)}(a) + \frac{(b - a)^n}{n!}f^{(n)}(c).$$

36 Rewrite the conclusion of Taylor's Theorem susbtituting $a + h$ for b.

Maclaurin's Theorem

37 By putting $a = 0$ and $h = x$ in qn 36 show that, for any function $f: [0, x] \to \mathbb{R}$ which is differentiable n times,

$$f(x) = f(0) + xf'(0) + \frac{x^2}{2}f''(0) + \ldots$$

$$+ \frac{x^{n-1}}{(n - 1)!}f^{(n-1)}(0) + \frac{x^n}{n!}f^{(n)}(\theta x),$$

for some $\theta \in (0, 1)$.

38 Use Maclaurin's Theorem to say what you can about $R_n(x)$ where

$$\exp(x) = 1 + x + \frac{x^2}{2!} + \ldots + \frac{x^{n-1}}{(n - 1)!} + R_n(x),$$

assuming that $\exp(0) = 1$ and $\exp'(x) = \exp(x)$.
Prove that, for any given x, $(R_n(x)) \to 0$ as $n \to \infty$.
Deduce that

$$\exp(x) = \sum_{n=0}^{\infty} \frac{x^n}{n!}.$$

Use computer graphics to observe the connection between the graphs of $y = 1$, $y = 1 + x$, $y = 1 + x + x^2, \ldots$, and $y = \exp(x)$.

39 Use Maclaurin's Theorem to say what you can about $R_{2n}(x)$ where

$$\sin x = x - \frac{x^3}{3!} + \ldots + (-1)^{n+1}\frac{x^{2n-1}}{(2n - 1)!} + R_{2n}(x).$$

Prove that, for any given x, $(R_{2n}(x)) \to 0$ as $n \to \infty$.

Deduce that

$$\sin x = \sum_{n=1}^{\infty} (-1)^{n+1} \frac{x^{2n-1}}{(2n-1)!}.$$

Observe the connection between the graphs you drew for qn 16 and this result.

40 Use Maclaurin's Theorem to say what you can about $R_{2n+1}(x)$ where

$$\cos x = 1 - \frac{x^2}{2!} + \ldots + (-1)^n \frac{x^{2n}}{(2n)!} + R_{2n+1}(x).$$

Prove that, for any given x, $(R_{2n+1}(x)) \to 0$ as $n \to \infty$.
Deduce that

$$\cos x = \sum_{n=0}^{\infty} (-1)^n \frac{x^{2n}}{(2n)!}.$$

Observe the connection between the graphs you drew for qn 16 and this result.

41 If $f(x) = \log(1 + x)$, use the result that $f'(x) = 1/(1 + x)$ from chapter 11 to show that

$$f^{(n)}(x) = (-1)^{n+1} \frac{(n-1)!}{(1+x)^n}.$$

Use Maclaurin's Theorem to say what you can about $R_n(x)$ where

$$\log(1 + x) = x - \frac{x^2}{2} + \frac{x^3}{3} + \ldots + (-1)^n \frac{x^{n-1}}{n-1} + R_n(x).$$

Prove that $(R_n(x)) \to 0$ as $n \to \infty$
   (i) when $0 \leqslant x < 1$;
   (ii) when $x = 1$;
   (iii) when $-\frac{1}{2} \leqslant x < 0$.
Deduce that

$$\log(1 + x) = \sum_{n=1}^{\infty} (-1)^{n+1} \frac{x^n}{n}, \quad \text{when } -\frac{1}{2} \leqslant x \leqslant 1.$$

In fact, the series expansion is valid for $-1 < x \leqslant 1$ as we will show in qns 46 and 11.38.

42 How do you know that the series $x - \frac{1}{2}x^2 + \frac{1}{3}x^3 - \ldots$ is not convergent when $x > 1$? Deduce that, in this case, the sequence $(R_n(x))$ in qn 41 is not a null sequence.

In qn 36,

$$f(a) + hf'(a) + \frac{h^2}{2!} f''(a) + \ldots + \frac{h^{n-1}}{(n-1)!} f^{(n-1)}(a)$$

is called the *polynomial expansion* of $f(a + h)$ and

$$R_n(h) = \frac{h^n}{n!} f^{(n)}(a + \theta h),$$

is called the *Lagrange form of the remainder*.
In qn 37,

$$f(0) + xf'(0) + \frac{x^2}{2!} f''(0) + \ldots + \frac{x^{n-1}}{(n-1)!} f^{(n-1)}(0)$$

is called the *polynomial expansion* of $f(x)$ and

$$R_n(x) = \frac{x^n}{n!} f^{(n)}(\theta x),$$

is called the *Lagrange form of the remainder*.
 (i) If we regard the polynomial expansions as the partial sums of a
     power series (known as the Taylor series in the case of qn 36 and
     as the Maclaurin series in the case of qn 37) the power series
     may be convergent or not. In any case Taylor's Theorem is still a
     valid theorem and the remainder term measures how good an
     approximation to the function is given by the polynomial. There
     is an example of a non-convergent Taylor series in qn 42.
 (ii) If the Taylor series is convergent and the sequence of remainders
     tends to 0 as n tends to infinity, then the Taylor series converges
     to the function as we have seen in questions 38–41. In fact,
     when the remainders tend to 0, the Taylor series is necessarily
     convergent.
(iii) Somewhat suprisingly, a Taylor series may be convergent *without*
     the remainders tending to zero. In this case, the Taylor series
     cannot converge to the function. There is an example of this in
     qn 44.

43 Write down the Maclaurin series for the function
   $f(x) = \surd(1 + x)$. Does it coincide with the binomial expansion of
   qn 5.93 for $a = \frac{1}{2}$? Show that the series converges when $x = 1$.
   How many terms of this series are needed before $\surd 2$ has been
   found correct to 1 place of decimals, that is, until the error is less
   than 0.05.

44 The function $f: \mathbb{R} \to \mathbb{R}$ defined by

$$f(x) = \begin{cases} e^{-1/x^2} & \text{when } x \neq 0, \\ 0 & \text{when } x = 0, \end{cases}$$

has the curious property that $f^{(n)}(0) = 0$ for every value of n.
[A proof is given in Scott and Tims p. 335.]
What is the Maclaurin series for this function?
Is there a non-zero value of x for which the remainders can form a null sequence?
Construct an infinity of different functions each of which has the same Maclaurin series.

A Taylor series is a valid description of the function from which it originates if and only if its sequence of remainders is null. It is so important to be able to prove that the sequence of remainders of a Taylor series is null that a number of different forms of the remainder have been devised since the Lagrange form of the remainder, which we have been using, is not always sufficient to decide the matter, as we found in qn 41.

(45) With the notation and conditions of qn 35 let

$$\frac{f(b) - f(a) - (b - a)f'(a) - \ldots - \dfrac{(b - a)^{n-1}}{(n - 1)!}f^{(n-1)}(a)}{b - a} = K.$$

Apply Rolle's Theorem to the function $F: [a, b] \to \mathbb{R}$ given by

$$F(x) = f(b) - f(x) - (b - x)f'(x) - \frac{(b - x)^2}{2!}f''(x) - $$

$$\ldots - \frac{(b - x)^{n-1}}{(n - 1)!}f^{(n-1)}(x) - K(b - x)$$

to show that, for some c with $a < c < b$,

$$K = \frac{(b - c)^{n-1}}{(n - 1)!}f^{(n)}(c).$$

So

$$f(b) = f(a) + (b - a)f'(a) + \frac{(b - a)^2}{2!}f''(a) + \ldots$$

$$+ \frac{(b - a)^{n-1}}{(n - 1)!}f^{(n-1)}(a) + \frac{(b - c)^{n-1}(b - a)}{(n - 1)!}f^{(n)}(c).$$

This last term is called *Cauchy's form of the remainder*.

(46) Use Cauchy's form of the remainder to prove that the Taylor series for $\log(1 + x)$ converges to the function when $-1 < x < 1$.

Summary

Rolle's Theorem
      qn 7   If $f\colon [a, b] \to \mathbb{R}$
            (i) is continuous on $[a, b]$
            (ii) is differentiable on (a, b), and
            (iii) $f(a) = f(b)$,
            then $f'(c) = 0$ for some c, with $a < c < b$.

An intermediate value theorem for derivatives
      qn 12   If $f\colon [a, b] \to \mathbb{R}$ is differentiable on $[a, b]$ and
            $f'(a) < k < f'(b)$,
            then $f'(c) = k$ for some c, with $a < c < b$.

Mean Value Theorem
      qn 7   If $f\colon [a, b] \to \mathbb{R}$ is
            (i) continuous on $[a, b]$, and
            (ii) differentiable on (a, b), then

$$\frac{f(b) - f(a)}{b - a} = f'(c) \quad \text{for some } c,$$

            with $a < c < b$.

Theorem   If $f\colon [a, b] \to \mathbb{R}$ satisfies the conditions of the Mean
  qn 15   Value Theorem and $f'(x) > 0$ for all x, with
            $a < x < b$, then f is strictly monotonic increasing.

Theorem   If $f\colon [a, b] \to \mathbb{R}$ satisfies the conditions of the Mean
  qn 17   Value Theorem and $f'(x) = 0$ for all x, with
            $a < x < b$ then f is constant.

Theorem   If $f\colon [a, b] \to \mathbb{R}$ satisfies the conditions of the Mean
  qn 18   Value Theorem and $|f'(x)| < 1$ for all x, with
            $a < x < b$, then $f(x) = x$ has exactly one solution,
            which is the limit of any sequence (a_n) defined by
            $a_{n+1} = f(a_n)$.

Theorem   If $f\colon [a, b] \to \mathbb{R}$ satisfies the conditions of the Mean
  qn 22   Value Theorem and if, for some c, with $a < c < b$,
            $f'(c) = 0$ and $f''(c) > 0$, then f has a local minimum
            at c.

Cauchy's Mean Value Theorem
      qn 25   If $f\colon [a, b] \to \mathbb{R}$ and $g\colon [a, b] \to \mathbb{R}$ are functions such
            that

(i) both are continuous on $[a, b]$,

(ii) both are differentiable on (a, b), and

(iii) $g'(x) \neq 0$ for any x, with $a < x < b$, then

$$\frac{f(b) - f(a)}{b - a} = \frac{f'(c)}{g'(c)}$$

for some c, with $a < c < b$.

de l'Hôpital's rule

qn 26 If $f: [a, b] \to \mathbb{R}$ and $g: [a, b] \to \mathbb{R}$ are functions such that

(i) both are continuous on $[a, b]$

(ii) both are differentiable on (a, b),

(iii) $f(a) = g(a) = 0$, and

(iv) $\lim\limits_{x \to a^+} \dfrac{f'(x)}{g'(x)} = l$,

then

$$\lim\limits_{x \to a^+} \frac{f(x)}{g(x)} = l.$$

Taylor's Theorem with Lagrange's form of the remainder,
or nth Mean Value Theorem

qn 35 If $f: [a, b] \to \mathbb{R}$

(i) has a continuous $(n - 1)$th derivative on $[a, b]$,

(ii) is differentiable n times on (a, b), then

$$f(b) = f(a) + (b - a)f'(a) + \frac{(b - a)^2}{2!}f''(a) +$$

$$\dots + \frac{(b - a)^{n-1}}{(n - 1)!}f^{(n-1)}(a) + \frac{(b - a)^n}{n!}f^{(n)}(c).$$

for some c, with $a < c < b$.

Taylor's Theorem with Cauchy's form of the remainder

qn 45 If $f: [a, b] \to \mathbb{R}$

(i) has a continuous $(n - 1)$th derivative on $[a, b]$,

(ii) is differentiable n times on (a, b), then

$$f(b) = f(a) + (b - a)f'(a) + \frac{(b - a)^2}{2!}f''(a) +$$

$$\dots + \frac{(b - a)^{n-1}}{(n - 1)!}f^{(n-1)}(a) + \frac{(b - c)^{n-1}(b - a)}{(n - 1)!}f^{(n)}(c).$$

for some c, with $a < c < b$.

Maclaurin's Theorem

qn 37   If $f: [0, x] \to \mathbb{R}$ is differentiable n times, then

$$f(x) = f(0) + xf'(0) + \frac{x^2}{2!} f''(0) + \dots$$

$$+ \frac{x^{n-1}}{(n-1)!} f^{(n-1)}(0) + \frac{x^n}{n!} f^{(n)}(\theta x),$$

for some θ, with $0 < \theta < 1$.

Definition   The power series

$$f(a + h) =$$

$$f(a) + hf'(a) + \frac{h^2}{2!} f''(a) + \dots + \frac{h^n}{n!} f^{(n)}(a) + \dots$$

is called the Taylor series of f at a.

Definition   The power series

$$f(x) = f(0) + xf'(0) + \frac{x^2}{2!} f''(0) + \dots + \frac{x^n}{n!} f^{(n)}(0) + \dots$$

is called the Maclaurin series for f.

Theorem   The Taylor series expansion of a function at a

qn 38   converges to $f(a + h)$ if and only if the difference between the nth partial sum of the series and $f(a + h)$ is a null sequence as $n \to \infty$; or, in other words, if the remainders form a null sequence.

Historical note

Early in the seventeenth century Cavalieri affirmed that there was always a tangent parallel to a chord, and this must count as an embryonic form of the Mean Value Theorem.

In 1691, Michel Rolle proved that between two adjacent roots of a polynomial $f(x)$ there was a root of $f'(x)$. His definition of $f'(x)$ was algebraic (so that, if $f(x) = x^n$, then $f'(x) = nx^{n-1}$) and his proof was like an application of the Intermediate Value Theorem in that he claimed that f' had a different sign at adjacent roots of f and therefore must be zero somewhere between the roots.

Lagrange was the first mathematician to bring inequalities into theorems and proofs about the foundations of calculus. In 1797 he argued (wrongly as it turned out) that every function had a power series expansion. He obtained $f(a + h) = f(a) + hf'(a) + hr(h)$ where $r(h)$ tends to 0 as h tends to 0, from his power series for f

and deduced firstly that functions with positive derivatives were increasing. He then deduced that $(f(a + h) - f(a))/h$ lay between the greatest and least values of $f'(x)$ on the interval $[a, a + h]$. Presuming the Intermediate Value Theorem gave him the Mean Value Theorem. He applied the same argument to obtain the Second and Third Mean Value Theorems. In 1823, Cauchy started from an ε–δ definition of derivative on an interval. He divided his interval $[a, b]$ into lengths shorter than δ and obtained a chain of inequalities of the form

$$f'(x_i) - \varepsilon \leqslant \frac{f(x_i) - f(x_{i-1})}{x_i - x_{i-1}} \leqslant f'(x_i) + \varepsilon.$$

The sum of the numerators is $f(b) - f(a)$ and the sum of the denominators is $b - a$, and as this forms a sort of mean of the quotients, it lies between the greatest and least of the derivatives. Now Cauchy claimed the Intermediate Value Theorem (assuming continuous derivatives) and obtained the Mean Value Theorem. Rolle's theorem and the Mean Value Theorem, as we know them today, depend on the results of Weierstrass which distinguish precisely between properties of continuous functions and properties of differentiable functions and therefore expose exactly what is assured according to the number of times a function may be differentiable. In a letter which Schwarz wrote to Cantor in 1870 (both Schwarz and Cantor had been students of Weierstrass) Schwarz proved that $f'(x) = 0 \Rightarrow f(x)$ is constant, and said that this was the foundation of the differential and integral calculus. In 1878, U. Dini published the first proof of Rolle's Theorem with Weierstrassian rigour.

De l'Hôpital's rule first appeared in his *Analyse des infiniments petits*, (1696) where de l'Hôpital singularly failed to give the credit for this result where it was due, namely to his teacher, Johann Bernoulli. In 1823, working with functions with continuous derivatives, Cauchy proved his extension of the Mean Value Theorem and deduced de l'Hôpital's rule from it.

The expansion of binomial, trigonometric and exponential expressions in the form of power series goes back to Newton and James Gregory (1669). Brook Taylor constructed the series which bears his name in 1712. His method was based on the Gregory–Newton construction of a polynomial from finite differences which appeared in Newton's *Principia* (1687), to which Taylor applied a limiting process. Maclaurin (1742) obtained his series by repeated differentiation of a presumed infinite series expansion, putting $x = 0$ at each stage. Lagrange (1797) having 'proved' that every function has a Taylor series expansion established the First, Second and Third Mean Value Theorems as we have described above. In each case he obtained bounds on the remainder (really on K, in qns 13, 31 and 33) after one, two and three terms of the Taylor series (respectively) and claimed, correctly, that his method extended to n terms giving the Lagrange form of the remainder as we know it today.

In 1822 Cauchy showed that all the derivatives of $\exp(-1/x^2)$ are zero at $x = 0$ so that this function does not have a Taylor series at this point. This undermined Lagrange's programme for a theory of analytic functions based entirely on Taylor series. Cauchy derived his form of the Taylor series remainder from the remainder in integral form, which he also established.

In Weierstrass' lectures in 1861 he claimed the Fundamental Theorem of Analysis to be that, when $f^{(i)}(x_0) = 0$ for $i = 1, 2, 3, \ldots, n - 1$,

$$f(x) - f(x_0) = \frac{(x - x_0)^n}{n!} \cdot f^{(n)}(x_0 + \theta(x - x_0)),$$

for some $\theta, 0 < \theta < 1$,

a form of Taylor's Theorem with Lagrange remainder which had been obtained by Cauchy in 1829.

Answers

1 Yes. Either a local maximum or a local minimum for the function.

2 A differentiable function is continuous and a continuous function on a closed interval is bounded and attains its bounds, from qns 7.31 and 7.34. So the function must have *either* a maximum *or* a minimum in the open interval (a, b), but not necessarily both.
If $f(a) = f(b)$ is not a maximum then the maximum occurs on the open interval. If $f(a) = f(b)$ is not a minimum then the minimum occurs on the open interval. If $f(a) = f(b)$ is both a maximum and a minimum, then the function is constant. Yes.

3 Yes, by qn 8.30

4 If $f(a) = f(b) = M > m$, there is a c with $f(c) = m$.
If $f(a) = f(b) = m < M$, then there is a c with $f(c) = M$.
If $f(a) = f(b) = M = m$, then $f(x) = M = m$ for all x, and so $f'(x) = 0$ from qn 8.6.

5 See figure 9.1.

6 Differentiable $\Rightarrow$ Continuous.
At a local maximum or minimum $f'(x) = 0$.

7 (i) $\Rightarrow f$ is bounded and attains its bounds.
(iii) $\Rightarrow$ when $f(a)$ is not a maximum, there is a maximum $f(c)$ with $a < c < b$; when $f(a)$ is not a minimum there is a minimum $f(c)$ with $a < c < b$; and when $f(a)$ is a maximum and a minimum $f(c)$ is a maximum and a minimum.
(ii) $\Rightarrow f'(c) = 0$.

8 See figure 9.2.

9 Apply Rolle's Theorem to the function
$x \mapsto a_0 x + a_1 x^2/2 + a_2 x^3/3 + \ldots + a_n x^{n+1}/(n + 1)$ on the interval $[0, 1]$.

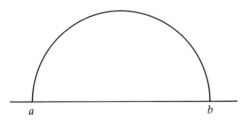

a b

Figure 9.1

Property (i)
fails

Property (ii)
fails

Property (iii)
fails

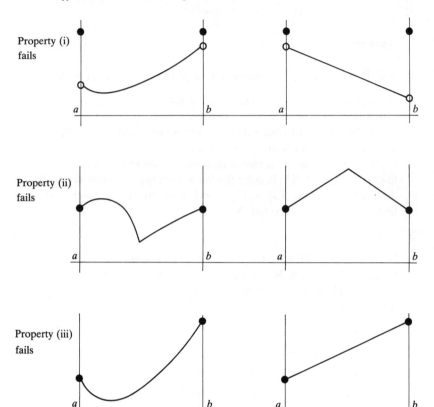

Figure 9.2

10 Apply Rolle's Theorem to $[a, b]$ and $[b, c]$, to find $f'(x) = 0$ with $a < x < b$ and $f'(y) = 0$ with $b < y < c$. Then apply Rolle's Theorem to the function f' on the interval $[x, y]$.

If an n-times differentiable function has $n + 1$ zeros, then between the extreme roots there is a point c at which $f^{(n)}(c) = 0$.

11 (i) Continuous and differentiable by qns 8.15 and 8.17 since the denominator is never 0. $f(0) = f(2) = \frac{1}{4}$. Rolle's Theorem fails since the domain is not an interval.

(ii) Polynomials continuous and differentiable. $g(1) = g(4) = 0$. g is bounded below but does not attain its lower bound. Domain not an interval, i.e. not complete.

12 (i) By qn 6.23. (ii) By qns 7.31, 7.32 and 7.34. (iii) By qns 8.10 and 8.11.

(iv) By qn 8.6 since two derivatives are unequal.
Apply Rolle's Theorem to g on $[a, b]$ then, for some c in (a, b), $g'(c) = 0$. The function f in qn 8.22 is differentiable at every point. For this function, $f'(x) = 2x \cdot \sin(1/x) - \cos(1/x)$, when $x \neq 0$, and $f'(0) = 0$. The derived function f' is discontinuous at 0. Examine the graph of f' on a graphics calculator to see how the intermediate value property for derivatives holds on any interval $[0, a]$. Near to zero, the function oscillates infinitely many times between positive and negative values.

13 (i) If the point (x, y) lies on the chord joining $(a, f(a))$ to $(b, f(b))$ then

$$\frac{y - f(a)}{x - a} = \frac{f(b) - f(a)}{b - a},$$

so

$$y = f(a) + \frac{f(b) - f(a)}{b - a}(x - a)$$

and

$$D(x) = f(x) - f(a) - \frac{f(b) - f(a)}{b - a}(x - a).$$

D statisfies the conditions for Rolle's Theorem on $[a, b]$.

$$D'(c) = 0 \Leftrightarrow f'(c) = \frac{f(b) - f(a)}{b - a}.$$

(ii) $F(a) = 0$ from the definition of K. $F(b) = 0$, trivially. Differentiability follows from qn 8.10. $F'(c) = 0 \Leftrightarrow f'(c) = K$. For some c in (a, b),

$$f'(c) = K = \frac{f(b) - f(a)}{b - a}.$$

14 If $b = a + h$, any c in (a, b) has the form $a + \theta h$.

15 Let $a \leqslant x < y \leqslant b$; then, for some c in the interval (x, y),

$$\frac{f(y) - f(x)}{y - x} = f'(c) > 0.$$

So $x < y \Leftrightarrow f(x) < f(y)$.

16 $f_1'(x) = 1 - \cos x > 0$ on $(0, 2\pi)$, so f_1 is strictly increasing by qn 15. Since $f_1(0) = 0$, $f_1(x) > 0$ on $(0, 2\pi)$.
$f_2'(x) = f_1(x) > 0$ on $(0, 2\pi)$, so f_2 is strictly increasing. Since $f_2(0) = 0$, $f_2(x) > 0$ on $(0, 2\pi)$. $f_3'(x) = f_2(x)$ and $f_4'(x) = f_3(x)$. The argument may be repeated.

17 Let $a < x \leq b$; then

$$\frac{f(x) - f(a)}{x - a} = f'(c) = 0 \quad \text{for some } c \text{ in } (a, x).$$

So $f(x) = f(a)$.
For non-constant f, the domain must be disconnected.

18 $\dfrac{f(y) - f(x)}{y - x} = f'(c).$

$|f'(c)| \leq L$ gives the result.

$$|a_{m+1} - a_2| = |f(a_m) - f(a_1)| \leq L \cdot |a_m - a_1| \leq L \cdot |b - a|.$$

$$|a_{m+2} - a_3| = |f(a_{m+1}) - f(a_2)| \leq L \cdot |a_{m+1} - a_2|$$
$$\leq L^2 \cdot |a_m - a_1| \leq L^2 \cdot |b - a|.$$

Further steps by induction. Now $0 < L < 1$, so (L^n) is a null sequence. So for sufficiently large n, $|a_{n+m} - a_{n+1}| < \varepsilon$, and (a_n) is a Cauchy sequence and so, by qn 4.57, is convergent.
A and B lie in $[a, b]$ by qn 3.66, since the interval is closed. $(a_n) \to A$ implies $(a_{n+1}) \to A$, so $(f(a_n)) \to A$. But f is continuous at A, so $f(A) = A$ and similarly $f(B) = B$.

$$|b_{n+1} - a_{n+1}| = |f(b_n) - f(a_n)| \leq L \cdot |b_n - a_n| \leq \ldots \leq L^n \cdot |b_1 - a_1|.$$

So $(b_n - a_n)$ is a null sequence and by qn 3.44 (ii) (a_n) and (b_n) have the same limit.

19 Let $f(x) = \cos x$. Then if $0 \leq x \leq \frac{1}{2}\pi$, $0 \leq \cos x \leq 1$.
$f'(x) = -\sin x$. So f is a contraction mapping on $[0, 1]$ and, by qn 18, $f(x) = x$ has a unique solution. Obviously no soution on $[1, \frac{1}{2}\pi]$.

20 If $1 \leq x \leq 2$, then $3 \leq x + 2 \leq 4$, so $\sqrt{3} \leq \sqrt{(x + 2)} \leq 2$ which implies $1 \leq \sqrt{(x + 2)} \leq 2$. Thus f has domain and co-domain $[1, 2]$.
$f'(x) = 1/2\sqrt{(x + 2)}$. $1 \leq x \leq 2 \Rightarrow \frac{1}{4} \leq f'(x) \leq 1/2\sqrt{3} \Rightarrow |f'(x)| < 1$. So f is a contraction mapping on $[1, 2]$. Therefore if $a_1 \in [1, 2]$, the sequence defined by $a_{n+1} = f(a_n)$ is convergent.

21 See figure 9.3.

22 Let $f''(c) = l > 0$. Then, for some δ,

$$0 < |x - c| < \delta \Rightarrow \left| \frac{f'(x) - f'(c)}{x - c} - l \right| < \tfrac{1}{2}l \Leftrightarrow \tfrac{1}{2}l < \frac{f'(x)}{x - c} < \tfrac{3}{2}l,$$

since $f'(c) = 0$.
Thus $c - \delta < x < c \Rightarrow f'(x) < 0$, and $c < x < c + \delta \Rightarrow f'(x) > 0$.

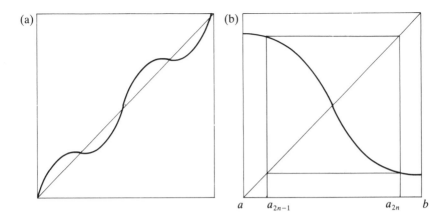

Figure 9.3

Now, by the Mean Value Theorem, for any x satisfying $c - \delta < x < c$,

$$\frac{f(x) - f(c)}{x - c} = f'(d)$$

for some $d \in (x, c)$, so $f'(d) < 0$ and $f(x) > f(c)$.
Likewise for $c < x < c + \delta$, $f(x) > f(c)$, so f has local minimum at c.

23 f_{n+1} is strictly monotonic decreasing by qn 15, and so has at most one root. $n + 1$ odd $\Rightarrow f_{n+1}$ has at least one root from qn 7.19 since x^{n+1} changes sign. $f'_{n+2}(x) = 0 \Leftrightarrow f_{n+1}(x) = 0$ and this occurs once, at $x = a \neq 0$.
For any value of x, $f''_{n+2}(x) = -f'_{n+1}(x) = f_n(x) > 0$.
So f_{n+2} is minimal at $x = a$ by qn 22.

$$f_{n+2}(a) = f_{n+1}(a) + \frac{a^{n+2}}{(n + 2)!} = \frac{a^{n+2}}{(n + 2)!} > 0,$$

since n is even and $a \neq 0$.
At its only minimum value f_{n+2} is positive, so the function is positive everywhere. Thus f_n positive implies f_{n+2} positive.
$f_2(x) = 1 - x + \frac{1}{2}x^2 = \frac{1}{2}[1 + (1 - x)^2] \geq \frac{1}{2}$. So by induction f_n is positive when n is even. We have also shown that, if this is the case, then f_{n+1} has a unique root.

24 If $0 < \theta < 1$ and $x < y$, then $x < \theta x + (1 - \theta)y < y$.
Let $\theta x + (1 - \theta)y = s$, then

$$\frac{f(s) - f(x)}{s - x} = f'(c) \text{ and } \frac{f(y) - f(s)}{y - s} = f'(d) \quad \text{for some } c \text{ and } d,$$

with $x < c < s < d < y$, by the Mean Value Theorem. Now $f''(x) > 0$ for all x, so f' is strictly monotonic increasing by qn 15, and so
$f'(c) < f'(d)$.
Substituting from the equations above gives the result. The graph is convex downwards.

25 Note that $g(b) \neq g(a)$ because, if not, Rolle's Theorem applied to g on $[a, b]$ would contradict $g'(x) \neq 0$.

26 For some c, $a < c < b$,

$$\frac{f(b)}{g(b)} = \frac{f'(c)}{g'(c)}.$$

First equation follows directly from this.
For the conditions we have established $0 < k < h$. So if $h < \delta$, then certainly $k < \delta$.
What this establishes is a convenient rule for finding difficult limits. If $f(a) = g(a) = 0$, then our ordinary procedures for finding $\lim_{x \to a} f(x)/g(x)$ may not be used, but we can try to find $\lim_{x \to a} f'(x)/g'(x)$, and, if we are successful, that is the answer we are looking for.

27 (i) $1/\sqrt{2}$. (ii) 1. (iii) 1, 1/2, 1/6. The first result in (iii) is of great importance and is variously justified in different treatments. Some make it plausible from geometric considerations and then use it to find the derivative of sine. In such a case, de l'Hôpital's rule may not be used, for that would make a circular argument. Some treatments define sine by a power series, and then de l'Hôpital's rule is needed for a sound argument. We will define sine analytically (but with geometric motivation) in chapter 11, and that provides a proper basis for the argument here.

28 The basis of the question is the application of Cauchy's Mean Value Theorem in case $f(a) = g(a) = 0$. In this case, for $b > a$, we have

$$\frac{f(b)}{g(b)} = \frac{f'(c)}{g'(c)} \quad \text{for some } c, \text{ with } a < c < b.$$

If, given B, there exists a δ such that

$$a < x < a + \delta \Rightarrow \frac{f'(x)}{g'(x)} > B,$$

then $a < y < a + \delta \Rightarrow \dfrac{f(y)}{g(y)} = \dfrac{f'(x)}{g'(x)} > B$

for some x, with $a < x < y < a + \delta$. The result is proved.

29 After applying de l'Hôpital's rule, use qn 8.23.
$f'(x) = 2x$ for $x \geq 0$, and $f'(x) = -2x$ for $x < 0$. This f' is not differentiable at 0. But the limit exists and equals 0.

30 For the first result see qn 8.21.
$f'(x)/g'(x) = 2x \sin(1/x) - \cos(1/x)$, of which the first component tends to 0 as in qn 8.21, but the second component oscillates near 0 like the function of qn 8.20.
Cauchy's Mean Value Theorem says that, if $a < b$, then there exists a c with $a < c < b$, such that . . . . . But if $b \to a$ there is nothing to indicate that the consequent cs for each b exhaust the possibilities.

31 $F(a) = 0$ from the definition of K. $F(b) = 0$, trivially. F is differentiable from qns 8.10 and 8.13.

32 $f(a + h) = f(a) + hf'(a) + \frac{1}{2}h^2 f''(a + \theta h)$, for some θ, $0 < \theta < 1$. Put $a = 0$ and $h = x$.

33 Let

$$\frac{f(b) - f(a) - (b - a)f'(a) - \frac{1}{2}(b - a)^2 f''(a)}{(b - a)^3} = K,$$

and let $F(x) = f(b) - f(x) - (b - x)f'(x) - \frac{1}{2}(b - x)^2 f''(x) - K(b - x)^3$.
The application of Rolle's Theorem to F gives $K = f'''(c)/6$ for some c, with $a < c < b$.

$$f(a + h) = f(a) + hf'(a) + (h^2/2!)f''(a) + (h^3/3!)f'''(a + \theta h).$$

36 For some θ, $0 < \theta < 1$,

$$f(a + h) = f(a) + hf'(a) + \frac{h^2}{2!}f''(a) + \ldots + \frac{h^{n-1}}{(n-1)!}f^{(n-1)}(a)$$

$$+ \frac{h^n}{n!}f^{(n)}(a + \theta h).$$

38 $R_n(x) = \dfrac{x^n}{n!} \exp(\theta x)$, for some θ, $0 < \theta < 1$.

The function exp is monotonic (see chapter 11) so if $k = \max(1, \exp(x))$, $|R_n(x)| \leq |x^n/n!| \cdot k$. Now $(|x^n/n!|)$ is a null sequence by qn 3.62(ii). So $(R_n(x)) \to 0$ as $n \to \infty$ by qn 3.28 (sandwich theorem). The nth partial sum of the power series $= \exp(x) - R_n(x) \to \exp(x)$ as $n \to \infty$.

39 If $f(x) = \sin x$, $f'(x) = \cos x = \sin(x + \frac{1}{2}\pi)$, so $f^{(n)}(x) = \sin(x + \frac{1}{2}n\pi)$. $R_{2n}(x) = (x^n/n!) \cdot \sin(\theta x + n\pi)$. So $|R_{2n}(x)| \leq |x^n/n!|$. This shows that $(R_{2n}(x)) \to 0$ as $n \to \infty$. As in qn 38, this enables us to prove that the series converges to the function for each value of x.

40 If $f(x) = \cos x$, $f'(x) = -\sin x = \cos(x + \frac{1}{2}\pi)$, so $f^{(n)}(x) = \cos(x + \frac{1}{2}n\pi)$. The argument proceeds as in qn 38.

41 $f''(x) = -1/(1 + x)^2$, $f'''(x) = 2/(1 + x)^3$ and, by induction, $f^{(n)}(x) = (-1)^{n+1}(n - 1)!/(1 + x)^n$. Thus

$$R_n(x) = \frac{x^n \cdot (-1)^{n+1}(n - 1)!}{n!(1 + \theta x)^n} = (-1)^{n+1}\frac{x^n}{n(1 + \theta x)^n}.$$

(i) For $0 \leqslant x < 1$, $|R_n(x)| \leqslant |x^n/n|$ and so $(R_n(x))$ is a null sequence, and the series converges to the function, as in qn 38.

(ii) $|R_n(1)| = 1/n(1 + \theta)^n < 1/n$, and again $(R_n(x))$ is a null sequence.

(iii) For $-\frac{1}{2} \leqslant x < 0$, $|x| \leqslant \frac{1}{2} < |1 + \theta x|$, so $|x/(1 + \theta x)| < 1$ and therefore $|R_n(x)| < 1/n$, so that $(R_n(x))$ is a null sequence.

42 From qn 5.78 (or 5.90) the radius of convergence is 1. The nth partial sum of series $= \log(1 + x) - R_n(x)$. If $(R_n(x))$ were a null sequence, the sequence of nth partial sums would converge.

43 The two series coincide. Circle of convergence has radius 1. If $x = 1$, the terms of the series alternate in sign, so the series will be convergent if the absolute value of the terms gives a monotonic null sequence.

These are $\frac{1}{1}, \frac{1}{2}, \frac{1}{2} \cdot \frac{1}{4}, \frac{1}{2} \cdot \frac{1}{4} \cdot \frac{3}{6}, \frac{1}{2} \cdot \frac{1}{4} \cdot \frac{3}{6} \cdot \frac{5}{8}, \frac{1}{2} \cdot \frac{1}{4} \cdot \frac{3}{6} \cdot \frac{5}{8} \cdot \frac{7}{10}, \ldots$

Denote these terms by a_1, a_2, a_3, etc., respectively. $a_3 = a_2/4$, $a_4 = a_2(3/4)/6 < a_2/6$, $a_5 = a_2(3/4)(5/6)/8 < a_2/8, \ldots$ and in general $a_n < a_2/(2n - 2)$. $(1/(2n - 2))$ is a null sequence, so $(a_2/(2n - 2))$ is also a null sequence and so the sequence we are concerned with is null by a sandwich theorem.

Each term is obtained from its predecessor by multiplication by a positive factor less than 1, so the sequence is monotonic.

For $n \geqslant 2$, |the remainder after the nth term| $= a_{n+1}/(1 + \theta)^{n-3/2} < a_{n+1}$, $a_4 = 0.06 \ldots$, $a_5 = 0.039 \ldots$. So four terms needed.

44 The Maclaurin series is $0 + 0 + 0 + \ldots$.

The remainder always equals the whole value of the function, so remainders are constant and not null.

For each choice of the constant k, the function $x \mapsto \sin x + k \cdot f(x)$ has the same Maclaurin series.

45 As in qn 35, $F(a) = F(b) = 0$, and F continuous on $[a, b]$ and differentiable on (a, b),

$$F'(c) = 0 \Leftrightarrow -\frac{(b - c)^{n-1}}{(n - 1)!}f^{(n)}(c) + K = 0.$$

46 Using the work of qn 41, Cauchy's form of the remainder

$$R_n(x) = \frac{(x - \theta x)^{n-1} x}{(n - 1)!} (-1)^{n+1} \frac{(n - 1)!}{(1 + \theta x)^n}$$

$$= (-1)^{n+1} x^n \left(\frac{1 - \theta}{1 + \theta x}\right)^{n-1} \cdot \frac{1}{1 + \theta x}.$$

Now $-1 < x \Rightarrow -\theta < \theta x \Rightarrow 0 < 1 - \theta < 1 + \theta x \Rightarrow (1 - \theta)/(1 + \theta x) < 1$.
So $-1 < x \Rightarrow |R_n(x)| < |x^n/(1 + \theta x)| < |x^n/(1 - |x|)|$. But, if $|x| < 1$, this last expression gives a null sequence, so $(R_n(x))$ is a null sequence.

10

Integration
The Fundamental Theorem of Calculus

Preliminary reading: Gardiner ch. III.3, Toeplitz ch. 2
Concurrent reading: Bryant ch. 5, Courant and John ch. 2.
Further reading: Spivak chs 13 and 14.

Areas with curved boundaries

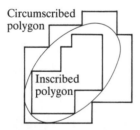

Figure 10.1

The idea of integration comes from the need to measure areas. The measurement of area is the measurement of the quantity of units of surface needed for an exact covering. From the use of a unit square comes the area of a rectangle as length times breadth. Since a triangle can be dissected and reassembled to form a rectangle, the area of a triangle can be calculated, and since any area bounded by a polygon can be dissected into triangles, the area of any polygon can also be calculated. However, when the boundary is curved, the measurement of area is more difficult. The areas of inscribed and circumscribed polygons provide lower and upper bounds on the area (see figure 10.1), and if the construction of inscribed and circumscribed polygons can be done progressively in such a way that the areas of inscribed polygons and the areas of circumscribed polygons tend to the same limit, then that common limit will be the area to be determined, provided the boundary is continuous.

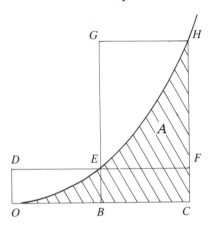

Figure 10.2

1 We start by investigating the shaded area, A, bounded by the parabola $y = x^2$, the x-axis and the line $x = a$, taking a to be positive. In figure 10.2 $O = (0, 0)$, $B = (\frac{1}{2}a, 0)$, $C = (a, 0)$, $D = (0, \frac{1}{4}a^2)$, $E = (\frac{1}{2}a, \frac{1}{4}a^2)$, $F = (a, \frac{1}{4}a^2)$, $G = (\frac{1}{2}a, a^2)$ and $H = (a, a^2)$.

Justify the inequalities

area $BEFC < A <$ area $ODEB +$ area $BGHC$,

and deduce that

$\frac{1}{8}a^3 < A < \frac{1}{8}a^3 + \frac{1}{2}a^3 = \frac{5}{8}a^3$.

By inscribing rectangles with bases $(\frac{1}{3}a, 0)(\frac{2}{3}a, 0)$ and $(\frac{2}{3}a, 0)(a, 0)$, in A, show that $5a^3/27 < A$, and, by covering A with rectangles with bases $(0, 0)(\frac{1}{3}a, 0)$, $(\frac{1}{3}a, 0)(\frac{2}{3}a, 0)$ and $(\frac{2}{3}a, 0)(a, 0)$ show that $A < 14a^3/27$. In the first case A was shown to lie within an interval of size $\frac{1}{2}a^3$, and in the second case A was shown to lie within an interval of size $\frac{1}{3}a^3$, so the gap was narrowed.

Now, for a positive integer n, inscribe rectangles with bases $(a/n, 0)(2a/n, 0)$, $(2a/n, 0)(3a/n, 0)$, $\ldots$, $((n-1)a/n, 0)(a, 0)$, in A, to show that

$$\frac{a^3}{n^3}\left(1^2 + 2^2 + \ldots + (n-1)^2\right) < A.$$

Use qn 1.1 to show that $\frac{1}{3}a^3(1 - 1/n)(1 - 1/2n) < A$.

Now, cover A with rectangles with bases $(0, 0)(a/n, 0)$, $(a/n, 0)(2a/n, 0)$, ..., $((n - 1)a/n, 0)(a, 0)$, to show that

$$A < \frac{a^3}{n^3}(1^2 + 2^2 + \ldots + n^2) = \tfrac{1}{3}a^3(1 + 1/n)(1 + 1/2n).$$

Deduce that $A = \tfrac{1}{3}a^3$ is the only value which satisfies the inequalities

$$\tfrac{1}{3}a^3(1 - 1/n)(1 - 1/2n) < A < \tfrac{1}{3}a^3(1 + 1/n)(1 + 1/2n)$$

for all values of n.

2   Let B be the area bounded by the curve $y = x^3$, the x-axis and the line $x = a$, where we take a to be positive. Using the method of qn 1, and appealing to qn 1.3(iii), prove that, for all values of the positive integer n,

$$\tfrac{1}{4}a^4(1 - 1/n)^2 < B < \tfrac{1}{4}a^4(1 + 1/n)^2.$$

Deduce that $B = \tfrac{1}{4}a^4$ is the only value which satisfies all of these inequalities.

Using this method to find the area bounded by the curve $y = x^k$, the x-axis and the line $x = a$ depends upon knowing a formula for the sum $\sum_{r=1}^{r=n} r^k$, which we may not have to hand.

3   Let C be the area between the curve $y = x^k$, the x-axis and the lines $x = 1$ and $x = a$, where we take a to be positive and greater than 1, and we suppose that k is a positive integer. Let $h = \sqrt[n]{a}$. By inscribing rectangles in C with bases $(1, 0)(h, 0)$, $(h, 0)(h^2, 0)$, ..., $(h^{n-1}, 0)(h^n, 0)$, prove that

$$(h - 1)\frac{h^{n(k+1)} - 1}{h^{k+1} - 1} < C,$$

using qn 1.3(vi).

Deduce that

$$\frac{a^{k+1} - 1}{h^k + h^{k-1} + \ldots + h + 1} < C.$$

From qn 3.47, prove that $(h) \to 1$ as $n \to \infty$, and deduce from qns 3.44(ii), 3.56 and 3.66 that

$$\frac{a^{k+1} - 1}{k + 1} \leqslant C.$$ I

By just covering C with rectangles with the same bases as the inscribed rectangles, prove that

$$C < (h - 1)h^k \frac{h^{n(k+1)} - 1}{h^{k+1} - 1}.$$

As above, prove that

$$C \leqslant \frac{a^{k+1} - 1}{k + 1}. \qquad\qquad\qquad \text{II}$$

Deduce the value of C from the two inequalities, I and II.

4 Where must the argument of qn 3 be modified to deal with the case $0 < a < 1$? Now let $a \to 0$, and determine the area bounded by the curve $y = x^k$, the x-axis and the line $x = 1$. Deduce the area bounded by the curve $y = x^k$, the x-axis and the line $x = a$, for any positive number a.

5 Show that the result of qn 3 holds when k is a negative integer different from -1.

6 Let D be the area bounded by the curve $y = 1/x$, the lines $x = 1$, and $x = a$ and the x-axis, where $a > 1$. Use the method of qn 3 to show that

$$\frac{n(\sqrt[n]{a} - 1)}{\sqrt[n]{a}} < D < n(\sqrt[n]{a} - 1).$$

From qn 4.40 we know that $\left(n(\sqrt[n]{a} - 1)\right)$ has a limit as $n \to \infty$. Use qn 3.47(iv), 3.56 and 3.66 to deduce that that limit is D.

Monotonic functions

7 (*Newton*, 1687) Let $f: [a, b] \to R$ be a function which is positive, monotonic increasing and continuous. Let $c = \frac{1}{2}(a + b)$.
   (i) Explain why

$$(b - a)f(a) \leqslant (c - a)f(a) + (b - c)f(c)$$
$$\leqslant (c - a)f(c) + (b - c)f(b)$$
$$\leqslant (b - a)f(b).$$

   (ii) Divide the interval $[a, b]$ into n equal parts each of length $(b - a)/n$. Let $I_n = $ the sum of the areas of the n rectangles inscribed under the graph of $y = f(x)$, with bases $(a, 0)(a + l/n, 0), (a + l/n, 0)(a + 2l/n, 0), \ldots,$ $(a + (n - 1)l/n, 0)(b, 0)$, where $l = b - a$; and let $C_n = $ the

sum of the areas of n rectangles just covering the graph of
$y = f(x)$, with bases $(a, 0)(a + l/n, 0)$,
$(a + l/n, 0)(a + 2l/n, 0), \ldots, (a + (n - 1)l/n, 0)(b, 0)$. Use
the argument at the beginning of this question to show that
$I_n \leqslant I_{2n} \leqslant C_{2n} \leqslant C_n$.

(iii) Prove that (I_{2^n}) is a monotonic increasing sequence.
Deduce that this sequence is convergent to a limit I, say.

(iv) Prove that (C_{2^n}) is a monotonic decreasing sequence.
Deduce that this sequence is convergent to a limit C, say.

(v) Prove that $C_n - I_n = ((b - a)/n)(f(b) - f(a))$, and deduce
that $(C_n - I_n)$ is a null sequence.

(vi) Use qn 3.44(iii) to show that $C = I$, so that we have found
a measure of the area bounded by the graph of $y = f(x)$,
the lines $x = a$, $x = b$ and the x-axis.

8   In qn 7, had f been monotonic decreasing, in what way would
the argument have been affected?

9   A famous example of a continuous function which is monotonic
decreasing is the function given by $f(x) = 1/x$ for positive values
of x. From qn 8, the area bounded by this graph, the x-axis and
the lines $x = a$ and $x = b$, for positive a and b, is well defined,
being the limit of the sum of the areas of inscribed rectangles.

(i) Show that if n rectangles with bases of equal width are
inscribed under this graph on the interval $[1, a]$, and that if
n rectangles with bases of equal width are inscribed under
this graph on the interval $[b, ba]$, then the area of the two
sets of n rectangles are equal.

(ii) Deduce from qn 8 that the area under the graph on $[1, a]$ is
equal to the area under the graph on $[b, ba]$.

(iii) Deduce further that the area under the graph on $[1, ab]$ is
equal to the sum of the areas under the graph on $[1, a]$ and
$[1, b]$, the essential property of the logarithm function.

(iv) Intepret this result in terms of the limits of qn 6.

The definite integral

It is now appropriate to recognise that the argument in qn 7 did
not require the function f to have positive values except for the
purpose of using the word 'area'. The inequalities of qn 7(i) are
independent of any notion of area.

In qn 7(ii) we can define

$$I_n = \sum_{i=1}^{i=n} \frac{l}{n} f\left(a + \frac{(i-1)l}{n}\right) \quad \text{and} \quad C_n = \sum_{i=1}^{i=n} \frac{l}{n} f\left(a + \frac{il}{n}\right),$$

without reference to area, and we obtain convergent sequences as before in qn 7(iii) and 7(iv), with equal limits, as in qn 7(vi). The common limit is called the *integral* of the function f on $[a, b]$, and is denoted by $\int_a^b f$. The symbol $\int$ is a large squashed S (for sum).

10 Give examples of monotonic functions f and g which are both continuous on $[a, b]$, for which the equation

$$\int_a^b (f + g) = \int_a^b f + \int_a^b g$$

would fail if the integral were always to denote the positive value of the area between the graph and the x-axis.

A further distinction between the notion of integral and the measurement of area under a graph seems natural when we recognise that the argument of qn 7, as modified before qn 10, does not depend on the continuity of the function f. An area must have a boundary, and in order to give *boundaries* to the areas under discussion in qn 7, the function f was taken to be continuous. The graph of a discontinuous function may have gaps in it and so does not bound an area. But the limiting arguments of qn 7 as modified before qn 10 hold for a monotonic function f even if f is discontinuous at many points of the interval $[a, b]$. So such a function can have a well-defined integral.

11 (*Dirichlet*, 1829) If even discontinuous functions can have integrals, are there any functions which cannot? Find the area of the smallest circumscribed rectangle and the largest inscribed rectangle for the area bounded by $x = 0$, the x-axis, $x = 1$ and the function defined by

$$f(x) = \begin{cases} 1 & \text{when } x \text{ is rational,} \\ 0 & \text{when } x \text{ is irrational;} \end{cases}$$

on the interval $[0, 1]$.

As we search for a definition of $\int_a^b f$, we will limit our attention to functions which are bounded, so that they may be covered above and below, by rectangles, and we seek a definition with the following two properties which we retain from our experience of finding areas with curved boundaries using inscribed and circumscribed polygons:

(i) if $m \leqslant f(x) \leqslant M$ on $[a, b]$, then $m(b - a) \leqslant \int_a^b f \leqslant M(b - a)$;

(ii) $\int_a^c f + \int_c^b f = \int_a^b f$, when these integrals exist.

Step functions

In qns 12–15, we construct a family of functions, called *step functions*, with convenient, and obvious, integrals, which we define in qn 15. These functions are not generally either positive or continuous, but they will be able to play the rôle of inscribed and circumscribed polygons for us in an analytical setting. The use of properties (i) and (ii) in qns 12–15 is to establish the reasonableness of the definition of the integral of a step function when it comes in qn 15.

12 If $f(x) = A$ on $[a, b]$, show that there is a unique definition of $\int_a^b f$ which is compatible with (i) above.

13 If $f(x) = A$ on $[a, c]$ and $f(x) = B$ on $(c, b]$, with $A \leqslant B$ use the equation

$$\int_a^b f = \int_a^c f + \int_c^{c+h} f + \int_{c+h}^b f$$

to show that

$$(c - a)A + hA + (b - c - h)B \leqslant \int_a^b f$$

$$\leqslant (c - a)A + hB + (b - c - h)B$$

so

$$(c - a)A + (b - c)B - h(B - A) \leqslant \int_a^b f$$

$$\leqslant (c - a)A + (b - c)B$$

and, if this is to hold for all h, however small, $\int_a^b f$ must be defined as $(c - a)A + (b - c)B$.
Is the same definition necessary if $A \geqslant B$?

14 Show that the value of the integral found in qn 13 must also be the value of the integral for the function, f, defined by

$f(x) = A$ on $[a, c)$,

$f(c) = C$,

$f(x) = B$ on $(c, b]$,

by using the equation $\int_a^b f = \int_a^{c-h} f + \int_{c-h}^{c+h} f + \int_{c+h}^b f$.

15 Let $a = x_0 < x_1 < x_2 < \ldots < x_n = b$, and let a real function f be defined on $[a, b]$ such that $f(x) = A_i$, for $x_{i-1} < x < x_i$, $i = 1, 2,$ $\ldots, n$; and $f(x_i) = B_i$, $i = 0, 1, 2, \ldots, n$.
Such a function is called a *step function*.
Use qn 14 to prove that if this step function has an integral compatible with properties (i) and (ii) before qn 12, then its value is

$$\sum_{i=1}^{i=n} (x_i - x_{i-1}) A_i.$$

We now adopt this result as the *definition* of the integral of a step function.

Lower integral and upper integral

16 Consider a function $f: [a, b] \to [m, M]$. Let $s: [a, b] \to \mathbb{R}$ be a step function such that $s(x) \leqslant f(x)$ for all values of x. Such a step function is called a *lower step function* for f.

We call $\int_a^b s$, a *lower sum* for f.
Why must $m(b - a)$ be a lower sum for f?
Why must every lower sum be $\leqslant M(b - a)$?
Since, for the given bounded function f, the set of all lower sums is non-empty and bounded above it has a supremum, which is called the *lower integral* for f, and is denoted by $\underline{\int}_a^b f$.

17 Every bounded function on a closed interval has a lower integral. Find the lower integral for the function in qn 11.

18 Consider a function $f: [a, b] \to [m, M]$. Let $S: [a, b] \to \mathbb{R}$ be a step function such that $S(x) \geqslant f(x)$ for all values of x. Such a step function is called an *upper step function* for f.
We call $\int_a^b S$, an *upper sum* for f.
Why must $M(b - a)$ be an upper sum for f?
Why must every upper sum be greater than or equal to $m(b - a)$?
Since, for the given bounded function f, the set of all upper sums is non-empty and bounded below it has an infimum, which is called the *upper integral* for f, and is denoted by $\overline{\int}_a^b f$.

19 Find the upper integral for the function in qn 11.

It is in proving the existence of upper and lower integrals for bounded functions that completeness is needed in building a theory of integration.

20 For a given bounded function $f: [a, b] \to \mathbb{R}$, let S be an upper step function, which is constant on the intervals (x_{i-1}, x_i) where $a = x_0 < x_1 < x_2 < \ldots < x_n = b$.
Let s be a lower step function which is constant on the intervals (y_{j-1}, y_j) where $a = y_0 < y_1 < y_2 < \ldots < y_m = b$.
Let $\{z_0, z_1, z_2, \ldots, z_l\} = \{x_0, \ldots, x_n\} \cup \{y_0, \ldots, y_m\}$
where $a = z_0 < z_1 < z_2 < \ldots, < z_l = b$.
Why are both S and s constant on each of the intervals (z_{k-1}, z_k)? Prove that $s(x) \leq S(x)$ on such an interval.
Does $\int_a^b S$ have the same value whether it is calculated on the intervals (x_{i-1}, x_i) or on the intervals (z_{k-1}, z_k)? What is the corresponding claim for $\int_a^b s$? Deduce that the lower sum from the lower step function $s \leq$ the upper sum from the upper step function S.
You have proved that every lower sum $\leq$ every upper sum.

21 Show that for any bounded function $f: [a, b] \to \mathbb{R}$, the lower integral must be less than or equal to the upper integral, that is $\underline{\int}_a^b f \leq \overline{\int}_a^b f$, by showing that if this were false, we could use the basic property of supremum and infimum (see qn 4.60) to contradict the result of qn 20.

Putting together the definitions of upper and lower integral with the results of qns 20 and 21, we obtain

$$\int_a^b s \leq \underline{\int}_a^b f \leq \overline{\int}_a^b f \leq \int_a^b S,$$

for all lower step functions s and all upper step functions S.

22 Verify that any number lying between the upper integral and the lower integral would satisfy the conditions we stated before qn 12 for the integral of the function.

The Riemann integral

Question 22 implies that no unique integral satisfying the conditions (i) and (ii), as stated before qn 12, can exist unless $\underline{\int}_a^b f = \overline{\int}_a^b f$, and then $\int_a^b f$ must equal the common value. When this is the case the bounded function f is defined to be *integrable* (or more precisely *Riemann integrable*) on $[a, b]$, and the common value of $\int_{\underline{a}}^{b} f$ and $\int_{a}^{\overline{b}} f$ is called the *Riemann integral* of the function.

When $\int_{\underline{a}}^{b} f < \int_{a}^{\overline{b}} f$, we say that the function f is *not integrable* on $[a, b]$. A function which is not bounded cannot have an integral in this sense.

23 Use the basic property of an infimum (see qn 4.60) and the analogous property of a supremum, to show that if the function f is integrable on $[a, b]$ (that is $\int_{\underline{a}}^{b} f = \int_{a}^{\overline{b}} f = \int_{a}^{b} f$) then it is possible to find an upper sum and a lower sum for f which are arbitrarily close to each other; that is, given $\varepsilon > 0$, it is possible to find an upper step function S and a lower step function s such that

$$\int_{a}^{b} S - \int_{a}^{b} s < \varepsilon.$$

24 (Converse of qn 23.) If for a function $f: [a, b] \to \mathbb{R}$ it is possible to find upper sums and lower sums whose differences are arbitrarily small, use the chain of inequalities before qn 22 to prove that the upper and lower integrals are equal, so that the function is integrable.

Question 24 provides the standard test for integrability. We proved in qns 12–15 that a step function was necessarily integrable. In qns 7 and 8 we gave an argument which showed that a bounded monotonic function was integrable even if it was not continuous.

25 Prove that the function $f: [0, 1] \to [0, 1]$, given by

$$f(1/n) = 1 \quad \text{when } n = 1, 2, 3, \ldots,$$
$$f(x) \ = 0 \quad \text{otherwise,}$$

is integrable.

Theorems on integrability

While obtaining the basic algebraic properties of integrals (in qns 26–31) it will be useful to have notation available that will not have to be repeatedly defined. We suppose that the integrable function $f: [a, b] \to \mathbb{R}$ has integral F, and, following qn 23, given $\varepsilon > 0$, f has a lower step function s with lower sum σ and an upper step function S with an upper sum Σ, such that $\Sigma - \sigma < \varepsilon$.

26 If, for $k > 0$, the function $k \cdot f$ is defined by

$k \cdot f: x \mapsto k \cdot f(x)$,

show that $k \cdot s$ is a lower step function and $k \cdot S$ is an upper step function for $k \cdot f$ and deduce from qn 24 that $k \cdot f$ is integrable, and has integral $k \cdot F$.

27 If the function $-f$ is defined by

$-f: x \mapsto -f(x)$,

show that $-S$ is a lower step function and $-s$ is an upper step function for $-f$, and deduce from qn 24 that $-f$ is integrable and has integral $-F$.

28 From qns 26 and 27 extend the result of qn 26 to show that the result in that question holds for any real value of k.

We suppose further that the integrable function $g: [a, b] \to \mathbb{R}$ has integral G and, following qn 23, given $\varepsilon > 0$, g has a lower step function s' with lower sum σ' and an upper step function S' with upper sum Σ', such that $\Sigma' - \sigma' < \varepsilon$.

29 If the function $f + g: [a, b] \to \mathbb{R}$ is defined by

$f + g: x \mapsto f(x) + g(x)$,

show that $s + s'$ is a lower step function for $f + g$ and that $S + S'$ is an upper step function for $f + g$, and that the difference between the upper and lower sums for these step functions is $(\Sigma - \sigma) + (\Sigma' - \sigma')$. Deduce that $f + g$ is integrable and has integral $F + G$.

Putting together the results of qns 28 and 29, we can show that, if $f: [a, b] \to \mathbb{R}$ is integrable with integral F and $g: [a, b] \to \mathbb{R}$ is integrable with integral G, then the function $k \cdot f + l \cdot g$ is integrable with integral $k \cdot F + l \cdot G$. This makes the obtaining of an integral a linear function of the family of integrable functions on the domain $[a, b]$.

30 Use qns 28 and 29 with qn 4 to determine the value of $\int_0^a f$ when:

(i) $f(x) = x^2 + x^3$;

(ii) $f(x) = 2x + 3x^2 + 4x^3$;

(iii) $f(x) = c_0 + c_1 x + c_2 x^2 + \ldots + c_n x^n$.

31 If the function f^+ is defined by

$$f^+(x) = \begin{cases} f(x) & \text{when } f(x) \geqslant 0, \\ 0 & \text{when } f(x) < 0, \end{cases}$$

prove that s^+ is a lower step function for f^+ and that S^+ is an upper step function for f^+. Then by considering s^+ and S^+ on the intervals on which both are constant, as in qn 20, show that the difference between the upper and lower sums which they give is less than or equal to $\Sigma - \sigma$. Deduce that, if f is integrable, then so is f^+.

32 If the function f^- is defined by

$$f^-(x) = \begin{cases} 0 & \text{when } f(x) > 0, \\ f(x) & \text{when } f(x) \leqslant 0, \end{cases}$$

show that $f^-(x) = -(-f)^+(x)$ for all values of x.
Deduce from qns 27 and 31 that, if f is integrable, then so is f^-.

33 If the function $|f|$ is defined by

$$|f|(x) = \begin{cases} f(x) & \text{when } f(x) \geqslant 0, \\ -f(x) & \text{when } f(x) < 0, \end{cases}$$

check that $|f| = f^+ - f^-$, and deduce that, if f is integrable, then so is $|f|$.

34 Give an example to show that $|f|$ may be integrable even when f is not.

35 Use the inequality

$$\left| \int_a^b f^+ + \int_a^b f^- \right| \leqslant \left| \int_a^b f^+ \right| + \left| \int_a^b f^- \right|,$$

which follows from the triangle inequality of qn 2.65 to prove that $|\int_a^b f| \leqslant \int_a^b |f|$, when f is integrable on $[a, b]$.

(36) If the function f^2 is defined by

$$f^2: x \mapsto (f(x))^2,$$

s is a non-negative lower step function for $|f|$ giving the lower sum σ, and S is an upper step function for $|f|$ giving the upper sum Σ, prove that s^2 is a lower step function for f^2 and that S^2 is an upper step function for f^2.
If the lower sum from s^2 is σ' and the upper sum from S^2 is Σ', and M is an upper bound for $|f(x)|$, and, without loss of generality, for $S(x)$ also, show that

$$S^2(x) - s^2(x) =$$

$$(S(x) + s(x))(S(x) - s(x)) \leqslant 2M(S(x) - s(x)),$$

and deduce that

$$\Sigma' - \sigma' \leqslant 2M(\Sigma - \sigma),$$

so that, if f is integrable, so is f^2.

(37) Use the equation $f \cdot g = \frac{1}{4}[(f + g)^2 - (f - g)^2]$, to prove that, if f and g are integrable, then so is $f \cdot g$.

(38) If f and g are integrable functions on $[a, b]$, explain why $k \cdot f + l \cdot g$ is also integrable on $[a, b]$, for any numbers k and l. Deduce that $(k \cdot f + l \cdot g)^2$ is integrable and that its integral is greater than or equal to 0.
Use qn 2.54 to show that

$$\left(\int_a^b f \cdot g\right)^2 \leqslant \left(\int_a^b f^2\right)\left(\int_a^b g^2\right).$$

39 (This is a hard question on which nothing further in this book depends, but it opens up the question as to what functions may be Riemann integrable in a most dramatic way.)
In qn 6.72 we examined the continuity of the function f defined on $[0, 1]$ by

$f(x) = 0$ when x is irrational or zero,

$f(p/q) = 1/q$ when p and q are non-zero integers with no common factor.

We found that this function was continuous at each irrational point and discontinuous at each rational point, except 0.
We will consider the integrability of this function on the interval $[0, 1]$ by seeking an upper sum and a lower sum which differ by less than $\varepsilon > 0$.

   (i) Identify a lower step function for which the lower sum = 0.
  (ii) Choose a positive integer m such that $1/m < \frac{1}{2}\varepsilon$. Show that there cannot be more than $\frac{1}{2}m(m - 1)$ points $x \in [0, 1]$ such that $f(x) > 1/m$.
 (iii) Suppose that there are exactly N such points, which we denote in order of magnitude by $c_1, c_2, c_3, \ldots, c_N$.
Find a value of δ such that an upper step function defined on the subdivision

$$0 \leqslant c_1 - \delta < c_1 + \delta < c_2 - \delta < c_2 + \delta <$$

$$\ldots < c_N + \delta \leqslant 1$$

by

$$S(x) = 1 \qquad \text{when } c_i - \delta < x < c_i + \delta,$$
$$S(x) = 1/m \quad \text{otherwise},$$

gives an upper sum $< \varepsilon$, where points are suppressed in the subdivision where there is overlap.

(iv) Deduce that the function *is* Riemann integrable on $[0, 1]$.

Integration and continuity

40 (i) Why is the function f defined by $f(x) = 1/x$ when $x \neq 0$ and $f(0) = 0$ not integrable on $[0, 1]$?

(ii) Give an example to show that a function which is not continuous at one point of a closed interval may none the less be integrable on that interval.

41 A real function f is continuous on $[a, b]$. How do you know that f is bounded and that therefore f has an upper and a lower integral?

42 (*Cauchy*, 1823) The real function f is continuous on $[a, b]$ and $a = x_0 < x_1 < x_2 < \ldots < x_n = b$.

(i) Must the function f be bounded and attain its bounds on $[x_{i-1}, x_i]$?

(ii) Let $m_i = \inf \{f(x) | \, x_{i-1} \leqslant x \leqslant x_i\}$,
and $M_i = \sup \{f(x) | \, x_{i-1} \leqslant x \leqslant x_i\}$.
Explain why $\sum_{i=1}^{i=n} m_i(x_i - x_{i-1})$ is a lower sum for f and why $\sum_{i=1}^{i=n} M_i(x_i - x_{i-1})$ is an upper sum for f.

(iii) Now suppose that the division of the interval $[a, b]$ has been into n equal lengths, so that $x_i - x_{i-1} = (b - a)/n$ for all values of $i = 1, 2, \ldots, n$. Check that in this case the difference between the upper sum and the lower sum

$$= \frac{b - a}{n} \sum_{i=1}^{i=n} (M_i - m_i).$$

If it were possible to prove that this may be made less than any given $\varepsilon > 0$, we would have succeeded in proving that f was integrable.

Now use the fact that a function which is continuous on a closed interval is uniformly continuous on that interval to show that, for any given $\varepsilon > 0$, there exists a $\delta > 0$ such that $|f(x) - f(y)| < \varepsilon/(b - a)$ when $|x - y| < \delta$.

Show how to choose n so that the difference of upper and lower sums previously calculated is less than ε.

This establishes that a continuous function on a closed interval is integrable.

43 Does the converse of the result at the end of qn 42 hold? If a function is integrable must it be continuous?

44 $\int_a^b f = 0$ and $f(x) \geq 0$ for all $x \in [a, b]$. If f is continuous, prove that $f(x) = 0$ for all $x \in [a, b]$. What if f were not necessarily continuous?

45 If f and g are integrable on $[a, b]$ and $\int_a^b (k \cdot f + l \cdot g)^2 = 0$, under what conditions could you deduce that $k \cdot f(x) = -l \cdot g(x)$ for all $x \in [a, b]$?

46   (i) If f is a continuous function on $[a, b]$, prove that

$$\frac{b - a}{n} \sum_{i=1}^{i=n} f\left(a + \frac{i}{n}(b - a)\right)$$

lies between the upper sum and the lower sum calculated in qn 42.

  (ii) Deduce that its limit as $n \to \infty$ is $\int_a^b f$.

  (iii) Check that the limit obtained coincides with the measurement of area in qns 1 and 2.

The notation for integrals, proposed by Leibniz, which is in standard use today, namely $\int f(x)\,dx$, is deliberately suggestive of the limit of the sum $\sum f(x_i)\delta x_i$, where δx_i denotes the difference $x_i - x_{i-1}$, and the limit is taken as the greatest of the $\delta x_i \to 0$.

47 Determine whether the functions f and g defined below are integrable on the interval $[0, 1]$.

$f(x) = x \sin(1/x)$ when $x \neq 0$, and $f(0) = 0$.

$g(x) = \sin(1/x)$ when $x \neq 0$, and $g(0) = 0$.

Mean Value Theorem for integrals

48 If f is a continuous function on $[a, b]$,

$m = \inf\{f(x)|a \leq x \leq b\}$ and $M = \sup\{f(x)|a \leq x \leq b\}$,

explain why $m(b - a) \leq \int_a^b f \leq M(b - a)$.
Deduce that, for some $c \in [a, b]$, $\int_a^b = f(c) \cdot (b - a)$.
This is called the *Mean Value Theorem for integrals*.

If $a = x_0 < x_1 < x_2 < \ldots < x_n = b$, and $x_i - x_{i-1} = (b-a)/n$, show that

$$\left(\left(\sum_{i=1}^{i=n} f(x_i)\right)/n\right) \to f(c) \text{ as } n \to \infty.$$

Integration on subintervals

49 If a function f is integrable on $[a, b]$ and $a \leqslant c < d \leqslant b$, prove that f is integrable on $[c, d]$.

50 If a function f is integrable on $[a, b]$ and $a < c < b$, prove that $\int_a^b f = \int_a^c f + \int_c^b f$. We started using this as an intuitively desirable property for integrals, but now that we have a formal definition of what an integral is we need to prove that it is a formal consequence of the definition.

51 Make a definition of $\int_a^a f$ for any function f, which is compatible with the result of qn 49 in case $c = a$ or $c = b$.
Define $\int_a^b f$ in terms of $\int_b^a f$ in a way which will extend the validity of the result in qn 50 whatever the order of the numbers a, b and c on the number line.

Indefinite integrals

52 Give a reason why the integer function defined by $f(x) = [x]$ is integrable on any closed interval.
Give a formula for $F(x) = \int_0^x f$
   (i) when $0 \leqslant x \leqslant 1$,
   (ii) when $1 \leqslant x \leqslant 2$,
   (iii) when $2 \leqslant x \leqslant 3$.
Is the function F continuous at $x = 1$ and $x = 2$?

53 The function f is given to be integrable on the interval $[a, b]$ and a function F is defined on the interval $[a, b]$ by

$$F(x) = \int_a^x f.$$

Prove that F is continuous at every point of $[a, b]$.
Identify a real number L such that $|F(x) - F(y)| \leqslant L \cdot |x - y|$ for all x and y in $[a, b]$.
The function F is called an *indefinite integral* for f.

54 (*Darboux*, 1875) Let f be a function which is integrable on $[a, b]$, and suppose there exists a function F such that $F'(x) = f(x)$ for all $x \in [a, b]$.
Suppose also that $a = x_0 < x_1 < x_2 < \ldots < x_n = b$ is a subdivision of $[a, b]$.

(i) How do you know that there exists a $c_i \in (x_i, x_{i-1})$ such that

$$\frac{F(x_i) - F(x_{i-1})}{x_i - x_{i-1}} = F'(c_i)?$$

(ii) By adding equations of the type

$$F(x_i) - F(x_{i-1}) = f(c_i)(x_i - x_{i-1}),$$

show that

$$F(b) - F(a) = \sum_{i=1}^{i=n} f(c_i)(x_i - x_{i-1}).$$

(iii) Deduce that $F(b) - F(a) = \int_a^b f$.

It is customary to write $F(b) - F(a) = [F(x)]_a^b$.
There was no requirement in qn 54 that f be continuous.

55 Apply qn 54 on $[0, x]$ to
$F(x) = x^2 \sin(1/x)$ for $x \neq 0$,
$F(0) = 0$.

In qn 54, two conditions were postulated, namely
(i) f is integrable on $[a, b]$, and
(ii) $F'(x) = f(x)$.
This second condition appears to imply that f has an anti-derivative as will be defined after qn 57, and therefore might be thought to imply that f is integrable, but this is not so. Conditions (i) and (ii) are independent, as qn 56 will show. There are even bounded derivatives which are not integrable as Volterra found in 1881.

56 Suppose $F(x) = x^2 \sin(1/x^2)$ when $x \neq 0$ and $F(0) = 0$. Check that F is differentiable on $[0, 1]$. Use computer graphics to examine the graph of F'. Show that F' is unbounded on $[0, 1]$ and so not integrable.

The Fundamental Theorem of Calculus

57 (*Cauchy*, 1823) Let f be a continuous function on $[a, b]$ and let

$$F(x) = \int_a^x f.$$

(i) Why, for some θ, $0 < \theta < 1$, must
$F(x + h) - F(x) = h \cdot f(x + \theta h)$?

(ii) Deduce that $\dfrac{F(x + h) - F(x)}{h} - f(x) = f(x + \theta h) - f(x)$.

(iii) Use the continuity of f to show that $F'(x) = f(x)$.
This result is the *Fundamental Theorem of Calculus*.

When f is continuous and $F'(x) = f(x)$ it is customary, following Leibniz, to write $F(x) = \int f(x)\, dx$ (echoing $\sum f(x_i)(x_i - x_{i-1})$) and F is called an *anti-derivative* for f.

58 If both F and G are anti-derivatives for f on a given interval, use the Mean Value Theorem to deduce that $F(x) - G(x)$ is independent of x. Use this to verify that
$F(b) - F(a) = G(b) - G(a)$.

As a consequence of qn 58, it follows that, if F is an anti-derivative of f, then every other anti-derivative of f has the form $x \mapsto F(x) + c$, for a constant real number c.

Integration by parts

59 If both the functions f and g have continuous derivatives, use the fact that $f \cdot g$ is an anti-derivative of $f \cdot g' + f' \cdot g$ to prove that

$$\int_a^b f' \cdot g = [f \cdot g]_a^b - \int_a^b f \cdot g'.$$

60 If the function f has a continuous second derivative on $[a, b]$, prove that

$$f(b) = f(a) + (b - a)f'(a) + \int_a^b (b - x)f''(x)\, dx.$$

61 (*Cauchy*, 1823) If the function f is infinitely differentiable on $\mathbb{R}$, prove that, for any real number b:

$$f(b) = f(0) + bf'(0) + \frac{b^2 f''(0)}{2!} + \ldots + \frac{b^n f^{(n)}(0)}{n!}$$

$$+ \frac{1}{n!}\int_0^b (b - x)^n f^{(n+1)}(x)\, dx.$$

If $f(x) = \sin x$ show that the last term of this expansion tends to 0 as $n \to \infty$ and deduce that

$$\sin x = \sum_{n=0}^{\infty} \frac{(-1)^n x^{2n+1}}{(2n + 1)!} \quad \text{for all values of } x.$$

Integration by substitution

62 If f is a continuous function on $[g(a), g(b)]$ with $F' = f$, and the function g has a continuous derivative on $[a, b]$, identify an anti-derivative for the function $(f \circ g) \cdot g'$.
Justify each of the equations

$$\int_{g(a)}^{g(b)} f = [F]_{g(a)}^{g(b)} \quad \text{and} \quad \int_{a}^{b} f(g(x)) \cdot g'(x) \, dx = [F(g(x))]_{a}^{b}.$$

What further condition must the function g satisfy if we are to be sure that all four expressions exist and are equal?

Leibniz' notation is helpfully suggestive here as it takes the form

$$\int f(g) \, dg = \int f(g(x)) \frac{dg}{dx} \, dx.$$

Improper integrals

63 If a and b are positive numbers, find the value of $\int_a^b f$ where $f(x) = 1/x^2$.

Determine $\lim_{b \to \infty} \int_a^b f$.

When a function f is integrable on the interval $[a, b]$ with integral I, and $I \to L$ as $b \to \infty$, we write $\int_a^{\infty} f = L$. When the limit exists, $\int_a^{\infty} f$ is called an *improper integral*.
In qns 5.53–5.58 there are theorems which show that the existence of an improper integral may be equivalent to the convergence of an infinite series.

64 Give a definition for $\int_{-\infty}^{b} f$ analogous to that above. Illustrate your definition with an example. When the limit exists, $\int_{-\infty}^{b} f$ is called an *improper integral*.

65 If a and b are positive numbers, find the value of $\int_a^b f$ where $f(x) = 1/\sqrt{x}$.

Determine $\lim\limits_{a \to 0^+} \int_a^b f$.

When a function f is integrable on the interval $[a, b]$ with integral I, and $I \to L$ as $a \to c^+$, we write $\int_c^b f = L$, even when f is not integrable on $[c, b]$. When this limit exists $\int_c^b f$ is called an *improper integral*.

66 Find $\displaystyle\int_{-1}^0 \frac{dx}{\sqrt{(1+x)}}$ as an improper integral.

If $-1 < a < 0$, show that $0 < \displaystyle\int_a^0 \frac{dx}{\sqrt{(1-x^2)}} < \int_a^0 \frac{dx}{\sqrt{(1+x)}}$.

Deduce that $\displaystyle\int_{-1}^0 \frac{dx}{\sqrt{(1-x^2)}}$ exists as an improper integral.

67 Give a definition for $\int_a^c f$ when f is integrable on $[a, b]$ but not on $[a, c]$, analogous to that above.
Illustrate your definition with an example.

68 Find $\displaystyle\int_0^1 \frac{dx}{\sqrt{(1-x)}}$ as an improper integral.

If $0 < a < 1$, show that $0 < \displaystyle\int_0^a \frac{dx}{\sqrt{(1-x^2)}} < \int_0^a \frac{dx}{\sqrt{(1-x)}}$.

Deduce that $\displaystyle\int_0^1 \frac{dx}{\sqrt{(1-x^2)}}$ exists as an improper integral.

69 A function f is defined for $x \geq 0$ by

$$f(x) = \frac{(-1)^n}{n+1} \quad \text{when } n \leq x < n + 1;\ n = 0, 1, 2, \ldots$$

Do either (or both) of

$$\int_0^\infty f \quad \text{or} \quad \int_0^\infty |f|$$

exist?

70 Find the limit of $\displaystyle\frac{1}{\sqrt{n}}\left(\frac{1}{1} + \frac{1}{\sqrt{2}} + \frac{1}{\sqrt{3}} + \ldots + \frac{1}{\sqrt{n}}\right)$ as $n \to \infty$.

Summary

Definition
qn 15
A function $s: [a, b] \to \mathbb{R}$ is called a *step function* if, for some subdivision

$$a = x_0 < x_1 < x_2 < \ldots < x_n = b,$$
$$s(x) = A_i \text{ when } x_{i-1} < x < x_i, i = 1, 2, \ldots, n;$$
$$s(x_i) = B_i \text{ when } i = 0, 1, \ldots, n.$$

The integral of this step function is

$$\int_a^b s = \sum_{i=1}^{i=n} A_i(x_i - x_{i-1}).$$

Definition
qns 16,
18
If $s: [a, b] \to \mathbb{R}$ is a step function and $f: [a, b] \to \mathbb{R}$ is a function, s is called a *lower step function* for f when $s(x) \leq f(x)$ for all values of x, and s is called an *upper step function* for f when $f(x) \leq s(x)$ for all values of x.

When s is a lower step function, its integral is called a *lower sum* for f. When s is an upper step function, its integral is called an *upper sum* for f.

Theorem
qns 16, 18
Any bounded function $f: [a, b] \to [m, M]$ has an upper sum $M(b - a)$ and a lower sum $m(b - a)$.

Theorem
qn 20
Each lower sum of a bounded function $f: [a, b] \to \mathbb{R}$ is less than or equal to each upper sum.

Definition
qns 16, 18
The supremum of the lower sums of a bounded function $f: [a, b] \to \mathbb{R}$ is called the *lower integral* of f and denoted by $\underline{\int}_a^b f$.

The infimum of the upper sums of f is called the *upper integral* of f and denoted by $\overline{\int}_a^b f$.

Theorem
qn 21
For a given bounded function $f: [a, b] \to \mathbb{R}$,

each lower sum $\leq$ the lower integral
$\leq$ the upper integral
$\leq$ each upper sum.

Definition
qn 22
A bounded function $f: [a, b] \to \mathbb{R}$ is said to be *Riemann integrable* when its upper integral is equal to its lower integral. Their common value is called the *Riemann integral* of the function and denoted by $\int_a^b f$.

Theorem
qns 23, 24
A function $f: [a, b] \to \mathbb{R}$ is Riemann integrable if and only if, given $\varepsilon > 0$, there exists an upper sum Σ and a lower sum σ, such that $\Sigma - \sigma < \varepsilon$.

Theorem
qns 7, 8, 10+
Every function which is monotonic on a closed interval is integrable.

Theorem qns 3, 4, 5	If $f(x) = x^n$ for n an integer $\neq -1$, then $\int_0^a f = a^{n+1}/(n+1)$.
Theorem qns 6, 9	If $f(x) = 1/x$, then $\int_b^{ba} f = \int_1^a f = \lim_{n \to \infty} n(\sqrt[n]{a} - 1)$.
Theorem qns 26, 27, 28	If $f: [a, b] \to \mathbb{R}$ has integral F, then $k \cdot f$ has integral $k \cdot F$ for any real number k.
Theorem qns 33, 35	If $f: [a, b] \to \mathbb{R}$ is integrable, then $|f|$ is also integrable, and $|\int_a^b f| \leq \int_a^b |f|$.
Theorem qns 29, 37	If $f: [a, b] \to \mathbb{R}$ has integral F and $g: [a, b] \to \mathbb{R}$ has integral G, then $f + g$ has integral $F + G$, and $f \cdot g$ is integrable.
Theorem qn 42	If $f: [a, b] \to \mathbb{R}$ is continuous, then f is integrable.

Mean Value Theorem for integrals

qn 48	If $f: [a, b] \to \mathbb{R}$ is continuous, then $\int_a^b f = f(c) \cdot (b - a)$ for some $c \in [a, b]$.
Theorem qn 49	If $f: [a, b] \to \mathbb{R}$ is integrable and $a \leq c < d \leq b$, then $f: [c, d] \to \mathbb{R}$ is integrable.
Theorem qn 50	If $f: [a, b] \to \mathbb{R}$ is integrable and $a < c < b$, then $\int_a^b f = \int_a^c f + \int_c^b f$.
Definition qn 51	$\int_a^a f = 0$ and $\int_b^a f = -\int_a^b f$.
Theorem qn 53	If $f: [a, b] \to \mathbb{R}$ is integrable, then the function F defined by $F(x) = \int_a^x f$, is continuous on $[a, b]$.
Theorem qn 54	If $f: [a, b] \to \mathbb{R}$ is integrable and $F'(x) = f(x)$, then $\int_a^b f = F(b) - F(a)$.

The Fundamental Theorem of Calculus

qn 57	If $f: [a, b] \to \mathbb{R}$ is continuous and $F(x) = \int_a^x f$, then $F'(x) = f(x)$.

Definition of improper integrals

qn 63	If $I(b) = \int_a^b f$ and $I(b) \to L$ as $b \to \infty$, then we write $\int_a^\infty f = L$.
qn 67	If $I(b) \to L$ as $b \to c^-$, then we write $\int_a^c f = L$.
qn 64	If $I(a) = \int_a^b f$ and $I(a) \to L$ as $a \to -\infty$, then we write $\int_{-\infty}^b f = L$.
qn 65	If $I(a) \to L$ as $a \to c^+$, then we write $\int_c^b f = L$.

Historical note

In Archimedes' book on the *Quadrature of the Parabola* (c. 250 B.C.) he finds the area bounded by a parabolic segment by inscribing a succession of triangles and summing their areas with a geometric series. In his later work *On Spirals* he included a summation of segments of areas which led to the same algebraic calculations and argument as in our qn 1. Cavalieri, a pupil of Galileo, published further studies of area in 1635 and 1647, in which by ingenious use of certain cases of the Binomial Theorem he obtained the area under $y = x^3$ and $y = x^4$, and suggested a formula for the area under $y = x^n$ for positive integral n. Proving this result and extending it to all rational $n \neq -1$ (as in qns 3–5) is commonly credited to Fermat who claimed the result in general form, about 1643, though Torricelli also obtained the result about this time. The analysis of the area under a rectangular hyperbola, as in qn 9, is due to Gregory of St Vincent (1647).

In a book written by Isaac Barrow, Newton's teacher, and published in 1669, while discussing continuous and monotonic distance–time and velocity–time graphs, Barrow described the Fundamental Theorem of Calculus. The description is geometrical (and therefore, from a modern viewpoint, hard to recognise) and the proof uses infinitesimals. Newton helped Barrow to write the book, and there was a copy of it in Leibniz' personal library. From his papers, we know that Newton came to understand the Fundamental Theorem in the year 1665–6. He considered two variables x and y (which he called fluents) changing with time. Their rates of change he denoted by $\dot{x}$ and $\dot{y}$ (which he called fluxions) and he obtained $\dot{A}/\dot{x} = y$, where A is the area under the graph of y against x. Newton's discovery of the inverse relationship between integration and differentiation, or as he would have put it, between the method of quadratures and the method of fluxions, transformed the study of areas and generated a range of powerful applications. Newton's readiness to differentiate and integrate power series term by term (without tests of convergence) brought a further host of significant results relating to the binomial series and trigonometric functions. Leibniz obtained Newton's fundamental theorem independently in 1675. In his *Principia* (1687), Newton included a proof that any monotonic function was integrable. His illustrations show that he was considering only differentiable monotonic functions, but all the functions he was concerned with were piecewise monotonic, so his proof dealt satisfactorily with his field of concern. Quite trivially, his proof may be extended to show that discontinuous monotonic functions are integrable, but this was not done until the integrability of such functions was considered in the nineteenth century.

The wide application of this result and the fact that, in the century following Newton and Leibniz, functions were considered to be infinitely differentiable meant that, during the eighteenth century, the study of integration consisted of the search for anti-derivatives. In 1822 Fourier's work on the conduction of heat raised the possibility of the integrability of discontinuous functions (obtained from trigonometic series) which he was not able to settle. In 1823 Cauchy considered the integration of continuous functions (a much more general family of functions than the eighteenth century had considered) and proved their integrability by defining the integral to be the limit of $\sum f(x_i)(x_i - x_{i-1})$ as the lengths of the subintervals $|x_i - x_{i-1}|$ tend to 0. Although he actually assumed uniform continuity for his functions, this enabled him to prove the Fundamental Theorem of the Calculus as we know it. In the same lectures, Cauchy obtained the integral form of the remainder for a Taylor series. Cauchy acknowledged that his definition of the integral could be applied to a function with a finite number of discontinuities. Cauchy's work did not answer Fourier's question as to which functions were possibly integrable. Dirichlet, who had had discussions with Fourier, and perhaps Cauchy, in Paris, published his famous example (qn 11) of a function which is *not* integrable in Cauchy's sense in 1829 and asserted in the same paper that a function was integrable in Cauchy's sense provided its discontinuities were nowhere dense.

Riemann tackled the subject of what functions could be integrable under the direction of Dirichlet, and in 1854 delivered a lecture in Göttingen (which was not published until 1868, after his death) in which the possibility of the integration of trigonometric series was considered to a depth that was to be definitive for the rest of the century. Riemann's definition of integral was the limit of $\sum f(t_i)(x_i - x_{i-1})$, where $x_{i-1} \leqslant t_i \leqslant x_i$, as the lengths of the subintervals $|x_i - x_{i-1}|$ tend to 0, and he constructed a function which was integrable in this sense but which was discontinuous on a set of points that was everywhere dense. We have given an example of such a function in qn 39 (using a construction due to Baire, 1899). Riemann's example was put together rather differently: let $g(x) = 0$ when $x - [x] = \frac{1}{2}$ and let $g(x) = x - [x + \frac{1}{2}]$ otherwise, then Riemann's function

$$f(x) = g(x) + \frac{g(2x)}{2^2} + \frac{g(3x)}{3^2} + \ldots + \frac{g(nx)}{n^2} + \ldots$$

which is discontinuous at every point $x = p/2q$, where p and q are odd numbers without a common divisor. The function f is none the less integrable because the series is uniformly convergent everywhere.

The treatment of the Riemann integral that we have given, with step functions and upper and lower integrals, is due to Darboux (1875).

Answers

1 *BEFC* is an inscribed polygon. *ODEGHC* is a circumscribed polygon.
Three inscribed rectangles of width $\frac{1}{3}a$: heights 0, $a^2/9$, $4a^2/9$.
Combined area $= \frac{1}{3}a(a^2/9 + 4a^2/9) = 5a^3/27$.
Three circumscribed rectangles of width $\frac{1}{3}a$: heights $a^2/9$, $4a^2/9$, a^2.
Combined area $= \frac{1}{3}a(a^2/9 + 4a^2/9 + a^2) = 14a^2/9$.
$5a^3/8 - a^3/8 = 4a^3/8 = \frac{1}{2}a^3$. $14a^3/27 - 5a^3/27 = 9a^3/27 = \frac{1}{3}a^3$.

$$\left(1 - \frac{1}{n}\right)\left(1 - \frac{1}{2n}\right) < \frac{A}{\frac{1}{3}a^3} < \left(1 + \frac{1}{n}\right)\left(1 + \frac{1}{2n}\right).$$

Each of the bounds tends to 1 as $n \to \infty$, so no number different from
$A = \frac{1}{3}a^3$ will satisfy all the inequalities.

3 Factorise the denominator and put $h^n = a$.

4 If $0 < a < 1$, then $0 < h < 1$, so for positive terms we must write $1 - h$
and $1 - a^{k+1}$, etc.

As $a \to 0$, $\dfrac{1 - a^{k+1}}{k + 1} \to \dfrac{1}{k + 1}$.

For $0 < a < 1$, area required $= \dfrac{1}{k + 1} - \dfrac{1 - a^{k+1}}{k + 1} = \dfrac{a^{k+1}}{k + 1}$.

For $1 < a$, area required $= \dfrac{a^{k+1} - 1}{k + 1} + \dfrac{1}{k + 1} = \dfrac{a^{k+1}}{k + 1}$.

5 Provided $k \neq -1$, the algebra of qn 3 still holds when k is a negative
integer. If $(h) \to 1$, then $(1/h) \to 1$ and $(1/h)^k \to 1$.

6 Let $h = \sqrt[n]{a}$. Inscribed rectangles have area

$$(h - 1)/h + (h^2 - h)/h^2 + \ldots + (h^n - h^{n-1})/h^n = n(h - 1)/h.$$

Circumscribed rectangles have area

$$(h - 1) \cdot 1 + (h^2 - h)/h + \ldots + (h^n - h^{n-1})/h^{n-1} = n(h - 1).$$

The sequence $\left(n(\sqrt[n]{a} - 1)\right)$ is monotonic decreasing from qn 2.48 so,
from qn 3.66, its limit $\leq D \leq$ its limit.

7 (i) Since $c = \frac{1}{2}(a + b)$, $b - c = c - a = \frac{1}{2}(b - a)$. Thus the first
inequality is equivalent to $f(a) \leq f(c)$; the third to $f(c) \leq f(b)$; and
the second to these two together.
(ii) Apply the argument of (i) to each of the n subintervals.

(iii) From (ii), (I_{2^n}) is monotonic increasing. It is bounded above by C_1.
So it is convergent by qn 4.34.

(iv) From (ii), (C_{2^n}) is monotonic decreasing. It is bounded below by I_1.
So it is convergent by qn 4.33.

(v) $C_2 - I_2 = (b - c)f(b) - (c - a)f(a) = \frac{1}{2}(b - a)(f(b) - f(a))$. Draw a figure for $C_n - I_n$.

(vi) $C_n - I_n \to C - I = 0$. $I_n < \text{area} < C_n$.

8 All the inequalities would have been reversed.

9 (i) n maximal rectangles with equal width inscribed in the area bounded by $y = 1/x$, $x = b$, the x-axis and $x = ba$ have area

$$\frac{ba - b}{n}\left(\frac{1}{b + (1/n)(ba - b)} + \frac{1}{b + (2/n)(ba - b)} + \dots \right.$$

$$\left. + \frac{1}{b + (n/n)(ba - b)}\right)$$

$$= \frac{a - 1}{n}\left(\frac{1}{1 + (1/n)(a - 1)} + \frac{1}{1 + (2/n)(a - 1)} + \dots \right.$$

$$\left. + \frac{1}{1 + (n/n)(a - 1)}\right)$$

which is the area of n maximal rectangles with equal width inscribed in the area bounded by $y = 1/x$, $x = 1$, the x-axis and $x = a$.

(ii) Since $y = 1/x$ is monotonic decreasing for positive x, the areas under the curve are well defined by qn 8. The two areas are each equal to the limit of the sums of the areas of the inscribed rectangles as $n \to \infty$.

(iii) Area under graph on $[1, ab]$
= area under graph on $[1, b]$ + area under graph on $[b, ab]$
= area under graph on $[1, b]$ + area under graph on $[1, a]$ by (ii).

(iv) $\lim n(\sqrt[n]{(ab)} - 1) = \lim n(\sqrt[n]{a} - 1) + \lim n(\sqrt[n]{b} - 1)$ as $n \to \infty$.
See also qn 4.40.

10 Let $f(x) = x^2$ and $g(x) = -x^2$: both functions are monotonic for positive x. On a positive interval $\int_a^b f + g = 0$, while $\int_a^b f = \int_a^b g \neq 0$ if all integrals were positive.

11 Inscribed area $= 0$. Circumscribed area $= 1$. Not equal, so no integral.

12 $A \leqslant f(x) \leqslant A$, so $A(b - a) \leqslant \int_a^b f \leqslant A(b - a)$ and $\int_a^b f = A(b - a)$.

13 The same definition is necessary when $A \geqslant B$.

14 Suppose that $A < B < C$ (the argument can be easily adapted for other orderings of these numbers).

$$A(c - h - a) + 2hA + B(b - c - h) < \int_a^b f$$

$$< A(c - h - a) + 2hC + B(b - c - h)$$

which gives

$$A(c - a) + B(b - c) - h(B - A) < \int_a^b f$$

$$< A(c - a) + B(b - c) + h(2C - B - A).$$

If this must hold for all h, however small, $\int_a^b f = A(c - a) + B(b - c)$.

15 The argument of qn 14 shows that a distinct value at an isolated point does not affect the value of the integral. Questions 12 and 13 establish the rest of the result.

16 Since $m \leqslant f(x)$, $s(x) = m$ defines a lower step function, so $m(b - a)$ is a lower sum.
Now $s(x) \leqslant f(x) \leqslant M$ for all x, so if $s(x) = A_i$ when $x_{i-1} < x < x_i$, $A_i \leqslant M$.
So $A_i(x_i - x_{i-1}) \leqslant M(x_i - x_{i-1})$, and $\int_a^b s \leqslant M(b - a)$.

17 Every lower step function $\leqslant 0$, so every lower sum $\leqslant 0$. But 0 is a lower sum, so the lower integral is 0.

18 Since $f(x) \leqslant M$, $S(x) = M$ defines an upper step function, so $M(b - a)$ is an upper sum.
$m \leqslant f(x) \leqslant S(x)$ for all x so, using the argument in qn 16,
$m(b - a) \leqslant \int_a^b S$.

19 Every upper step function $\geqslant 1$, so every upper sum $\geqslant 1$. But 1 is an upper sum, so the upper integral is 1.

20 Since $z_k = x_i$ for some i or y_j for some j, the open interval (z_{k-1}, z_k) is contained both within an interval of the type (x_{i-1}, x_i), so that S is constant on (z_{k-1}, z_k), and within an interval of the type (y_{j-1}, y_j), so that s is constant on (z_{k-1}, z_k).
If $z_{k-1} < x < z_k$, then $s(x) \leqslant f(x) \leqslant S(x)$, so $s(x) \leqslant S(x)$ on each of the open intervals (z_{k-1}, z_k). So, calculated on the z-intervals, $\int_a^b s \leqslant \int_a^b S$.
If $(x_{i-1}, x_i) = (z_{k-p}, z_k)$, then $(x_i - x_{i-1})A = (z_k - z_{k-p})A$
$= (z_k - z_{k-1})A + (z_{k-1} - z_{k-2})A + \ldots + (z_{k-p+1} - z_{k-p})A$, and so $\int_a^b S$

has the same value whichever intervals it is defined on. Likewise for $\int_a^b s$. So $\int_a^b s \leqslant \int_a^b S$ for any upper and lower step functions S and s for the function f.

21 Suppose $\overline{\int}_a^b f < \underline{\int}_a^b f$ and that $\underline{\int}_a^b f - \overline{\int}_a^b f = \varepsilon > 0$. Then there is an upper step function S and a lower step function s such that

$$\overline{\int}_a^b f \leqslant \int_a^b S < \overline{\int}_a^b f + \tfrac{1}{2}\varepsilon = \underline{\int}_a^b f - \tfrac{1}{2}\varepsilon < \int_a^b s \leqslant \underline{\int}_a^b f.$$

This gives $\int_a^b S < \int_a^b s$, which contradicts qn 20.

22 A constant function is a step function and therefore the inequality $\int_a^b s \leqslant \underline{\int}_a^b f \leqslant \overline{\int}_a^b f \leqslant \int_a^b S$ guarantees the first condition before qn 12 for any supposed 'integral' lying between the upper and the lower integral. The second condition also holds because it holds for step functions.

23 Since $\int_a^b f = \sup \int_a^b s$, there is a lower step function s such that $\int_a^b f - \tfrac{1}{2}\varepsilon < \int_a^b s \leqslant \int_a^b f$ and, since $\int_a^b f = \inf \int_a^b S$, there is an upper step function S such that $\int_a^b f \leqslant \int_a^b S < \int_a^b f + \tfrac{1}{2}\varepsilon$.

24 If the difference between the upper and lower integrals were ε, then to find an upper step function S and a lower step function s such that $\int_a^b S - \int_a^b s < \varepsilon$ would contradict the chain of inequalities before qn 22.

25 If we consider upper and lower step functions on the intervals $(0, 1/n)$, $(1/n, 1/(n-1))$, $\ldots$, $(\tfrac{1}{4}, \tfrac{1}{3})$, $(\tfrac{1}{3}, \tfrac{1}{2})$, $(\tfrac{1}{2}, 1)$; then we have an upper sum $= 1/n$ and a lower sum $= 0$. Since these are arbitrarily close, the function is integrable, and has integral 0.

26 $s(x) \leqslant f(x) \leqslant S(x) \Rightarrow k \cdot s(x) \leqslant k \cdot f(x) \leqslant k \cdot S(x) \Rightarrow k\sigma$ is a lower sum and $k\Sigma$ is an upper sum for $k \cdot f$. $k \cdot \Sigma - k \cdot \sigma < k \cdot \varepsilon$, so the difference between an upper sum and a lower sum may be made arbitrarily small and so $k \cdot f$ is integrable. Clearly the integral is $k \cdot F$.

27 $s(x) \leqslant f(x) \leqslant S(x) \Rightarrow -S(x) \leqslant -f(x) \leqslant -s(x) \Rightarrow -S$ is a lower step function and $-s$ is an upper step function for $-f$. So $-\Sigma$ is a lower sum and $-\sigma$ is an upper sum for $-f$. $(-\sigma) - (-\Sigma) = \Sigma - \sigma < \varepsilon$. Clearly the integral is $-F$.

28 From qn 26, $k \cdot f$ is integrable so, from qn 27, $-k \cdot f$ is integrable, with integral $-k \cdot F$.

29 $s(x) \leqslant f(x) \leqslant S(x)$ and $s'(x) \leqslant g(x) \leqslant S'(x)$
$\Rightarrow s(x) + s'(x) \leqslant f(x) + g(x) \leqslant S(x) + S'(x)$.
Now $s + s'$ is a step function on the union of the subintervals for s and s', in the sense of qn 20, and is therefore a lower step function for $f + g$. Moreover, $\int_a^b (s + s') = \int_a^b s + \int_a^b s'$ using the definition of the integral of a

step function on the union of the subintervals. So $\sigma + \sigma'$ is a lower sum for $f + g$, and likewise $\Sigma + \Sigma'$ is an upper sum.
$(\Sigma + \Sigma') - (\sigma + \sigma') = (\Sigma - \sigma) + (\Sigma' - \sigma') < 2\varepsilon$, which may be made arbitrarily small, so $f + g$ is integrable.
$\sigma \leqslant F \leqslant \Sigma$ and $\sigma' \leqslant G \leqslant \Sigma' \Rightarrow \sigma + \sigma' \leqslant F + G \leqslant \Sigma + \Sigma'$.
so the integral is $F + G$.

30 $\int_0^a f = a^{k+1}/(k + 1)$ from qn 4.
   (i) $\frac{1}{3}a^3 + \frac{1}{4}a^4$ from qn 29.
   (ii) $a^2 + a^3 + a^4$ from qns 28 and 29.
   (iii) $c_0 a + c_1 a^2/2 + c_2 a^3/3 + \ldots + c_n a^{n+1}/(n + 1)$.

31 $s^+(x) > 0 \Rightarrow 0 < s^+(x) = s(x) \leqslant f(x) = f^+(x)$.
$s^+(x) = 0 \Rightarrow s^+(x) = 0 \leqslant f^+(x)$.
Likewise $f(x) > 0 \Rightarrow 0 < f^+(x) = f(x) \leqslant S(x) = S^+(x)$.
$f(x) \leqslant 0 \Rightarrow f^+(x) = 0 \leqslant S^+(x)$.
So $s^+(x) \leqslant f^+(x) \leqslant S^+(x)$.
Evidently since s and S are step functions, so are s^+ and S^+. So we have established that s^+ is a lower step function and S^+ is an upper step function for f^+.
Now clearly, if $S^+(x) = 0$, then $s^+(x) = 0$, so
$S(x) - s(x) \geqslant S^+(x) - s^+(x) \geqslant 0$. Thus the difference between the upper sums and lower sums from S^+ and $s^+ \leqslant \Sigma - \sigma < \varepsilon$. So the difference between upper and lower sums may be made arbitrarily small and so f^+ is integrable.

32 $f^-(x) = 0 \Leftrightarrow f(x) \geqslant 0 \Leftrightarrow -f(x) \leqslant 0 \Leftrightarrow (-f)^+(x) = 0$.
$f^-(x) < 0 \Leftrightarrow f(x) < 0 \Leftrightarrow -f(x) > 0 \Rightarrow (-f)^+(x) = -f(x)$
$\Rightarrow -(-f)^+(x) = f(x)$.
f integrable $\Rightarrow -f$ integrable by qn 27.
$-f$ integrable $\Rightarrow (-f)^+$ integrable by qn 31 $\Rightarrow f^-$ integrable by qn 27.

33 Examine $f(x) > 0, = 0, < 0$, to show $|f| = f^+ - f^-$.
Integrability of $|f|$ follows from qns 31, 32, 27 and 29.

34 On the interval $[0, 1]$ consider $f(x) = 1$ when x is irrational and $f(x) = -1$ when x is rational.

35 From the definitions, $f = f^+ + f^-$, so the left-hand side of the inequality $= |\int_a^b f|$.
$|f| = f^+ - f^- = |f^+| + |f^-|$, so $\int_a^b |f| = \int_a^b |f^+| + \int_a^b |f^-|$
$= |\int_a^b f^+| + |\int_a^b f^-|$.

36 $0 \leqslant s(x) \leqslant |f(x)| \Rightarrow 0 \leqslant s^2(x) \leqslant |f(x)|^2 = f^2(x)$.
Now s^2 is constant on any interval on which s is constant, so s^2 is a lower step function for f^2.

If $|f(x)| \leqslant M$, then $s(x) \leqslant M$ and we may choose $S(x) \leqslant M$, so $s(x) + S(x) \leqslant 2M$.

Since $\Sigma - \sigma$ may be made arbitrarily small, so may $2M(\Sigma - \sigma)$ and hence $\Sigma' - \sigma'$. Thus the integrability of f implies the integrability of f^2.

37 Use qns 29, 36, 27, and 26.

38 Use qns 26, 29 and 36. 0 is a lower sum.

$$\int_a^b (k \cdot f + l \cdot g)^2 \geqslant 0, \text{ so } k^2 \int_a^b f^2 + 2kl \int_a^b f \cdot g + l^2 \int_a^b g^2 \geqslant 0.$$

Now put $A = \int_a^b f^2$, $B = 2 \int_a^b f \cdot g$ and $C = \int_a^b g^2$, then $A \geqslant 0$.

$A \geqslant 0$ and $Ak^2 + Bkl + Cl^2 \geqslant 0 \Rightarrow A^2k^2 + ABkl + ACl^2 \geqslant 0$
$\Leftrightarrow (Ak + \frac{1}{2}Bl)^2 + (AC - \frac{1}{4}B^2)l^2 \geqslant 0$. Since this is true for all k and l.
$AC - \frac{1}{4}B^2 \geqslant 0$, so $\left(\int_a^b f^2\right)\left(\int_a^b g^2\right) \geqslant \left(\int_a^b f \cdot g\right)^2$.

39 (i) $s(x) = 0$.
  (ii) $f(x) > 1/m \Leftrightarrow x = p/q$, where $q < m$. So $m - 1$ possible values of q. Now $1 \leqslant p \leqslant q$, so at most $(m - 1) + (m - 2) + \ldots + 2 + 1$ $= \frac{1}{2}m(m - 1)$ possible values of x.
  (iii) Take $\delta < \frac{1}{4}\varepsilon/N$, then upper sum $\leqslant 2\delta N + (1/m)(1 - 2\delta N)$
    $< \frac{1}{2}\varepsilon + \frac{1}{2}\varepsilon = \varepsilon$.
    To avoid overlap, take $\delta < \min \{\frac{1}{2}(c_i - c_{i-1}), \frac{1}{4}\varepsilon/N\}$, then the formula for the upper sum is exact.
  (iv) Lower sum $= 0$, upper sum $< \varepsilon$. Function integrable, integral $= 0$.

40 (i) The function is not bounded above, so there is no upper integral.
  (ii) $\int_{1/2}^{3/2} [x] = \frac{1}{2}$.

41 By qns 7.31, 7.32 and 7.34.

42 (i) Yes, as in qn 41.
  (ii) $s(x) = m_i$ when $x_{i-1} \leqslant x < x_i$, gives a lower step function, and $S(x) = M_i$ when $x_{i-1} \leqslant x < x_i$, gives an upper step function.
  (iii) Choose n so that $(b - a)/n < \delta$, then

$$\frac{b - a}{n} \sum_{i=1}^{i=n} (M_i - m_i) < \frac{b - a}{n} \cdot \frac{n\varepsilon}{b - a} = \varepsilon.$$

43 No. Step functions and monotonic functions are integrable whether they are continuous or not.

44 If $f(x) \geqslant 0$, $s(x) = 0$ is a lower step function and so 0 is a lower sum. Since $\int_a^b f = 0$, there must be arbitrarily small upper sums. However, if $f(c) > 0$ for some c, there is a neighborhood of c on which $|f(x) - f(c)| < \frac{1}{2}f(c)$. So for some δ, $|x - c| < \delta \Rightarrow f(x) > \frac{1}{2}f(c)$. So any upper sum $> 2\delta \cdot \frac{1}{2}f(c)$, which is not arbitrarily small. This contradiction shows that $f(x) = 0$ for all x. In qns 25 and 42, discontinuous f satisfies the conditions.

45 The conclusion holds if both f and g are continuous functions, using qn 44. Otherwise we could take f as in qn 25 and $g(x) = 0$.

46 (i) $m_i = \inf \{f(x) | x_{i-1} \leqslant x \leqslant x_i\}$,
and $M_i = \sup \{f(x) | x_{i-1} \leqslant x \leqslant x_i\}$. So $m_i \leqslant f(x_i) \leqslant M_i$, and $m_i(x_i - x_{i-1}) \leqslant f(x_i)(x_i - x_{i-1}) \leqslant M_i(x_i - x_{i-1})$.
Thus

$$\text{lower sum} \leqslant \sum_{i=1}^{i=n} f(x_i)(x_i - x_{i-1}) \leqslant \text{upper sum}.$$

$x_i = a + (i/n)(b - a)$ and $x_i - x_{i-1} = (b - a)/n$.

(ii) We proved in qn 42 that the difference between these upper and lower sums may be made arbitrarily small for sufficiently large n, and it therefore follows that each of these tends to the integral as $n \to \infty$.
Denoting the sum in question by Σ, lower sum $\leqslant \Sigma \leqslant$ upper sum $\Rightarrow$ $0 \leqslant \Sigma -$ lower sum $\leqslant$ upper sum $-$ lower sum, and now, by a sandwich theorem, Σ tends to the integral as $n \to \infty$.

(iii) This summation gives a minimal upper sum for the given subdivision when the function is monotonic increasing.

47 The function f is continuous on $[0, 1]$ by qn 6.49 and is thus integrable. $s(x) = -1$ is a lower step function and $S(x) = 1$ is an upper step function for g. Consider step functions on $[0, 1/n]$ and $[1/n, 1]$. The function is continuous on $[1/n, 1]$ and therefore there are arbitrarily close upper and lower sums. On $[0, 1/n]$ the difference between upper and lower sums is $2/n$ which may be arbitrarily small. So the function is integrable by qn 24.

48 See qns 16 and 18. Since f is continuous, $(b - a)f$ is continuous. The minimum value of this function is $m(b - a)$ and the maximum value of this function is $M(b - a)$. Since the integral lies between these values, or at one of them, the integral is equal to $(b - a)f(c)$ for some $c \in [a, b]$ by the Intermediate Value Theorem.

Divide the result of qn 46(ii) by $(b - a)$.

49 Since f is integrable on $[a, b]$, there exist step functions s and S giving upper and lower sums which are arbitrarily close. If c and d were not part of the subdivisions for these step functions, introduce these points into the subdivision for each step function. Then the difference between the upper and lower sums on $[c, d] \leqslant$ the difference between the upper and lower sums on $[a, b]$, so f is integrable on $[c, d]$.

50 If f is a step function the result is obvious. To obtain the result for any integrable f, apply this result to upper and lower step functions for f.

51 To retain the equation for qn 50, we must define $\int_a^a f = 0$ and $\int_b^a f = -\int_a^b f$.

52 The function f is monotonic.

$$F(x) = \begin{cases} 0 & \text{when } 0 \leqslant x \leqslant 1, \\ x - 1 & \text{when } 1 \leqslant x \leqslant 2, \\ 2x - 3 & \text{when } 2 \leqslant x \leqslant 3. \end{cases}$$

F is continuous at $x = 1$ and 2, by considering limits from above and below at each of these points.

53 Since f is integrable, f is bounded. Let $m \leqslant f(x) \leqslant M$.
$F(c + h) - F(c) = \int_c^{c+h} f$, so for positive h,
$mh \leqslant F(c + h) - F(c) \leqslant Mh$.
Thus $\lim_{h \to 0^+} \big(F(c + h) - F(c)\big) = 0$, so $\lim_{h \to 0^+} F(c + h) = F(c)$.
Likewise for the limit from below. Then use qn 6.89.
$L = \max\{|m|, |M|\}$.

54 (i) By the Mean Value Theorem applied to F.
(iii) Every summation of this kind lies between an upper sum and a lower sum evaluated on the given subdivision. Since the subdivision is arbitrary, the only number satisfying this condition for all possible upper step functions and all possible lower step functions is the integral itself.

55 $F'(x) = 2x \sin(1/x) - \cos(1/x)$ when $x \neq 0$, and
$F'(0) = 0$ from qn 8.22.
Note that F' is not continuous at 0. None the less, F' is integrable by qn 29, because $x \sin(1/x)$ gives a continuous function, which is necessarily integrable, and $\cos(1/x)$ gives an integrable function like g in qn 47.
Thus $\int_0^x F' = x^2 \sin(1/x)$ provided $x \neq 0$.

56 When $x \neq 0$, $F'(x) = 2x \sin (1/x^2) - (2/x) \cos (1/x^2)$, which is unbounded near $x = 0$.

$$\frac{F(x) - F(0)}{x - 0} = x \sin \frac{1}{x^2}, \quad \text{so } -|x| \leqslant \frac{F(x) - F(0)}{x - 0} \leqslant |x|,$$

and $F'(0) = 0$.

57 (i) By the Mean Value Theorem for integrals, qn 48.

(iii) Since F is continuous at x, $f(x + \theta h) \to f(x)$ as $h \to 0$.

So

$$\frac{F(x + h) - F(x)}{h} \to f(x) \quad \text{as } h \to 0,$$

and this implies that $F'(x) = f(x)$.

58 $F'(x) = G'(x) = f(x) \Rightarrow (F - G)'(x) = 0 \Rightarrow (F - G)(x) = \text{constant by}$ qn 9.17.
$\Rightarrow (F - G)(b) = (F - G)(a) \Rightarrow F(b) - F(a) = G(b) - G(a)$.
Since $(F - G)(x) = \text{constant}$, $G(x) = F(x) + c$.

59 $(f \cdot g)' = f' \cdot g + f \cdot g'$, so, since $f' \cdot g + f \cdot g'$ is continuous, the Fundamental Theorem of Calculus gives $[f \cdot g]_a^b = \int_a^b (f' \cdot g + f \cdot g')$
$= \int_a^b f' \cdot g + \int_a^b f \cdot g'$.

60 $\int_a^b f''(x)(b - x)dx = [f'(x)(b - x)]_a^b - \int_a^b f'(x)(-1)dx$

$= -f'(a)(b - a) + \int_a^b f'(x)dx = -f'(a)(b - a) + f(b) - f(a)$.

61 Apply qn 59 to obtain

$$\int_0^b (b - x)^n f^{(n+1)}(x)dx = [f^{(n)}(x)(b - x)^n]_0^b - \int_0^b -n(b - x)^{n-1} f^{(n)}(x)dx$$

$$= -b^n f^{(n)}(0) + n \int_0^b (b - x)^{n-1} f^{(n)}(x)dx.$$

Qn 59 must be applied n times in all.
If $f(x) = \sin x$, then $f'(x) = \cos x = \sin (x + \frac{1}{2}\pi)$,
so $f^{(n)}(x) = \sin (x + \frac{1}{2}n\pi)$.

So $\left| \frac{1}{n!} \int_0^b (b - x)^n f^{(n+1)}(x)dx \right| \leqslant \frac{|b|^n}{n!} |\sin (c + \frac{1}{2}(n + 1)\pi)| \leqslant \frac{|b|^n}{n!}$.

The last term $\to 0$ as $n \to \infty$ from qn 3.62(ii).
So $(\sin x - \text{partial sum of the first } n \text{ terms}) \to 0$. So the series tends to $\sin x$.

62 $(F \circ g)' = (F' \circ g) \cdot g' = (f \circ g) \cdot g'$. The first equation comes from the application of the Fundamental Theorem to $F' = f$ on $[g(a), g(b)]$. The second equation comes from the application of the Fundamental Theorem to $(F \circ g)' = (f \circ g) \cdot g'$ on $[a, b]$.

All four expressions exist and are equal provided f is continuous on $g([a, b])$, which follows, for example, if g' is positive on $[a, b]$.

63 Using the Fundamental Theorem, the integral $= -1/b + 1/a$. As $b \to \infty$, integral $\to 1/a$.

64 If $I(a) = \int_a^b f$ exists for a negative and unbounded below, and
$$\lim_{a \to -\infty} I(a) = L.$$
then we write $\int_{-\infty}^b f = L$. Same example as qn 63.

65 Integral $= 2\sqrt{b} - 2\sqrt{a}$. $\lim_{a \to 0^+} (2\sqrt{b} - 2\sqrt{a}) = 2\sqrt{b}$.

66 $\lim_{a \to -1^+} [2\sqrt{(1 + x)}]_a^0 = 2$.
$$-1 < x \leqslant 0 \Rightarrow 1/\sqrt{2} < 1/\sqrt{(1 - x)} \leqslant 1 \Rightarrow 0 < 1/\sqrt{(1 - x^2)}$$
$$\leqslant 1/\sqrt{(1 + x)}.$$
Now $I(a) = \int_a^0 \dfrac{dx}{\sqrt{(1 - x^2)}}$ increases as $a \to -1^+$, but is bounded above by 2.
Let $\sup \{I(a)| -1 < a < 0\} = L$, then as $a \to -1^+$, $I(a) \to L$ and
$$\int_{-1}^0 \frac{dx}{\sqrt{(1 - x^2)}} = L.$$

67 If $I(b) = \int_a^b f$ and $\lim_{b \to c^-} I(b) = L$, then we write $\int_a^c f = L$.
$$\int_{-1}^a \frac{dx}{\sqrt{(-x)}} = [-2\sqrt{(-x)}]_{-1}^a = -2\sqrt{(-a)} + 2 \to 2 \text{ as } a \to 0^-.$$

68 $[-2\sqrt{(1 - x)}]_0^a = -2\sqrt{(1 - a)} + 2 \to 2$ as $a \to 1^-$.
$$0 \leqslant x < 1 \Rightarrow 1/\sqrt{(1 + x)} \leqslant 1 \Rightarrow 1/\sqrt{(1 - x^2)} \leqslant 1/\sqrt{(1 - x)}.$$
Thus $0 < \int_0^a \dfrac{dx}{\sqrt{(1 - x^2)}} < \int_0^a \dfrac{dx}{\sqrt{(1 - x)}}$ for $0 < a < 1$.
Now $I(a) = \int_0^a \dfrac{dx}{\sqrt{(1 - x^2)}}$ increases as $a \to 1^-$, and is bounded above by 2.
So $\int_0^1 \dfrac{dx}{\sqrt{(1 - x^2)}}$ exists, as in qn 66.

69 $\int_0^N f = 1 - \frac{1}{2} + \frac{1}{3} - \frac{1}{4} + \ldots + (-1)^{N-1}/N$. This series is convergent by the alternating series test (qn 5.60) to $\log 2$, by qn 9.41.
$\int_0^N |f| = 1 + \frac{1}{2} + \frac{1}{3} + \frac{1}{4} + \ldots + 1/N$. This harmonic series is divergent, qn 5.27.

70 $\displaystyle\sum_{r=2}^{r=n+1} \frac{1}{\sqrt{r}} < \int_1^{n+1} \frac{dx}{\sqrt{x}} < \sum_{r=1}^{r=n} \frac{1}{\sqrt{r}}.$

Integral $= 2\sqrt{(n+1)} - 2$, using the Fundamental Theorem.

$$2\sqrt{(n+1)} - 2 < \sum_{r=1}^{r=n} \frac{1}{\sqrt{r}} < 2\sqrt{(n+1)} - 1 - \frac{1}{(n+1)}.$$

Now divide through by $\sqrt{n}$ and obtain the limit 2 on both sides as $n \to \infty$. See qn 5.56.

11

Indices and circle functions

For an elegant treatment of exponential and logarithmic functions starting from the counter-intuitive definition $\log x = \int_1^x dt/t$ see Hardy, or for a presentation of Hardy's treatment using a problem sequence see Quadling. For an alternative treatment starting from the definition

$$\exp(x) = 1 + x + \frac{x^2}{2!} + \frac{x^3}{3!} + \dots$$

see Burkill.

Exponential and logarithmic functions

Although we have used logarithms and exponentials in this book, they have not played a formal rôle in defining any of the concepts of analysis.

We now have the results we need to define these functions and to establish their properties.

Positive integers as indices

Definition
For all $x \in \mathbb{R}$, $x^1 = x$ and $x^{n+1} = x^n x$; $n \in \mathbb{N}$.
For $x \neq 0$, $x^0 = 1$ and $x^{-n} = 1/x^n$.

1 For $m, n \in \mathbb{N}$, prove by induction on n that

(i) $x^{m+n} = x^m x^n$,

(ii) $x^{mn} = (x^m)^n$, and

(iii) $(xy)^n = x^n y^n$.

Properties (i) and (ii) are called the *laws of indices* though they are in fact theorems. It will be necessary to establish that these two laws of indices still hold as we extend the family of numbers which may be used as indices.

2 For m, $n \in \mathbb{N}$, prove that when $x > 1$, $m < n \Leftrightarrow x^m < x^n$, from order axioms in chapter 2.
Deduce that, when $0 < x < 1$, $m < n \Leftrightarrow x^m > x^n$.

3 By appeal to chapter 6, show that, for $n \in \mathbb{N}$, the function $f: \mathbb{R}^+ \to \mathbb{R}^+$ given by $f(x) = x^n$ is strictly increasing, continuous and unbounded above.

Positive rationals as indices

4 With f as in qn 3, show that $f^{-1}: \mathbb{R}^+ \to \mathbb{R}^+$ exists by the Intermediate Value Theorem, and is strictly increasing, continuous and unbounded above.

Definition
$f^{-1}(x)$ (as in qn 4) is denoted by $x^{1/n}$.

Definition
For $x > 0$ and n, $m \in \mathbb{Z}^+$, $x^{n/m} = (x^n)^{1/m}$.

5 (i) For x, $y > 0$, prove that $(xy)^{1/n} = x^{1/n} y^{1/n}$.
   (ii) By induction on n prove that $x^{n/m} = (x^{1/m})^n$.
   (iii) Show that $x^{1/mn} = (x^{1/m})^{1/n}$. Use (ii) and the fact that the function $x \mapsto x^{mn}$ is a bijection of $\mathbb{R}^+$.

6 For $x > 0$ and r, $s \in \mathbb{Q}^+$, prove the two laws of indices
   (i) $x^{r+s} = x^r x^s$ and (ii) $x^{rs} = (x^r)^s$. Build on qns 1 and 5.

7 Prove that, when $x > 1$, and r, $s \in \mathbb{Q}^+$, $r < s \Leftrightarrow x^r < x^s$.
Deduce that, when $0 < x < 1$, and r, $s \in \mathbb{Q}^+$, $r < s \Leftrightarrow x^r > x^s$.

Rational numbers as indices

Definition
For $x > 0$ and $r \in \mathbb{Q}^+$, $x^0 = 1$ and $x^{-r} = 1/x^r$.

8 For $x > 0$ and r, $s \in \mathbb{Q}$, prove the two laws of indices
   (i) $x^{r+s} = x^r x^s$ and (ii) $x^{rs} = (x^r)^s$.

9 Prove that, when $x > 1$, and r, $s \in \mathbb{Q}$, $r < s \Leftrightarrow x^r < x^s$.
Deduce that, when $0 < x < 1$, and r, $s \in \mathbb{Q}$, $r < s \Leftrightarrow x^r > x^s$.

Definition
For $a > 1$, define $A: \mathbb{Q} \to \mathbb{R}^+$ by $A(x) = a^x$.

Use computer graphics to examine the graph of A for $a = 1.5$, 2 and 3.

For qns 10–28, we will always assume that $a > 1$.

10 Verify from qn 3.47 that $(A(1/n)) \to 1 = A(0)$ as $n \to \infty$. Deduce that $(A(-1/n)) \to 1 = A(0)$ as $n \to \infty$. Use the fact that A is strictly increasing, established in qn 9, to show that A is continuous at 0.

11 Prove that A is continuous at $q \in \mathbb{Q}$ by considering that

$$A(x) - A(q) = A(q)(a^{x-q} - 1) \to 0 \text{ as } x \to q.$$

So far we have only given a meaning to rational indices, and found that for such indices the function $x \mapsto a^x$ is continuous. Because the rational numbers are dense on the real line we can complete the definition of this function on $\mathbb{R}$ simply by insisting that it shall be continuous. Questions 12–18 provide the tools for showing that this extension may only be done in one way and for finding the derivative of the resulting function.

Definition
Keeping $a > 1$, define $D: \mathbb{Q}\backslash\{0\} \to \mathbb{R}^+$ by

$$D(x) = \frac{a^x - 1}{x}.$$

Use computer graphics to examine the graph of D for $a = 1.5$, 2 and 3.

12 Why is D continuous where it has been defined?

13 By considering qn 2.48(iii), show that, when $m, n \in \mathbb{Z}^+$, $m < n \Leftrightarrow D(m) < D(n)$. Use the idea of qn 2.48(iv) to deduce that, when $r, s \in \mathbb{Q}^+$, $r < s \Leftrightarrow D(r) < D(s)$.

We proved in qn 4.40(i) that $(D(1/n))$ is convergent as $n \to \infty$.

Definition
Let $(D(1/n)) \to L(a)$ as $n \to \infty$.

14 Use the fact that D is strictly increasing for $x > 0$ to show that

$$\lim_{x \to 0^+} D(x) = L(a).$$

15 Use the fact that, for non zero x, $D(-x) = D(x)/A(x)$, to show that

$$\lim_{x \to 0^-} D(x) = L(a).$$

16 How should $D(0)$ be defined so that D is continuous on $\mathbb{Q}$?

17 Find a positive number M such that

$$\frac{a^x - a^y}{x - y} < M,$$

provided $0 \leqslant y < x \leqslant c$.

Real numbers as indices

18 How does the condition in qn 17 guarantee that A may be extended in a unique way to a continuous function on $[0, \infty)$?

19 How can you be sure that A may also be extended in a unique way to a continuous function on $(-\infty, 0]$?
This gives us a continuous function $A: \mathbb{R} \to \mathbb{R}^+$, and at last we have a well-defined meaning for irrational indices.

20 For $a > 1$ and $x, y \in \mathbb{R}$, prove the two laws of indices

(i) $a^{x+y} = a^x a^y$ and (ii) $a^{xy} = (a^x)^y$.

For (ii), first establish the result for $y = n \in \mathbb{N}$, then for $y = 1/n$, and then for $y = m/n \in \mathbb{Q}$, before attempting $y \in \mathbb{R}$.

21 Show that A is strictly increasing on $\mathbb{R}$.

Since $A: \mathbb{R} \to \mathbb{R}^+$ is strictly increasing it is a bijection and has a unique inverse: $\mathbb{R}^+ \to \mathbb{R}$ called the logarithm to the base a.

Definition
$A^{-1}(x) = \log_a x$.

22 Prove that, when $X, Y > 0$, $\log_a X + \log_a Y = \log_a XY$.
Deduce that $\log_a 1/X = -\log_a X$.

23 (i) For $x \neq 0$, check that D is continuous at x if and only if A is continuous at x. See qn 12.
(ii) Use the argument of qn 21 to show that D is strictly increasing on $\mathbb{R}^+$.

(iii) If (x_n) is a null sequence of positive terms, and q_n is a rational number lying between x_n and $2x_n$, use the fact that $(D(q_n)) \to L(a)$ and $(D(\frac{1}{2}q_n)) \to L(a)$ to prove that $(D(x_n)) \to L(a)$ and deduce that $\lim_{x\to 0^+} D(x) = L(a)$.

(iv) By an argument like that of qn 15 show that $\lim_{x\to 0^-} D(x) = L(a)$.

(v) If $D(0) = L(a)$, must D be continuous on $\mathbb{R}$?

24 Prove that $A'(x) = A(x) \cdot L(a)$ for $x \in \mathbb{R}$.

Natural logarithms

25 From qn 10.6, we know that $D(1/n)$ is an upper sum and $D(-1/n)$ is a lower sum for $\int_1^a dx/x$.
Deduce that

$$L(a) = \int_1^a \frac{dx}{x}, \quad \text{for } a > 1.$$

26 Why must $L(a) \to 0$ as $a \to 1^+$, and $L(a) \to \infty$ as $a \to \infty$? Why is L strictly increasing, continuous and $L'(x) = 1/x$.

27 Why must there be a number $e > 1$ such that $L(e) = 1$? Check that $L(2) < 1$, so that $2 < e$.

Definition
When $a = e$, write $A(x) = E(x) = \exp(x)$.

28 Prove that $E'(x) = E(x)$. What is the Taylor series for $E(x)$?

If we now define A for $0 < a < 1$, we can use the fact that $1 < 1/a$ to apply the results of qn 10 onwards. With this condition, in the analogues of qns 10 and 21, A is then strictly decreasing. The analogue of qn 13 holds for negative integers and negative rationals and the analogues of qns 14 and 15 transpose the originals. The analogue of qn 16 holds, but for the analogues of qns 17 and 18 an interval $[-c, 0]$ must be used and the result extended from the negative reals to the positive reals. The analogues of qns 20, 23 and 24 hold, but for qn 25 have $D(-1/n)$ as an upper sum and $D(1/n)$ as a lower sum; so we get $L(a)$ as before when $0 < a$ and $L(a) \to -\infty$ as $a \to 0^+$.

It is trivial to define $A(x) = 1$ when $a = 1$, and then $L(1) = 0$.

Now L is defined on the domain $\mathbb{R}^+$ and is strictly increasing and continuous.

29 By differentiating the function $f = L \circ E$, show that $L(E(x)) = x$ for $x \in \mathbb{R}$.

30 Since L and E are bijections, $E(L(x)) = x$ for $x \in \mathbb{R}^+$. Deduce that $L(x) = \log_e x$.

The function L is called the natural logarithmic function and is usually denoted in school texts by ln. When $\log x$ is written without an explicitly named base, in university texts, the logarithm to the base e is meant. The function A is an exponential function, but when *the* exponential function is referred to, it is E that is meant.

31 Provided that a, $b > 0$, show that

(i) $a^x = (e^{\log a})^x = e^{x \log a}$.

(ii) $\log a^x = x \cdot \log a$.

(iii) $\log b = \log_a b \cdot \log a$.

Exponential and logarithmic limits

32 Use de l'Hôpital's rule to find the limit of

$$\frac{\log (1 + ax)}{x} \quad \text{as } x \to 0^+.$$

Deduce that $\big(n \log (1 + a/n)\big) \to a$ as $n \to \infty$.
Use the continuity of E to show that $\big((1 + a/n)^n\big) \to e^a$ as $n \to \infty$.

33 Construct an argument similar to that of qn 32 to show that

$$\lim_{n \to \infty} \left(1 - \frac{a}{n}\right)^{-n} = e^a.$$

34 The function f is defined on $\mathbb{R}^+$ by $f(x) = x^a$ where a is a real number different from 0. Prove that $f'(x) = a \cdot x^{a-1}$.

35 Use the ratio test to prove that $(n^a e^{-n}) \to 0$ as $n \to \infty$.
Deduce that $x^a e^{-x} \to 0$ as $x \to \infty$. Prove that $e^x/x^a \to \infty$ as $x \to \infty$ and illustrate this with graph drawing facilities on a computer.
Prove that, when $a > 0$, $\log x/x^a \to 0$ as $x \to \infty$.

36 Investigate the function given by $f(x) = x^x$, defined for positive x. Show that f is continuous throughout its domain. Find the minimum value of f. Show that f is monotonic decreasing on the domain $0 < x < 1/e$. Show that $\big(f(1/n)\big) \to 1$ as $n \to \infty$. Deduce that $\lim_{x \to 0^+} f(x) = 1$.

37 What is the Maclaurin series for $E(x)$? Does it converge to the value of the function for all values of x?

38 Use the equation

$$1 + (-t) + (-t)^2 + \ldots + (-t)^{n-1} = \frac{1 - (-t)^n}{1 - (-t)},$$

which is valid for all t unless $t = -1$, and the Fundamental Theorem of Calculus, to show that

$$\log(1 + x) = x - \frac{x^2}{2} + \frac{x^3}{3} - \ldots$$

$$+ (-1)^{n-1} \frac{x^n}{n} + (-1)^n \int_0^x \frac{t^n}{1 + t}\, dt,$$

provided $-1 < x$.

For any positive number K, show that, when $-K/(K + 1) < x$,

$$\left| (-1)^n \int_0^x \frac{t^n}{1 + t}\, dt \right| \le (K + 1) \int_0^x |t|^n dt = \frac{(K + 1)|x|^{n+1}}{(n + 1)}.$$

Check that this last expression gives a null sequence when $|x| \le 1$ and determine for what values of x the power series developed here is a valid expansion of $\log(1 + x)$.

Circular or trigonometric functions

The origins of the functions sine, cosine, tangent, cotangent, secant and cosecant are geometric. A particle, P, moves around the circumference of a circle with centre O and radius 1. The *angle* through which the radius turns is the length of arc which is traced out by P on the circumference. If the particle moves anti-clockwise from the point A to the point B, and the arc length from A to B is x, then the perpendicular from B to OA has length 'sine of x'. The cosine of x is the sine of the complementary angle. The tangent of x is the length along the tangent at B from B to the point where it meets OA produced. The cotangent of x is the tangent of the complementary angle.

If we only have the axioms for the real numbers, how do we define these functions? If logic was all that mattered, we could ignore the geometric origins of the functions and define sine and cosine by their power series: this is the procedure adopted by Burkill. It is also possible to define the sine function by an infinite product or to

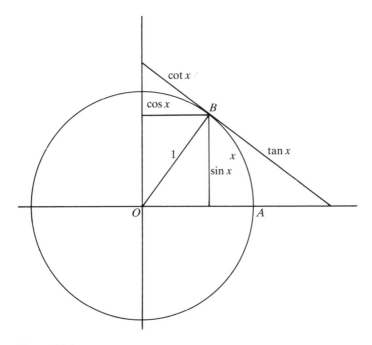

Figure 11.1

develop the circular functions from the definition

$$\arctan x = \int_0^x \frac{dt}{1 + t^2}:$$

this is the procedure adopted by Hardy.

But if the geometric origins of these functions are to be respected we must develop a formal definition of angle either from the notion of the area of a sector of a unit circle (the procedure adopted by Spivak) or from a formal definition of arc length. This is what is done in qns 39–48. You may, if you wish, skip to the definition of circular arc length following qn 48 and explore the intervening problems when your curiosity is aroused.

Length of a line segment

39 Give an algebraic formula for the non-negative function $f: [-1, 1] \to \mathbb{R}$ whose graph will appear as a semicircle with centre at the origin and radius 1.

40 If the distance between two points (x, y) and (a, b) is defined to be $\sqrt{((x - a)^2 + (y - b)^2)}$, show that this distance

$$\leqslant |x - a| + |y - b|.$$

What is the distance between the two points $(a, f(a))$ and $(b, f(b))$?

41 Why is the function f, of qn 39, continuous on $[-1, 1]$ and differentiable on $(-1, 1)$?

42 Apply the Mean Value Theorem to f to show that the distance between the points of qn 40 is equal to

$$\frac{b - a}{\sqrt{(1 - c^2)}}$$

for some c with $-1 \leqslant a < c < b \leqslant 1$.

Arc length

43 For any subdivision $a = x_0 < x_1 < x_2 < \ldots < x_n = b$, show that the *polygonal arc length* of the function f on $[a, b]$, defined by

$$\sum_{i=1}^{i=n} \sqrt{((x_i - x_{i-1})^2 + (f(x_i) - f(x_{i-1}))^2)}$$

can only increase if an additional point, or finite set of points, is added to the subdivision. Use qn 2.56.

44 Use qn 40 to show that any polygonal arc length for all or part of the function of qn 39 is less than or equal to 4.
Deduce that the polygonal arc length of this function has a supremum. The supremum for polygonal arc length on the interval $[a, b]$ is called the *arc length* of the function on $[a, b]$.

45 By applying qn 42, show that any polygonal arc length on the interval $[a, b]$ for the function of qn 39 has the value

$$\sum_{i=1}^{i=n} \frac{x_i - x_{i-1}}{\sqrt{(1 - c_i^2)}},$$

with x_i as in qn 43, for some c_is, with $x_{i-1} \leqslant c_i \leqslant x_i$.

46 If we define the function $g: (-1, 1) \to \mathbb{R}$ by

$$g(x) = \frac{1}{\sqrt{(1 - x^2)}}$$

show that any polygonal arc length on the interval $[a, b]$ for the

function f of qn 39 is greater than a lower sum for the function g on the interval $[a, b]$, and less than an upper sum for this function provided $-1 < a < b < 1$. Use qn 10.42.
Deduce that the lower integral $\int_a^b g \leq$ arc length on $[a, b]$.

47 Let $-1 < a = x_0 < x_1 < x_2 < \ldots < x_n = b < 1$ and
$a = y_0 < y_1 < y_2 < \ldots < y_m = b$ be two subdivisions of the interval $[a, b]$, and let the union of these two subdivisions be
$a = z_0 < z_1 < z_2 < \ldots < z_l = b$.
Use qn 43 to show that
the polygonal arc length of f with the 'x' subdivision
$\quad\quad \leq$ the polygonal arc length of f with the 'z' subdivision.

Explain why
the upper sum for g on the 'y' subdivision
$\quad\quad \geq$ the upper sum for g on the 'z' subdivision.
Deduce from qn 46 that
any polygonal arc length of f on $[a, b]$
$\quad\quad \leq$ any upper sum for g on $[a, b]$.

Deduce that the arc length on $[a, b] \leq$ the upper integral $\overline{\int}_a^b g$.

48 From qns 46 and 47 and the definition of the Riemann integral we know that the arc length of f on $[a, b]$ is equal to $\int_a^b g$.
Use qn 10.66 and 10.68 to show that the integral exists as an improper integral even when $a = -1$ or $b = 1$ or both.

Arc cosine

We now *define* an angle function (or arc length function) $A: [-1, 1] \to \mathbb{R}$ by

$$A(y) = \int_y^1 \frac{dx}{\sqrt{(1 - x^2)}}.$$

49 Say why the function A is
(i) continuous,
(ii) monotonic decreasing,
(iii) differentiable on $(-1, 1)$ with $A'(y) = \dfrac{-1}{\sqrt{(1 - y^2)}}$.

50 Say why

(i) $A(1) = 0$,
(ii) $A(-1) = 2A(0)$.
(iii) A is a bijection $[-1, 1] \to [0, A(-1)]$.

The function A is usually called arccos, and sometimes $\cos^{-1}$. We may now *define* $A(-1) = \pi$ and A^{-1} as the cosine function, so that cosine is a continuous and monotonic decreasing bijection with domain $[0, \pi]$ and range $[-1, 1]$. On the same domain we define the sine function by

$$\sin x = \sqrt{(1 - \cos^2 x)}.$$

Cosine and sine

51 Find the values of $\cos x$ and $\sin x$ when $x = 0, \frac{1}{2}\pi$ and π.

52 Use the equation $A(\cos x) = x$ to prove that $\cos' x = -\sin x$ for $0 < x < \pi$. Use the definition of sine to prove that $\sin' x = \cos x$ for $0 < x < \pi$.

53 Sketch the graphs of cosine and sine on the domain $[0, \pi]$. For $\pi < x \le 2\pi$, *define* $\cos x = \cos(2\pi - x)$ and $\sin x = -\sin(2\pi - x)$. Sketch the graphs of cosine and sine on the domain $[0, 2\pi]$. Verify that $\sin^2 x + \cos^2 x = 1$ on $[0, 2\pi]$, and that $\cos' x = -\sin x$ on $(\pi, 2\pi)$.

54 Prove that $\cos' \pi = -\sin \pi = 0$, by applying the Mean Value Theorem to $(\cos x - \cos \pi)/(x - \pi)$, so that $\cos' x = -\sin x$ on $(0, 2\pi)$. Prove likewise that $\sin' \pi = \cos \pi = -1$.

For any integer k, we now *define* sine and cosine for $2k\pi \le x < 2(k + 1)\pi$ by

$$\cos x = \cos(x - 2k\pi)$$
$$\text{and } \sin x = \sin(x - 2k\pi).$$

55 Prove that $\sin^2 x + \cos^2 x = 1$ for all real x.
Prove that $\cos' x = -\sin x$ and $\sin' x = \cos x$, except possibly when $x = 2k\pi$.
Use the method of qn 54 to prove that $\cos' 0 = -\sin 0 = 0$ and $\sin' 0 = \cos 0 = 1$, so that the formulae for the derived functions hold for all values of x.

56 Define $f(x) = \sin(a - x) \cdot \cos x + \cos(a - x) \cdot \sin x$. Prove that $f'(x) = 0$ for all x. Use the Mean Value Theorem to prove that $f(x) = \sin a$. Deduce the formula for $\sin(x + y)$.

57 Use the method of qn 56 to prove that
$\cos(x + y) = \cos x \cdot \cos y - \sin x \cdot \sin y$.

58 Prove that $\cos 2x = \cos^2 x - \sin^2 x$

$$= 2\cos^2 x - 1$$

$$= 1 - 2\sin^2 x.$$

Prove also that $\sin 2x = 2\sin x \cdot \cos x$.

Tangent

59 Define $\tan x = \sin x/\cos x$, except where $x = (k + \frac{1}{2})\pi$, and prove that $\tan' x = 1/\cos^2 x = \tan^2 x + 1$. Show that the tangent function is a monotonic increasing bijection $(-\frac{1}{2}\pi, \frac{1}{2}\pi) \to \mathbb{R}$.

60 Define arctan: $\mathbb{R} \to (-\frac{1}{2}\pi, \frac{1}{2}\pi)$ as the function inverse to tan, so that $\arctan(\tan x) = x$. Explain why arctan is differentiable and why $\arctan' x = 1/(1 + x^2)$.

61 Use the equation

$$1 + (-t^2) + (-t^2)^2 + \ldots + (-t^2)^{n-1} = \frac{1 - (-t^2)^n}{1 - (-t^2)},$$

which is valid for all t, and the Fundamental Theorem of Calculus, to show that

$$\arctan x = x - \frac{x^3}{3} + \frac{x^5}{5} - \ldots$$

$$+ (-1)^{n-1} \frac{x^{2n-1}}{2n - 1} + (-1)^n \int_0^x \frac{t^{2n}}{1 + t^2} \, dt.$$

Prove that

$$\left| (-1)^n \int_0^x \frac{t^{2n}}{1 + t^2} \, dt \right| \leqslant \int_0^x |t|^{2n} \, dt = \frac{|x|^{2n+1}}{2n + 1}.$$

Check that this last expression gives a null sequence when $|x| \leqslant 1$ and determine for what values of x the power series developed here is a valid expansion of $\arctan x$.

62 Prove that

$$\tfrac{1}{4}\pi = 1 - \frac{1}{3} + \frac{1}{5} - \frac{1}{7} + \ldots + (-1)^{n-1} \frac{1}{2n - 1} + \ldots$$

Summary

Exponential and logarithmic functions

Definitions	$x^1 = x$ and $x^{n+1} = x^n x$ for $n \in \mathbb{N}$.
Theorem	$f: \mathbb{R}^+ \to \mathbb{R}^+$ given by $f(x) = x^n$ is a continuous
qns 3, 4	bijection and strictly increasing.
	$f^{-1}: \mathbb{R}^+ \to \mathbb{R}^+$ given by $f^{-1}(x) = \sqrt[n]{x} = x^{1/n}$ is a
	continuous bijection and strictly increasing.
Definition	$x^{n/m} = (x^n)^{1/m}$, for $m, n \in \mathbb{N}$.
Theorem	$f: \mathbb{R}^+ \to \mathbb{R}^+$ given by $f(x) = x^q$ for $q \in \mathbb{Q}^+$ is a
qn 7	continuous bijection and strictly increasing.
Definition	When $x \neq 0$, $x^0 = 1$ and $x^{-q} = 1/x^q$ for $q \in \mathbb{Q}$.
Theorem	When $1 < a$, $A: \mathbb{Q} \to \mathbb{R}^+$ given by $A(x) = a^x$ is
qns 10, 11	continuous and strictly increasing.
	When $0 < a < 1$, A is strictly decreasing.
Theorem	$D: \mathbb{Q}\backslash\{0\} \to \mathbb{R}$ given by
qns 12, 13	$$D(x) = \frac{a^x - 1}{x}$$

is continuous.

If $1 < a$, D is increasing on $\mathbb{Q}^+$.

If $0 < a < 1$, D is increasing on negative $\mathbb{Q}$.

Theorem	If $c > 0$, $A: x \mapsto a^x$ is uniformly continuous on
qns 17, 18	$[0, c] \cap \mathbb{Q}$ and so may be extended in a unique way
	to a continuous function on $[0, c]$ and so on $[0, \infty)$.
Theorem	$A: x \mapsto a^x$ may be extended in a unique way to a
qn 19	continuous bijection $\mathbb{R} \to \mathbb{R}^+$, which is increasing
	when $1 < a$ and decreasing when $0 < a < 1$.
Theorem	$a^{x+y} = a^x a^y$ and $a^{xy} = (a^x)^y$,
qn 20	when $0 < a$ and $x, y \in \mathbb{R}$.
Definition	If $0 < a$ and $a^x = y$, then $\log_a y = x$.
Theorem	For $x, y > 0$, $\log_a x + \log_a y = \log_a xy$.
qn 22	
Theorem	$D: x \mapsto (a^x - 1)/x$ may be extended in a unique way
qns 14, 15,	to a continuous function on $\mathbb{R}$.
16, 23	
Theorem	$D(x) \to \int_1^a dy/y$ as $x \to 0$.
qns 14, 15,	
25	
Theorem	If $0 < a$ and $A(x) = a^x$ then $A'(x) = A(x) \cdot \int_1^a dy/y$.
qns 24, 25	
Definition	$\int_1^e dx/x = 1$, $E(x) = e^x = \exp(x)$.

Theorems $E'(x) = E(x),\ E(\int_1^a dx/x) = a,\ \int_1^a dx/x = \log_e a.$
qns 28, 29, 30
Definition $\log x = \log_e x$
Theorem When $a > 0$, $a^x = e^{x \log a}$; $\log a^x = x \log a.$
qn 31
Theorem $\lim\limits_{n\to\infty} \left(1 + \dfrac{a}{n}\right)^n = e^a.$
qn 32

Circular or trigonometric functions

Theorem The arc length on the unit circle from $(1, 0)$ to
qn 48 $(x, \sqrt{(1 - x^2)})$, $-1 \leqslant x \leqslant 1,$

$$A(x) = \int_x^1 \frac{dy}{\sqrt{(1 - y^2)}}.$$

Theorem The function $A : [-1, 1] \to [0, A(-1)]$ is continuous,
qn 49 decreasing and

$$A'(x) = \frac{-1}{\sqrt{(1 - x^2)}} \text{ on } (-1, 1).$$

$A(-1) = 2A(0).$
Definition $A(-1) = \pi.\ A^{-1} = \text{cosine}.$
Theorem $\cos : [0, \pi] \to [-1, 1]$
qn 51 $\cos 0 = 1,\ \cos \tfrac{1}{2}\pi = 0,\ \cos \pi = -1.$
Definition $\sin x = \sqrt{(1 - \cos^2 x)}$ on $[0, \pi].$
$\cos x = \cos(2\pi - x)$ on $(\pi, 2\pi].$
$\sin x = -\sin(2\pi - x)$ on $(\pi, 2\pi].$
$\cos x = \cos(x - 2k\pi)$ on $[2k\pi, 2(k + 1)\pi].$
$\sin x = \sin(x - 2k\pi)$ on $[2k\pi, 2(k + 1)\pi].$
Theorem $\sin 0 = 0,\ \sin \tfrac{1}{2}\pi = 1,\ \sin \pi = 0.$
qns 52, 55, $\cos' x = -\sin x,\ \sin' x = \cos x.$
56, 57 $\sin^2 x + \cos^2 x = 1.$
$\sin(x + y) = \sin x \cdot \cos y + \cos x \cdot \sin y.$
$\cos(x + y) = \cos x \cdot \cos y - \sin x \cdot \sin y.$
Definition $\tan x = \sin x/\cos x$ provided $x \neq (k + \tfrac{1}{2})\pi.$
Theorem $\tan' x = 1 + \tan^2 x.$
qn 59 $\tan : (-\tfrac{1}{2}\pi, \tfrac{1}{2}\pi) \to \mathbb{R}$ is a continuous bijection and
strictly increasing.

Historical note

In 1614, John Napier published tables matching n with
$10^7(1 - 10^{-7})^n$ for $n = 0, 1, \ldots, 100$ and used these to interpolate in

other similar tables to seven significant figures. Napier called n the *logarithm* of its matching number. Such logarithms do not exhibit the additive property of our logarithms but they provide the advantages of standard logarithms for the purpose of division. Napier's achievement is the more remarkable when one realises that the modern notation for exponents was developed some twenty years later. Briggs had extensive discussions with Napier and then used an algebraic transformation of Napier's tables to construct tables in which log 1 was 0 and log 10 was 1. Although these were in fact common logarithms to the base 10, it was not recognised at that time that they were exponents. These tables were published in 1619.

In 1647, Gregory of St Vincent showed that the additive property of logarithms applied to areas under the rectangular hyperbola $xy = 1$.

In his *De Analysi* (written in 1669 and circulated among friends, but not published until 1734) Newton had used term-by-term integration of the series for $1/(1 + x)$ (which he obtained by long division) to find the series for $\log(1 + x)$, and then developed a method of repeated successive approximations to calculate the inverse function of a power series. This enabled him to construct the series for $\exp(x) - 1$. Newton did not use the terminology of logarithms or exponentials. In 1668 Mercator published the series for $\log(1 + x)$ in his *Logarithmotechnia* and referred to areas under the rectangular hyperbola as *natural* logarithms. Leibniz described the integral $\int dx/x$ as a logarithm in 1676.

Negative and fractional exponents were first proposed by Wallis in 1656 and first used by Newton in his discussion of the Binomial Theorem in 1669. The first overt claim of the connection between exponents and logarithms was made by Wallis in 1685 in relation to the terms of a geometric progression.

The first claim that the common logarithm of x was the power to which 10 had to be raised to obtain x was made in tables published in England in 1742. In Euler's *Introductio in Analysin Infinitorum* published in 1748, he explicitly defined $\log_a x = y$ where $a^y = x$. He used infinitesimals and infinite numbers to great effect. For an infinitesimal ε, he let $(a^\varepsilon - 1)/\varepsilon = k$ (which depends on a) and called the value of a which makes $k = 1$, e (the first letter of his name). By combining infinitesimal and infinite numbers $\varepsilon N = x$, he obtained a^x as the exponential series $\exp(kx)$. The conventional exponential series follows. The original equation giving k, was rearranged to give $(1 + x/N)^N = e^x$. Further ingenious manipulation led to the result that

$$\log_a x = (N/k)(x^{1/N} - 1),$$

and he then used the binomial series to obtain the series for $\log(1 + x)$.

In 1821, Cauchy, presuming that a^x was well defined, proved that every continuous function f satisfying

$$f(x) \cdot f(y) = f(x + y)$$

was of the form $f(x) = a^x$ for some a. He used the Cauchy product to show that the exponential power series belonged to this family of functions, and defining $e = \sum 1/n!$ had thereby proved that the exponential power series was equal to e^x. Cauchy also used the Binomial Theorem to prove that, as $a \to 0$, $(1 + ax)^{1/a} \to e^x$. In 1881, A. Harnack defined a^x for rational x and extended his definition to irrational x by affirming the continuity of the new function.

In order to extend exponentials to complex numbers, Euler (1748) had defined $\exp(z)$ by means of the exponential series $1 + z + z^2/2! + \ldots$ and in the late nineteenth century it was recognised that, if this definition was used for a real variable, a formal definition of a^x as $\exp(x \log a)$ avoids some of the difficulties with which Cauchy had struggled.

Newton's discovery of the sine series depended on his previous discovery of the Binomial Theorem for rational index. He derived the series for arcsine from the term-by-term integration of the power series for $\sqrt{(1 - x^2)}$ and then used his repeated approximation method to construct the series for $\sin x$. This method is based on the formula $\frac{1}{2}r^2\theta$ for the area of a circular sector and appears in Newton's notes from 1666. In *De Analysi* Newton's method for these series was based on a calculation of arc length, which is closely related to the treatment in this chapter. Again, it was Euler who recognised that an extension of the trigonometric functions to complex variables required a formal definition of these functions by series, and he obtained the equation $e^{ix} = \cos x + i \sin x$ in 1748.

In 1821, Cauchy proved that a continuous solution of the functional equation $f(x + y) + f(x - y) = 2f(x)f(y)$ must take the form $\cos ax$ or $\cosh ax$ with the cosine occurring when $f(x) < 1$ for some x.

In the late nineteenth century it was recognised that, if Weierstrass' arithmetisation of analysis was to be carried through, a non-geometric definition of these functions was required for real variables. In 1880 Thomae gave an analytic definition of cosine using Cauchy's equation above. The series definition (as for complex variables) was also available for this purpose, and the alternative definition with an integral (which we have used) was commended by Felix Klein in 1908.

Answers

1   (i) The definition gives the basis for the induction.
Then $x^{m+n+1} = x^{m+n}x = x^m x^n x = x^m x^{n+1}$.
(ii) The definition gives the basis for the induction.
$x^{m(n+1)} = x^{mn+m} = x^{mn}x^m = (x^m)^n x^m = (x^m)^{(n+1)}$.
(iii) $(xy)^{n+1} = (xy)^n(xy) = x^n y^n xy = x^n xy^n y = x^{n+1}y^{n+1}$.

2   $m < n \Leftrightarrow 0 < n - m \Leftrightarrow 1 < x^{n-m} \Leftrightarrow x^m < x^n \Leftrightarrow 1/x^n < 1/x^m$.

3   See qn 2.35 for strict increase, qn 6.28 for continuity. $1 < x \leqslant x^n$ shows unboundedness above.

4   See qn 7.27.

5   (i) Use 1(iii) and the fact that $x \mapsto x^n$ is a bijection.
(ii) $(x^{1/m})^{(n+1)} = (x^{1/m})^n x^{1/m} = (x^{n/m})x^{1/m}$
$= (x^n)^{1/m}x^{1/m} = (x^n x)^{1/m} = (x^{n+1})^{1/m}$.

6   Let $r = p/m$ and $s = q/n$, where $m, n, p, q \in \mathbb{Z}^+$.
(i) $x^{r+s} = x^{(pn+qm)/mn} = (x^{1/mn})^{(pn+qm)} = (x^{1/mn})^{pn}(x^{1/mn})^{qm}$
$= x^{pn/mn}x^{qm/mn} = x^r x^s$
(ii) $x^{rs} = x^{pq/mn} = (x^{pq})^{1/mn} = (x^{p/m})^{q/n}$, using qn 5(ii) and 5(iii).

7   $r < s \Leftrightarrow 0 < s - r \Leftrightarrow 1 < x^{s-r} \Leftrightarrow x^r < x^s \Leftrightarrow 1/x^s < 1/x^r$.

8   The argument of qn 6 applies. For negative r and s, work with reciprocals.

9   The argument of qn 7 applies.

10   $A(-x) = 1/A(x)$. For some N, $m > N \Rightarrow |A(\pm 1/m) - 1| < \varepsilon$, and because A is strictly increasing, $-1/(N + 1) < x < 1/(N + 1) \Rightarrow |A(x) - 1| < \varepsilon$. Thus $A(x) \to 1 = A(0)$ as $x \to 0$.

11   $x \to q \Rightarrow x - q \to 0 \Rightarrow A(x - q) \to 1 \Rightarrow A(x - q) - 1 \to 0$
$\Rightarrow A(x) - A(q) \to 0 \Rightarrow A(x) \to A(q)$.

12   By qns 6.23 and 6.54.

13   Let $r = p/m$ and $s = q/n$, where $m, n, p, q \in \mathbb{Z}^+$.
$r < s \Rightarrow pn < qm \Rightarrow D(pn) < D(qm)$. Now put $b = a^{mn}$, and we get $D(r) < D(s)$ but with b for a in the definition of D.

14   If $m > N \Rightarrow |D(1/m) - L(a)| < \varepsilon$, then because D is strictly increasing on $\mathbb{Q}^+$, $0 < x < 1/(N + 1) \Rightarrow |D(x) - L(a)| < \varepsilon$.

15 Apply qn 6.93 from above.

16 $D(0) = L(a)$ by qn 6.89.

17 Expression $= A(y)D(x - y) < A(c) \cdot D(c) = M$, since both A and D are strictly increasing on $\mathbb{Q}^+ \cup \{0\}$.

18 A is uniformly continuous on $\mathbb{Q} \cap [0, c]$ by qn 7.42 and therefore may be extended to a continuous function in a unique way by qn 7.47. This extension may be made at any point $x > 0$ by selecting $c > x$.

19 $A(-x) = 1/A(x)$. Since $A(x) > 0$ for all x, the continuity of A at x is equivalent to the continuity of A at $-x$ from qn. 6.52.

20 If x and y are rational numbers, these results were obtained in qn 8. Let (x_n) be a sequence of rational numbers tending to x (possibly irrational) and let (y_n) be a sequence of rational numbers tending to y (also possibly irrational), then $(x_n + y_n) \to x + y$ and $(x_n y_n) \to xy$ by qn 3.44. By the continuity of A, $(A(x_n)) \to A(x)$, $(A(y_n)) \to A(y)$, $(A(x_n + y_n)) \to A(x + y)$, $(A(xy_n)) \to A(xy)$ and, by qn 8, $A(x_n + y_n) = A(x_n) \cdot A(y_n)$, so by qn 3.44(iv), $A(x + y) = A(x) \cdot A(y)$. So we have the first law. For the second law, first establish that it holds for $y = n \in \mathbb{N}$ by induction. Then establish that it holds for $y = 1/n$ using the bijection of qn 4. Then establish that it holds for $y = m/n$, a positive rational, using the two previous cases and extend this to negative rational y from the definitions. We now have the equipment to claim that $A(xy_n) = (A(x))^{y_n}$. Now $(A(xy_n)) \to A(xy)$ as we saw before and $((A(x))^{y_n}) \to (A(x))^y$, by the continuity of A, substituting a^x for a. So $A(xy) = (A(x))^y$.

21 We know that A is strictly increasing on $\mathbb{Q}$ and that A is continuous on $\mathbb{R}$. Now use qns 7.5 and 7.7 to show that, for irrational r,

$$A(r) = \sup \{A(q) | q < r, q \in \mathbb{Q}\} = \inf \{A(q) | q > r, q \in \mathbb{Q}\},$$

(it is necessary to consider rational sequences tending to r from above and below), so that, if $p, q \in \mathbb{Q}$ and $p < r < q$, then $A(p) < A(r) < A(q)$. Now, for any two distinct real numbers r and s, with $r < s$, there exists a rational number q such that $r < q < s$, and so $A(r) < A(q) < A(s)$, and A is strictly increasing on $\mathbb{R}$.

22 Since A is a bijection $\mathbb{R} \to \mathbb{R}^+$, there is a unique x such that $A(x) = X$ and a unique y such that $A(y) = Y$. From qn 20(i), $A(x + y) = XY$, so $A^{-1}(XY) = x + y$, and $\log_a XY = \log_a X + \log_a Y$.

23 (iii) $0 < \frac{1}{2}q_n < x_n < q_n$, and D strictly increasing, implies
$D(\frac{1}{2}q_n) < D(x_n) < D(q_n)$. From qn 14, $(D(\frac{1}{2}q_n)) \to L(a)$ and
$(D(q_n)) \to L(a)$. The result comes from qn 3.44(vi).
(v) From (i) and qn 6.89.

24 $(A(x') - A(x))/(x' - x) = A(x) \cdot D(x' - x) \to A(x) \cdot L(a)$ as $x' \to x$.

25 Both $(D(1/n))$ and $(D(-1/n)) \to L(a)$ as $n \to \infty$.

26 $0 < L(a) < a - 1$, so $L(a) \to 0$ as $a \to 1^+$.

For $a \geqslant 2$, $L(a) > \displaystyle\sum_{n=2}^{n=[a]} \frac{1}{n} \to \infty$ as $a \to \infty$, by qn 5.27.

L is strictly increasing because $x > 0 \Rightarrow 1/x > 0$.
L is continuous by qn 10.53. $L'(x) = 1/x$ by the Fundamental Theorem
of Calculus.

27 Since L is continuous and takes values from small positive to large
positive on $(1, \infty)$, L takes the value 1 by the Intermediate Value
Theorem for some value of a in this domain. On $(1, 2)$, $1/x < 1$, so
$L(2) < 1$.

28 Follows from $L(e) = 1$ and qn 24. Taylor series from qn 9.38.

29 $f'(x) = L'(E(x)) \cdot E'(x) = (1/E(x)) \cdot E(x) = 1$. Thus $f(x) = x + c$. But
$f(0) = 0$, so $c = 0$.

30 Since $E = L^{-1}$, $E(L(x)) = x$, and $L(x) = \log_e x$.

31 (i) First equality by definition of log. Second equality from second law
of indices.
(ii) The logarithm of the equality in (i). Use qn 29.
(iii) $E(\log b) = b$. $E(\log_a b \cdot \log a) = A(\log_a b)$ by (i), $= b$ by definition.

32 The function is well defined provided $|x| < |1/a|$.

Limit $= \displaystyle\lim_{x \to 0^+} \frac{a}{1 + ax} = a$. Then consider $x = 1/n$.

34 $f(x) = E(a \log x)$. $f'(x) = E'(a \log x) \cdot (a/x) = f(x)(a/x)$.

35 Let $f(x) = x^a e^{-x}$. $f(n + 1)/f(n) = (1 + 1/n)^a/e \to 1/e$ as $n \to \infty$. By qn
3.59, $(f(n)) \to 0$ as $n \to \infty$.
$f'(x) = f(x)(a/x - 1) < 0$ when $x > a$. So f is strictly decreasing when
$x > a$. Since the sequence $\to 0$, the function $\to 0$ as $x \to \infty$. So
$1/f(x) \to \infty$.

Let $g(x) = (\log x)/x^a$, then $g'(x) = \dfrac{1 - a \log x}{x^{a+1}} < 0$ when $x > e^{1/a}$.

Let $h(n) = g(e^n)$, then $h(n + 1)/h(n) = (1 + 1/n)/e^a \to 1/e^a$ as $n \to \infty$. So, by qn 3.59, $(h(n)) \to 0$ and since g is strictly decreasing, $g(x) \to 0$ as $x \to \infty$.

36 $f(x) = E(x \log x)$. f is continuous since log and E are continuous. $f'(x) = E'(x \log x) \cdot (\log x + x \cdot 1/x) = 0$ when $\log x = -1$ or $x = 1/e$. On $(0, 1/e)$, $f'(x) < 0$. $f(1/n) = \sqrt[n]{(1/n)} = 1/\sqrt[n]{n} \to 1$ by qn 4.41. Since f is strictly decreasing on this domain, $f(x) \to 1$ as $x \to 0^+$.

37 See qn 9.38.

38 Use qn 10.62 (integration by substitution) to show that

$$\int_1^{1+x} \frac{dt}{t} = \int_0^x \frac{dt}{1 + t}.$$

$$\int_0^x 1 + (-t) + \ldots + (-t)^{n-1} dt = \int_0^x \frac{dt}{1 - (-t)} - \int_0^x \frac{(-t)^n dt}{1 - (-t)}.$$

$-K/(K + 1) < t \Rightarrow 1/(K + 1) < t + 1 \Rightarrow 1/(t + 1) < K + 1$

$\Rightarrow |t|^n/(t + 1) < |t|^n (K + 1)$.

Also $\dfrac{(K + 1)|x|^n}{n + 1} \leqslant \dfrac{K + 1}{n + 1}$ if $|x| \leqslant 1$.

But K is constant, so $((K + 1)/(n + 1)) \to 0$ as $n \to \infty$.
Expansion is valid when $-1 < x \leqslant 1$.

39 $f(x) = \sqrt{(1 - x^2)}$.

40 Squaring both sides of the inequality makes the comparison trivial.
$\sqrt{((a - b)^2 + (f(a) - f(b))^2)}$.

41 f is continuous by qns 6.29, 7.27 and 6.40.
$f'(x) = -x/\sqrt{(1 - x^2)}$ on $(-1, 1)$.

42 $f(b) - f(a) = f'(c) \cdot (b - a)$, so

$$\sqrt{((b - a)^2 + (f(b) - f(a))^2)} = (b - a)\sqrt{(1 + (f'(c))^2)}$$

$$= \frac{b - a}{\sqrt{(1 - c^2)}}.$$

43 Adding a point to the subdivision replaces one part of the polygonal arc length by two new ones using the original end points, this increases the polygonal arc length by the triangle inequality.

44 On $[-1, 0]$ f is strictly increasing, so the polygonal arc length
$\leqslant \sum |x_i - x_{i-1}| + \sum |f(x_i) - f(x_{i-1})| \leqslant 1 + 1$.
Likewise on $[0, 1]$ f is strictly decreasing and again the polygonal arc

length is bounded above by 2. So all polygonal arc lengths, by qn 43, are bounded above by 4.

46 The function g is bounded on $[a, b]$ and so integrable. Working on the subdivision $-1 < a = x_0 < x_1 < x_2 < \ldots < x_n = b < 1$ and letting

$$m_i = \inf \{g(x)|\ x_{i-1} \leqslant x \leqslant x_i\},\ \text{and}$$

$$M_i = \sup \{g(x)|\ x_{i-1} \leqslant x \leqslant x_i\},$$

from qn 45, $\sum m_i(x_i - x_{i-1}) \leqslant$ polygonal arc length of f for this subdivision $\leqslant \sum M_i(x_i - x_{i-1})$.
So sup (lower sums) $\leqslant$ sup (polygonal arc lengths) = arc length.
So lower integral of g on $[a, b] \leqslant$ arc length of f on $[a, b]$.

47 The z-subdivision is the same as the x-subdivision but with some additional intervening points. The additional points increase the polygonal arc length by qn 43.
The z-subdivision is the same as the y-subdivision but with some additional intervening points, so if $y_{j-1} = z_k$, then $y_j = z_{k+p}$ for some p, so $[z_{s-1}, z_s] \subseteq [y_{j-1}, y_j]$ for $s = k + 1, \ldots, k + p$.
So $\sup \{g(x)|\ z_{s-1} \leqslant x \leqslant z_s\} \leqslant \sup \{g(x)|\ y_{j-1} \leqslant x \leqslant y_j\}$
for $s = k + 1, \ldots, k + p$.
Polygonal arc length on x-subdivision
 $\leqslant$ polygonal arc length on z-subdivision
 $\leqslant$ upper sum on z-subdivision (using qn 45)
 $\leqslant$ upper sum on y-subdivision.
So every polygonal arc length $\leqslant$ every upper sum.
So sup (polygonal arc lengths) $\leqslant$ every upper sum.
So arc length is a lower bound to the upper sums.
So arc length $\leqslant$ inf (upper sums) = upper integral.

48 Lower integral $\leqslant$ arc length $\leqslant$ upper integral (from qns 46 and 47). But g is bounded and continuous on $[a, b]$, so g is integrable on $[a, b]$, and its upper integral and lower integral are equal. So arc length = integral.

49 (i) A is continuous by qn 10.53.
(ii) If $-1 \leqslant t < s \leqslant 1$, then $A(s) - A(t) = \int_s^1 - \int_t^1 = \int_s^t = -\int_t^s < 0$.
(iii) A is differentiable by the Fundamental Theorem, qn 10.57.

50 (i) By qn 10.51.
(ii) $A(0) = \displaystyle\int_0^1 \frac{dx}{\sqrt{(1 - x^2)}}$.

Putting $u = -x$,

$$\int_{-1}^0 \frac{dx}{\sqrt{(1 - x^2)}} = \int_1^0 \frac{-du}{\sqrt{(1 - u^2)}} = A(0).$$

(iii) *A* is monotonic decreasing and so a bijection. Domain is given. The range is bounded by the values of the function at the extremities of the domain.

51 $A(1) = 0$, so $\cos 0 = 1$ and $\sin 0 = 0$.
$A(-1) = \pi$, so $\cos \pi = -1$ and $\sin \pi = 0$.
$A(0) = \frac{1}{2}A(-1) = \frac{1}{2}\pi$, so $\cos \frac{1}{2}\pi = 0$ and $\sin \frac{1}{2}\pi = 1$.

52 $A(\cos x) = x$ and cos is differentiable by 8.42.
By qn 8.40, $A'(\cos x) \cdot \cos' x = 1$, so $\left(-1/\sqrt{(1 - \cos^2 x)}\right) \cdot \cos' x = 1$.
$\cos' x = -\sin x$.

$$\sin x = \sqrt{(1 - \cos^2 x)}, \text{ so } \sin' x = \frac{\frac{1}{2}(-2\cos x)(-\sin x)}{\sqrt{(1 - \cos^2 x)}} = \frac{\cos x \cdot \sin x}{\sin x}.$$

54 The function cos is continuous on $[0, \pi]$ and differentiable on $(0, \pi)$ so we may apply the Mean Value Theorem and get

$$\frac{\cos x - \cos \pi}{x - \pi} = \cos' c = -\sin c \quad \text{for some } c, \; x < c < \pi.$$

As $x \to \pi$, $c \to \pi$, so $-\sin c \to 0$ and $\cos' \pi = 0$.
So we have $\cos' x = -\sin x$ on $(0, 2\pi)$.
Similarly $\sin' \pi = \cos \pi = -1$.

56 $f'(x) = -\cos(a - x)\cos x - \sin(a - x)\sin x$
$\qquad\qquad + \sin(a - x)\sin x + \cos(a - x)\cos x$
$\qquad = 0.$

From qn 9.17, f is constant. But $f(0) = \sin a$. Now put $x + y = a$ to obtain $\sin(x + y) = \sin x \cdot \cos y + \cos x \cdot \sin y$.

58 Apply qns 56 and 57 putting $y = x$, and use $\sin^2 x + \cos^2 x = 1$.

59 $\tan' x = \dfrac{\cos x \cdot \sin' x - \sin x \cdot \cos' x}{\cos^2 x} = \dfrac{\cos^2 x + \sin^2 x}{\cos^2 x} = 1 + \tan^2 x.$

The tan function is continuous on $(-\frac{1}{2}\pi, \frac{1}{2}\pi)$ and has positive derivative and is therefore strictly increasing, and so a bijection.
As $x \to \frac{1}{2}\pi^-$, $\cos x \to 0$, so $1/\cos x \to +\infty$; also $\sin x \to 1$, so $\tan x \to \infty$.
Similarly as $x \to -\frac{1}{2}\pi^+$, $\tan x \to -\infty$.

60 arctan is differentiable by qn 8.42.

$\arctan(\tan x) = x \Rightarrow \arctan'(\tan x)\tan' x = 1 \Rightarrow$

$$\arctan'(\tan x) = \frac{1}{1 + \tan^2 x}$$

61 Power series valid for $-1 \leqslant x \leqslant 1$.

62 Put $x = 1$ in the series.
Let $\arctan 1 = X$, then $\tan X = 1$, so $\sin X = \cos X$ and $\cos 2X = 0$ from qn 58. Thus $2X = \frac{1}{2}\pi$ and $X = \frac{1}{4}\pi$.

12

Sequences of functions

For this chapter, more than any other, the availability of a computer-assisted graph drawing package is essential, and the preliminary exercise (i)–(x) is dependent on this.

Concurrent reading: Burkill and Burkill ch. 5, Spivak ch. 23.
Further reading: Rudin ch. 7.

Sequences of functions are often considered in a second course of analysis. The difficulties encountered are not like those in earlier analysis since the apparatus of limits has already been developed. Difficulties arise from having two variables of unsymmetric status and having to think of keeping one constant and letting the other vary according to context. Before starting the questions in this chapter it is necessary to have some intuitive feel for the subject matter, and to have observed, in graphical form, how a sequence of functions may behave. To do this, use graph drawing facilities on a computer to illustrate the functions f_n for $n = 1, 2, 5,$ and 10, where

(i) $f_n(x) = x + 1/n$,

(ii) $f_n(x) = x/n$,

(iii) $f_n(x) = \dfrac{nx}{x + n}$ for $x \in \mathbb{R}^+$,

(vi) $f_n(x) = \dfrac{1}{x^2 + n^2}$,

(v) $f_n(x) = \dfrac{1}{1 + n^2 x^2}$,

(vi) $f_n(x) = \dfrac{nx}{1 + n^2 x^2}$,

(vii) $f_n(x) = \dfrac{x}{1 + n^2 x^2}$,

(viii) $f_n(x) = \dfrac{1}{(1 + x^2)^n}$,

(ix) $f_n(x) = x^n$

(x) $f_n(x) = \dfrac{x^n}{1 + x^{2n}}$,

The first thing to look for in these graphs, are the values of x for which the sequence of real numbers $(f_n(x))$ is convergent. One does this by thinking of x as constant and letting $n \to \infty$. Having obtained a limit function f in this way, one then compares the graphs of f_n and f as a whole, looking for their greatest distance apart and thinking of n as constant.

Pointwise limit functions

1 Sketch the graphs of the functions given by

$$f_1(x) = \frac{1}{1 + x^2}, \quad f_2(x) = \frac{1}{4 + x^2}, \quad \text{and } f_3(x) = \frac{1}{9 + x^2},$$

superimposing the diagrams on the same axes, using appropriate computer software.

For a constant real number x, find $\displaystyle\lim_{n\to\infty} \frac{1}{n^2 + x^2}$.

Because this limit is well defined for each real number x, we have a *limit function* $f: \mathbb{R} \to \mathbb{R}$ for the sequence of functions (f_n) defined by

$$f_n(x) = \frac{1}{n^2 + x^2}.$$

The limit function in this case is given by $f(x) = 0$, for all x.

2 Sketch the graphs of the functions given by

$$f_n(x) = (\sin x)/n, \quad \text{for } n = 1, 2 \text{ and } 3,$$

using computer software.

Is the limit $\lim_{n\to\infty} \sin x/n$ well defined, for each real number x?

Since $\lim_{n \to \infty} f_n(x) = 0$ for all x in this case, the function f defined by $f(x) = 0$ is the limit function for the sequence (f_n).

In qns 1 and 2, the limit functions were constant giving the same value for each x. In qns 3, 4 and 5, the values of the limit are not the same for all values of x.

3 If $f_n(x) = \dfrac{1}{1 + n^2 x^2}$,

find the limit function for the sequence (f_n) on the domain $\mathbb{R}$.

4 If $f_n(x) = x^n$, find the limit function for the sequence (f_n) on the domain $[0, 1]$.

5 If

$$f_n(x) = \frac{x^n - 1}{x^n + 1},$$

find the limit function for the sequence (f_n) on the domain $[0, \infty)$.

Because these limit functions are determined point by point for each x of the domain of the functions, they are called *pointwise limit functions*.

A formal definition runs like this: if a sequence of real functions (f_n), $f_n : A \to \mathbb{R}$, have the same domain A and, for each $x \in A$, the limit $\lim_{n \to \infty} f_n(x)$ exists, then a function $f : A \to \mathbb{R}$ may be defined by $f(x) = \lim_{n \to \infty} f_n(x)$.
This f is the *pointwise limit function* for the sequence (f_n). It is important to stress that pointwise limit functions are found by keeping x constant and letting $n \to \infty$.

It is somewhat disconcerting to find that even though the functions in the sequence are continuous it is possible for the pointwise limit function to be discontinuous.

This chapter investigates the context in which a property which is common to all the functions of a sequence is retained in the limit function. The possibility that continuity may be lost in the limit shows how significant this question can be. We will distinguish carefully between the cases where the pointwise limit function is continuous, and the cases where it is not.

6 In each of qns 1, 2, 3, 4 and 5, attempt to find a value of x such that

(i) $|f_{10}(x) - f(x)| > \frac{1}{2}$,

(ii) $|f_{100}(x) - f(x)| > \frac{1}{2}$.

7 Continuing the investigation started in qn 6, show that, for the functions of qn 1,
$|f_n(x) - f(x)| \leq 1/n^2$ for all x in the domain.

8 Show that, for the functions of qn 2,
$|f_n(x) - f(x)| \leq 1/n$ for all x in the domain.

9 For the functions of qn 3, find values of x for which
$|f_n(x) - f(x)| > \frac{1}{2}$.

10 For the functions of qn 4, find values of x for which
$|f_n(x) - f(x)| > 0.99$.

11 For the functions of qn 5, find values for x for which
$|f_n(x) - f(x)| > \frac{1}{2}$.

One way of distinguishing between the cases in qns 1 and 2 on the one hand and qns 3, 4 and 5 on the other is to imagine the graph of the pointwise limit functions f as an arm onto which sleeves of various diameters are pulled. In qns 1 and 2, however narrow the sleeves get the sequence of functions is eventually wholly within them. But in qns 3, 4 and 5 there are sleeves which the functions in the sequence never get wholly inside.

12 For the functions in qn 3, find
$\sup \{|f_n(x) - f(x)|: x \in \mathbb{R}\}$.

13 For the functions in qn 4, find
$\sup \{|f_n(x) - f(x)|: x \in [0, 1]\}$.

14 For the functions in qn 5, find
$\sup \{|f_n(x) - f(x)|: x \in [0, \infty)\}$.

Uniform convergence

When a sequence of functions, (f_n), with each $f_n: A \to \mathbb{R}$, has a pointwise limit function $f: A \to \mathbb{R}$, and, for any $\varepsilon > 0$, the graphs of the functions f_n are eventually inside a sleeve of radius ε about the graph of the limit function f, then, for sufficiently large n,

$f(x) - \varepsilon < f_n(x) < f(x) + \varepsilon$ for all values of x.

In this case we say that the sequence (f_n) *converges uniformly* to f.
[Think of a large fixed n, and check on the sleeve property by letting x vary right across the domain.]

We can also express the uniform convergence of (f_n) to f by saying that

$$\left(\sup\{|f_n(x) - f(x)|: x \in A\}\right) \to 0 \text{ as } n \to \infty,$$

or by saying that, given $\varepsilon > 0$, there exists an n_0 such that

$\sup\{|f_n(x) - f(x)|: x \in A\} < \varepsilon$, when $n > n_0$.

15 Use qns 7 and 8 to prove that the sequences of functions (f_n) in qns 1 and 2 converge uniformly to their respective pointwise limit functions f.

16 Use qns 12, 13 and 14 to prove that the sequences of functions (f_n) in qns 3, 4 and 5 do not converge uniformly to their respective pointwise limit functions.

17 The convergence of a sequence of functions, uniform or otherwise, may depend on the domain of those functions.
Examine the convergence of the sequence of functions given by
$f_n(x) = x/n$
(i) on the domain $[-a, a]$,
(ii) on the domain $\mathbb{R}$.

18 Examine the convergence of the sequence of functions in qn 4
(i) on the domain $[0, a]$, where $0 < a < 1$, and
(ii) on the domain $[0, 1)$.

19 The convergence of a seemingly well-behaved sequence of functions may fail to be uniform.
Examine the convergence of the sequence of functions given by
$f_n(x) = nx/(1 + n^2x^2)$ on the domain $\mathbb{R}$.
Draw the graphs of f_n for $n = 1, 2, 3, 5, 10$.

This particular example is a useful corrective to a wrong impression that might be gleaned from qns 1, 2, 3, 4 and 5. For in those questions there was uniform convergence to constant functions, and non-uniform convergence to non-constant functions. Question 19 shows that this coincidence was accidental.

Uniform convergence and continuity

Having defined uniform convergence with the intention of giving a condition that would make a sequence of continuous functions converge to a continuous limit, we need to see whether the condition we have given is sufficient to guarantee this in every case.

20 We suppose that the sequence of functions (f_n) converges uniformly to the function f and that each of the functions f_n is continuous on its domain A. We seek to prove that f is continuous on A or, in other words, that, for each $a \in A$,
$\lim_{x \to a} f(x) = f(a)$.
Or, again, given $\varepsilon > 0$, there exists a δ such that
$|f(x) - f(a)| < \varepsilon$ when $|x - a| < \delta$.
We break down $f(x) - f(a)$ into manageable pieces.
$f(x) - f(a) = f(x) - f_n(x) + f_n(x) - f_n(a) + f_n(a) - f(a)$
so $|f(x) - f(a)| \leq |f(x) - f_n(x)| + |f_n(x) - f_n(a)|$
$\qquad \qquad \qquad + |f_n(a) - f(a)|$.

For what reason must both $|f(x) - f_n(x)|$ and $|f_n(a) - f(a)|$ be less than $\frac{1}{3}\varepsilon$ for sufficiently large n?
Keeping to one of these sufficiently large ns, for what reason is it possible to find a δ such that $|f_n(x) - f_n(a)| < \frac{1}{3}\varepsilon$ when $|x - a| < \delta$?
Now complete the proof.

21 We have proved that the uniform limit of a sequence of continuous functions is continuous.
By considering qns 17(ii) and 19, show that the converse of this theorem is false: namely that a non-uniform limit of a sequence of continuous functions may also be continuous.

22 If (f_n) is a sequence of continuous functions which converges uniformly to the function f, and a is a point in the domain of these functions, justify each step of the following argument.

$$\lim_{n \to \infty} \lim_{x \to a} f_n(x) = \lim_{n \to \infty} f_n(a)$$
$$= f(a)$$
$$= \lim_{x \to a} f(x)$$
$$= \lim_{x \to a} \lim_{n \to \infty} f_n(x).$$

23 Illustrate the dependence of the argument in qn 22 on uniform convergence by showing how it would fail for the sequence of functions in qn 3 if we were to take $a = 0$.

Uniform convergence and integration

We have shown that the uniform limit of a sequence of continuous functions is continuous. Because of qn 10.42 this allows us to claim in certain cases that the uniform limit of a sequence of integrable functions is integrable.

24 What further condition must the functions of qn 20 satisfy if their integrability is to be claimed?
Even when the domain A of the functions in qn 20 is a closed interval, it is still an open question as to whether the limit of the integrals is equal to the integral of the limit function.
Use the inequality $|\int f_n - \int f| = |\int (f_n - f)| \leqslant \int |f_n - f|$, to prove that

$$\lim_{n \to \infty} \int f_n = \int f, \quad \text{or} \quad \lim_{n \to \infty} \int f_n = \int (\lim_{n \to \infty} f_n).$$

We have shown that the limit of an integral is equal to the integral of the limit in a context of uniform convergence but have not shown that there was any need for uniform convergence for this result.

25 Let $f_n(x) = nx(1 - x^2)^n$ on the domain $[0, 1]$.
Verify that the pointwise limit function is the zero function throughout this domain.
Prove that $\int f_n = n/(2n + 2)$ on $[0, 1]$ and deduce that

$$\lim_{n \to \infty} \int f_n \neq \int \lim_{n \to \infty} f_n.$$

Is the convergence of the sequence (f_n) uniform? Use question 24. Use computer software to draw the graphs of some functions from this sequence.
What is $\lim_{n \to \infty} \sup \{|f_n(x) - f(x)|: 0 \leqslant x \leqslant 1\}$?

26 Show that the result of qn 24 may *not* be extended to improper integrals by attempting to apply it to the sequence of functions defined by

$$f_n(x) = \begin{cases} (n - |x|)/n^2 & \text{when } x \in [-n, n], \\ 0 & \text{otherwise.} \end{cases}$$

Determine the pointwise limit function of the sequence (f_n). Prove that the convergence is uniform, and evaluate
$$\int f_n \text{ and } \int f \text{ on the domain } \mathbb{R}.$$

If we ask whether convergent sequences of integrable functions which are not necessarily continuous must converge to an integrable limit we would have part of an answer if we could construct a sequence of integrable functions which converge to Dirichlet's function given by

$$f(x) = \begin{cases} 1 & \text{when } x \text{ is rational, and} \\ 0 & \text{when } x \text{ is irrational,} \end{cases}$$

since this is an example of a function which is not Riemann integrable on any interval.

27 We construct a sequence of functions on the closed interval $[a, b]$. There is a countable infinity of rational numbers in this interval so there is a sequence in $[a, b]$, which we denote by x_1, $x_2, x_3, \ldots, x_n, \ldots$, which contains them all.
On the closed interval $[a, b]$ we define the function f_n by

$$f_n(x) = \begin{cases} 1 & \text{when } x = x_1, x_2, \ldots, x_n, \text{ and} \\ 0 & \text{otherwise.} \end{cases}$$

Illustrate f_1, f_2 and f_3 on a graph. Are these functions integrable? Is every member of the sequence (f_n) integrable? What is the pointwise limit function?

28 In qn 27, evaluate $|f_n(x_{n+1}) - f(x_{n+1})|$.
Determine whether the convergence of the sequence in qn 27 is uniform.

Having found that the pointwise limit of integrable functions may not be integrable, we now ask whether the uniform limit of a sequence of integrable functions must be integrable.

29 On the domain $[a, b]$ we let (f_n) be a sequence of integrable functions which tends uniformly to the function f. The condition of uniform convergence, that for sufficiently large n, $|f_n(x) - f(x)| < \varepsilon_1$, for any positive ε_1, allows us to set an upper bound and a lower bound on the function f, since
$$f_n(x) - \varepsilon_1 < f(x) < f_n(x) + \varepsilon_1, \quad \text{for all } x.$$

Now for any subdivision of the domain whatsoever, show how an upper sum for f_n may be used to construct an upper sum for f. Likewise show how, for the same subdivision, the lower sum for f_n may be used to construct a lower sum for f.
How close together can you be sure the upper and lower sums for f must be? For given $\varepsilon > 0$, can you choose n sufficiently large to make $2\varepsilon_1(b - a) < \frac{1}{2}\varepsilon$?
And can you then choose a subdivision for which the difference between the upper and lower sums for $f_n < \frac{1}{2}\varepsilon$? What then follows about the difference between the upper and lower sums for f for this subdivision of the domain $[a, b]$? Deduce that f is integrable on $[a, b]$.

30 If, on the domain $[a, b]$, the sequence of integrable functions (f_n) converges uniformly to the function f, use the argument of qn 24 to prove that the limit as $n \to \infty$ of $\int f_n$ is equal to $\int f$.

31 Let (f_n) be a sequence of functions which are integrable on the domain $[a, b]$ and which converge uniformly. Use an argument like that of qn 24 to prove that the sequence of functions (F_n) defined by $F_n(x) = \int_a^x f_n$ is uniformly convergent on the domain $[a, b]$.

Uniform convergence and differentiation

32 Investigate the convergence of the sequence of functions defined by $f_n(x) = x/(1 + nx^2)$ on $\mathbb{R}$. If f is the pointwise limit function, show that $\sup \{|f_n(x) - f(x)|: x \in \mathbb{R}\} = 1/2\sqrt{n}$, and deduce that the convergence is uniform.

Prove that $f_n'(x) = \dfrac{1 - nx^2}{(1 + nx^2)^2}$,

and deduce that $\lim\limits_{n \to \infty} f_n'(0) \neq f'(0)$.

Is $\lim\limits_{n \to \infty} \int_0^1 f_n = \int_0^1 f$? Use qn 24.

Question 32 shows that differentiability is less well behaved under uniform convergence than is integrability.

33 Let $f_n(x) = (\cos nx)/n$.
Find the pointwise limit function for the sequence (f_n) and
determine whether the convergence is uniform on $\mathbb{R}$.
Determine the function $f'_n(x)$. Is there a pointwise limit function
for the sequence (f'_n)? Consider values of $x \neq 0$. Determine the
value of $\int_0^x f_n$. Is $\lim_{n \to \infty} \int_0^x f_n = \int_0^x \lim_{n \to \infty} f_n$?

From qns 32 and 33 it is clear that, even when the sequence (f_n)
converges uniformly to f and every function f_n is differentiable, the
sequence (f'_n) need not be convergent and when it is convergent it
may not be uniformly convergent. If we are to formulate a theorem
about the uniform convergence of a sequence of derivatives, the
uniform convergence of the sequence of derivatives will have to be
assumed. Then, if the functions under discussion are all continuous,
the Fundamental Theorem of Calculus may enable us to use the good
behaviour of integrals under uniform convergence to claim results
about derivatives.

34 We suppose that we have a sequence of differentiable functions
(f_n) on an interval $[a, b]$ converging to the pointwise limit
function f.
We further suppose
(a)   that the sequence (f'_n) consists of continuous functions, and
(b)   that the sequence $(f'_n) \to \varphi$ uniformly.
Give reasons for each of the following propositions.
  (i)   The function φ is continuous.
  (ii)   Each function f'_n and the function φ are integrable on $[a, b]$.

  (iii) $\int_a^x \varphi = \lim_{n \to \infty} \int_a^x f'_n$.

  (iv) $= \lim_{n \to \infty} (f_n(x) - f_n(a))$

  (v) $= f(x) - f(a)$.
  (vi) $\varphi = f'$
  (vii) $(f_n) \to f$ uniformly on $[a, b]$.

The conditions of qn 34 appear to be quite restrictive. But we
need to apply it, typically, to prove theorems about power series and,
in this case, when the functions concerned are polynomials, some of
the conditions we have used are obvious. One of the issues at stake is
when we may integrate or differentiate a power series term by term
while still guaranteeing its continuity or differentiability. Our
theorems relating integration and differentiation to uniform
convergence will enable us to decide this matter.

Uniform convergence of power series

A series of functions is said to be uniformly convergent when its sequence of partial sums is uniformly convergent.

35 Let $e_n(x) = 1 + x + x^2/2! + \ldots + x^n/n!$.

  (i) Why is the sequence $(e_n(x))$ convergent for each x?

    We define $e(x) = \lim_{n \to \infty} e_n(x)$.

    If $-a \leqslant x \leqslant a$, then $|x^n/n!| \leqslant a^n/n!$.

 (ii) Prove that $|e_{n+m}(x) - e_n(x)| \leqslant e_{n+m}(a) - e_n(a)$

(iii) Let $m \to \infty$ in (ii) and prove that
$$|e(x) - e_n(x)| \leqslant e(a) - e_n(a) \quad \text{for all } x \text{ in } [-a, a].$$

(iv) Prove that $(e_n) \to e$ uniformly on $[-a, a]$.

 (v) Use qn 20 to prove that the function e is continuous on $[-a, a]$.

(vi) Use qn 34 to prove that $e'(x) = e(x)$.

The earlier results of this chapter lead to the proof of the uniform convergence of the sequence (e_n) in qn 35 on the basis of the inequality $|x^n/n!| \leqslant a^n/n!$ and the convergence of $\sum a^n/n!$. We isolate this claim in the basic theorem about the uniform convergence of series.

36 *The Weierstrass M-test*

We let $f_N(x) = \sum_{n=1}^{n=N} u_n(x)$ for $x \in A \subseteq \mathbb{R}$, and assume

  (i) for each n, and for all $x \in A$, there exists M_n such that $|u_n(x)| \leqslant M_n$, and

 (ii) $\sum M_n$ is convergent.

We investigate the convergence of the sequence (f_n).

Explain why the series $\sum u_n(x)$ is absolutely convergent for all $x \in A$. Deduce that the sequence (f_n) is pointwise convergent to a function f, say.

Prove that

$$\left| \sum_{n=N+1}^{n=N+m} u_n(x) \right| \leqslant \sum_{n=N+1}^{n=N+m} M_n,$$

and deduce that

$$|f_{N+m}(x) - f_N(x)| \leqslant \sum_{n=1}^{n=N+m} M_n - \sum_{n=1}^{n=N} M_n,$$

so that, letting $m \to \infty$,

$$|f(x) - f_N(x)| \le \sum_{n=1}^{\infty} M_n - \sum_{n=1}^{n=N} M_n.$$

Now deduce that (f_n) converges uniformly.

37 Use the Weierstrass M-test to prove that each of the following series is uniformly convergent on the domain given.

(i) $\sum x^n/n^2$ on the domain $[-1, 1]$,

(ii) $\sum (\sin nx)/2^n$ on the domain $\mathbb{R}$,

(iii) $\sum 1/(n^2 + x^2)$ on the domain $\mathbb{R}$,

(iv) $\sum x/(x^2 + n^2)$ on the domain $[-a, a]$,

(v) $\sum (x^2 + n)/(x^2 + n^3)$ on the domain $[-a, a]$,

(vi) $\sum (n + 1)x^n$ on the domain $[-a, a]$, where $0 < a < 1$,

(vii) $\sum x^n(1 - x)/n$ on the domain $[0, 1]$.

38 Let $f_n(x) = 1 + x + x^2 + \ldots + x^{n-1}$.
For $-1 < x < 1$, find the pointwise limit function f of the sequence (f_n).
For x in this range, prove that $|f_n(x) - f(x)| = |x|^n/(1 - x)$.
Deduce that the convergence of the sequence (f_n) is not uniform on the domain $(-1, 1)$.
Prove that the convergence of the sequence (f_n) is uniform on the domain $[-a, a]$, provided $0 < a < 1$.

39 Prove that each of the following series is not uniformly convergent on the domain given.

(i) $\sum (x^2 + n)/(x^2 + n^3)$ on the domain $\mathbb{R}$,

(ii) $\sum (n + 1)x^n$ on the domain $(-1, 1)$.

The heart of the proof lies in finding a value of x which shows that the sum of the series from n to ∞ cannot be made arbitrarily small by choice of n, however large.
Compare these results with qn 37(v) and 37(vi).

It is in fact quite straightforward to show that any power series is uniformly convergent within a closed interval inside its circle of convergence.

40 Suppose R is the radius of convergence of the power series $\sum a_n x^n$. Show that if $0 < a < R$ then the series is uniformly convergent on the interval $[-a, a]$ by choosing a real number r such that $a < r < R$, and using the Weierstrass M-test with $M_n = |a_n r^n|$. Deduce that the limit function must be continuous on this domain.

41 Use qn 24 to show that if a power series is integrated term by term then the integral of the limit function is equal to the limit of the power series obtained by term-by-term integration, on an interval $[0, x]$ inside its circle of convergence.
Is the radius of convergence of the integrated series the same as the radius of convergence of the original power series? See qn 5.102.

42 If a power series is differentiated term by term show that the radius of convergence of the differentiated series is the same as the radius of convergence of the original series. See qn 5.102. Use qn 34 to show that the limit function is differentiable and that the derivative of the limit function is equal to the limit of the power series obtained by term-by-term differentiation, within a closed interval inside its circle of convergence.

43 By differentiating the series for $1/(1 - x)$ m times, establish the binomial theorem for negative integral index.

44 By integrating the power series

$$1 + x + x^2 + \ldots + x^n + \ldots$$

within its circle of convergence, prove that

$$\log(1 - x) = -x - x^2/2 - \ldots - x^n/n - \ldots$$

and determine the radius of convergence of this power series.

The Binomial Theorem for any real index

45 A function f is defined for $-1 < x < 1$, by

$$f(x) = \sum_{n=0}^{\infty} \binom{a}{n} x^n.$$

Here a is a real number which is neither zero nor a positive integer. See qn 5.93. Prove that $(1 + x)f'(x) = af(x)$.
Use the Mean Value Theorem to prove that the function g defined on $(-1, 1)$ by $g(x) = (1 + x)^{-a} f(x)$ is constant. Since $g(0) = 1$, prove that $f(x) = (1 + x)^a$.

A function which is continuous everywhere and differentiable nowhere

46 The function $g: \mathbb{R} \to \mathbb{R}$ is given by

$$g(x) = \begin{cases} 1 + x, & \text{when } -2 < x \leqslant 0, \\ 1 - x, & \text{when } 0 < x \leqslant 2 \\ g(x + 4), & \text{for all } x. \end{cases}$$

(i) Sketch the graph of g. Verify that $|g(x)| \leqslant 1$ for all x. Verify that g is continuous on $\mathbb{R}$.

(ii) Sketch the graph of g_1 given by $g_1(x) = g(2^2 x)$,

and of g_2 given by $g_2(x) = g(2^{2^2} x)$.
Define the function g_n by $g_n(x) = g(2^{2^n} x)$ and the function f by $f(x) = \sum_{n=1}^{\infty} 2^{-n} g_n(x)$.

(iii) Use the Weierstrass M-test to show that this series is uniformly convergent to f and deduce that f is continuous at every point.
Now let $h(a, k) = \pm 1/2^{2^k}$, where the sign is chosen so that a and $a + h(a, k)$ lie on the same linear segment of g_k. Clearly $h(a, k) \to 0$ as $k \to \infty$.

(iv) Show that $g_n(a + h(a, k)) - g_n(a) = 0$ when $n > k$.

(v) Show that $|g_n(a + h(a, k)) - g_n(a)| = 1$ when $n = k$.

(vi) Show that $|g_n(a + h(a, k)) - g_n(a)| \leqslant 1/2^{2^{k-1}}$ when $n < k$.

(vii) $\left| \sum_{n=1}^{k-1} 2^{-n} \big(g_n(a + h(a, k)) - g_n(a) \big) \right|$

$$\leqslant \frac{(\frac{1}{2} + \frac{1}{4} + \ldots + 1/2^{k-1})}{2^{2^{k-1}}} < \frac{1}{2^{2^{k-1}}}.$$

(viii) $|f(a + h(a, k)) - f(a)| \geqslant 1/2^k - 1/2^{2^{k-1}}.$

(ix) $\left| \dfrac{f(a + h(a, k)) - f(a)}{h(a, k)} \right|$

$$\geqslant 2^{2^k - k} - 2^{2^{k-1}} = 2^{2^{k-1}} (2^{2^{k-1} - k} - 1)$$

(x) By letting $k \to \infty$, prove that f is not differentiable at a.

Summary

Definition (f_n) is a sequence of real functions, $f_n: A \to \mathbb{R}$, all of
qns 5, 6   which have the same domain A. If for each $x \in A$,

the limit $\lim_{n\to\infty} f_n(x)$ exists, then a function $f\colon A \to \mathbb{R}$ may be defined by $f(x) = \lim_{n\to\infty} f_n(x)$. This f is called the *pointwise limit function* of the sequence (f_n).

Definition   The sequence of functions, (f_n), has the pointwise
qns 14, 15   limit function $f\colon A \to \mathbb{R}$.
If $\left(\sup\{|f_n(x) - f(x)|\colon x \in A\}\right) \to 0$ as $n \to \infty$, we say that the sequence (f_n) *converges uniformly* to f.

Theorem   If the sequence of functions (f_n) converges uniformly
qn 20   to the function $f\colon A \to \mathbb{R}$ and each of the functions f_n is continuous on its domain A, then f is continuous on A.

Theorem   If the sequence of functions (f_n) converges uniformly
qns 29, 30   to the function $f\colon [a, b] \to \mathbb{R}$ and each of the functions f_n is integrable on $[a, b]$ then f is integrable on $[a, b]$ and $\lim_{n\to\infty} \int_a^b f_n = \int_a^b f$.

Theorem   If
qn 34
    (i) the sequence of functions (f_n) converges pointwise to the function $f\colon A \to \mathbb{R}$;
    (ii) each of the functions f_n has a continuous derivative on its domain A;
    (iii) the sequence (f_n') converges uniformly to $\varphi\colon A \to \mathbb{R}$;
    then $\varphi = f'$ and (f_n) converges uniformly to f on A.

The Weierstrass M-test

qn 36   A sequence of functions (f_n) is defined by the partial sums of a series:

$$f_N(x) = \sum_{n=1}^{n=N} u_n(x).$$

If there exist real numbers M_n such that
    (i) $|u_n(x)| \leq M_n$, and
    (ii) $\sum M_n$ is convergent,
then (f_n) is uniformly convergent.

Theorem   A power series is uniformly convergent on any
qn 40   closed interval inside its circle of convergence.
Theorem   If $f(x) = \sum_{n=0}^{\infty} a_n x^n$ has radius of convergence R,
qns 41, 42   then

$$f'(x) = \sum_{n=1}^{\infty} na_n x^{n-1} \text{ and } \int_0^x f = \sum_{n=0}^{\infty} \frac{a_n x^{n+1}}{n+1}$$

when $|x| < R$.

The Binomial Theorem

qn 45 $(1 + x)^a = \sum_{n=0}^{\infty} \binom{a}{n} x^n$,   for all real a,

provided $-1 < x < 1$.

Historical note

Newton and his contemporaries differentiated and integrated power series, term by term, without reference to their circle of convergence. During the eighteenth century it was presumed that the limit of a sequence of continuous functions was continuous. In 1821, Cauchy believed that he had a proof of this and in 1823 a proof that a sequence of integrable functions was integrable. However, he had not identified the distinctive features of uniform convergence and his proofs were fallacious. Fourier had produced counter-examples showing that a sequence of continuous functions might have a discontinuous limit and these were known to Dirichlet writing in 1829. In 1848 Seidel correctly analysed the defect in Cauchy's proofs, but his discussion of uniform convergence was clumsy by comparison with that of Weierstrass in his lectures in Berlin in the 1860s, to whom we owe the theorems of this chapter. Weierstrass was able to prove that every continuous function was the uniform limit of a sequence of polynomials and believed that this theorem legitimated the study of continuous but non-differentiable functions which he had initiated.

Answers

1 $\lim_{n\to\infty} 1/(n^2 + x^2) = 0$ for all x.

3 $f(0) = 1$, $f(x) = 0$ when $x \neq 0$.

4 $f(x) = 0$ when $0 \leqslant x < 1$, $f(1) = 1$.

5 $f(x) = -1$ when $0 \leqslant x < 1$, $f(1) = 0$, $f(x) = 1$ when $1 < x$.

6 Not possible with qns 1 or 2.

Question 3 (i) $x^2 < 1/10^2$, (ii) $x^2 < 1/100^2$.

Question 4 (i) $1/\sqrt[10]{2} < x < 1$, (ii) $1/\sqrt[100]{2} < x < 1$.

Question 5 (i) $1/\sqrt[10]{3} < x < 1$ or $1 < x < \sqrt[10]{3}$,

(ii) $1/\sqrt[100]{3} < x < 1$ or $1 < x < \sqrt[100]{3}$.

7 $1/(n^2 + x^2) \leqslant 1/n^2$.

8 $|(\sin x)/n| \leqslant 1/n$.

9 $0 < |x| < 1/n$.

10 $\sqrt[n]{(0.99)} < x < 1$.

11 $1/\sqrt[n]{3} < x < 1$ or $1 < x < \sqrt[n]{3}$.

12 $f_n(x) \to 1$ as $x \to 0$, so

$|f_n(x) - f(x)| \to 1 = \sup\{|f_n(x) - f(x)|: x \in \mathbb{R}\}$.

13 $f_n(x) \to 1$ as $x \to 1^-$, so

$|f_n(x) - f(x)| \to 1 = \sup\{|f_n(x) - f(x)|: 0 \leqslant x \leqslant 1\}$.

14 $f_n(x) \to 0$ as $x \to 1^-$, so

$|f_n(x) - f(x)| \to 1 = \sup\{|f_n(x) - f(x)|: 0 \leqslant x \leqslant 1\}$.

$f_n(x) \to 0$ as $x \to 1^+$, so

$|f_n(x) - f(x)| \to 1 = \sup\{|f_n(x) - f(x)|: 1 \leqslant x\}$.

17 The pointwise limit function is $f(x) = 0$ whatever the domain.
(i) On $[-a, a]$, $|f_n(x) - f(x)| \leqslant |a|/n \to 0$ as $n \to \infty$.
(ii) For given n, $|f_n(x) - f(x)|$ is unbounded for $x \in \mathbb{R}$.
The critical question is whether you can use n to control the greatest
values of $|f_n(x) - f(x)|$.

18 (i) On $[0, a]$, $|f_n(x) - f(x)| \leq a^n \to 0$ as $n \to \infty$.
(ii) On $[0, 1)$, $f_n(x) \to 1$ as $x \to 1^-$, so
$|f_n(x) - f(x)| \to 1 = \sup\{|f_n(x) - f(x)|: 0 \leq x < 1\}$.
This is a tricky and uncomfortable example, but it shows what a subtle business uniform convergence may be.

19 The pointwise limit function is given by $f(x) = 0$.
$\sup\{|f_n(x) - f(x)|: x \in \mathbb{R}\} = \frac{1}{2}$ irrespective of n.

20 By the uniform convergence of $(f_n) \to f$,
$\left(\sup\{|f_n(x) - f(x)|: x \in A\}\right) \to 0$ as $n \to \infty$.
So for some n_0, $|f(x) - f_n(x)|$ and $|f_n(a) - f(a)| \leq \frac{1}{3}\varepsilon$ for $n > n_0$.
Since each f_n is continuous a suitable δ may be found.
Now, if $|x - a| < \delta$, $|f(x) - f(a)| < \varepsilon$ since we know the three inequalities hold for some value of n. Thus $\lim_{x \to a} f(x) = f(a)$, and f is continuous at a.

22 First equality by continuity of f_n at a.
Second equality by definition of pointwise limit function.
Third equality by uniform convergence and qn 20.
Fourth equality by definition of pointwise limit function.

23 In qn 3,
$\lim_{n \to \infty} \lim_{x \to 0} f_n(x) = \lim_{n \to \infty} 1 = 1$.

$\lim_{x \to 0} \lim_{n \to \infty} f_n(x) = \lim_{x \to 0} 0 = 0$.

24 $A = [a, b]$. For sufficiently large n, $|f_n - f| < \varepsilon$,
So $\int |f_n - f| < \varepsilon(b - a)$. Use qn 10.35.

25 $\lim_{n \to \infty} \int_0^1 f_n = \frac{1}{2}$. $\int_0^1 \lim_{n \to \infty} f_n = 0$. $f_n'\left(1/\sqrt{(2n + 1)}\right) = 0$.

$\sup\{|f_n(x) - f(x)|: 0 \leq x \leq 1\} = \dfrac{n}{\sqrt{(1 + 2n)} \cdot \left(1 + 1/(2n)\right)^n} \to \infty$
$$\text{as } n \to \infty.$$

26 The pointwise limit function is given by $f(x) = 0$.
$\int f_n = 1$. $\int f = 0$.

27 The function f_n is integrable as in qn 10.14. The pointwise limit function is Dirichlet's function.

28 $|f_n(x_{n+1}) - f(x_{n+1})| = 1$. Convergence not uniform.

29 Let S_n be an upper sum for f_n. Then $S_n + \varepsilon_1(b - a)$ is an upper sum for f. Let s_n be a lower sum for f_n. Then $s_n - \varepsilon_1(b - a)$ is a lower sum for f. It is possible to take n sufficiently large so that $\varepsilon_1 < \varepsilon/4(b - a)$. Since f_n is integrable there are upper and lower sums such that $S_n - s_n < \frac{1}{2}\varepsilon$ from 10.23. Then $\left(S_n + \varepsilon_1(b - a)\right) - \left(s_n - \varepsilon_1(b - a)\right) < \varepsilon$. So f is integrable by qn 10.24.

30 The argument in qn 24 does not appeal to continuity.

31 Let $(f_n) \to f$, then by qn 30,

$$\lim_{n \to \infty} F_n(x) = \int_a^x \lim_{n \to \infty} f_n = \int_a^x f = F(x) \text{ (say)}.$$

$$F_n(x) - F(x) = \int_a^x (f_n - f) \text{ and, by qn 10.35,}$$

$$|F_n(x) - F(x)| \leqslant \int_a^x |f_n - f|.$$

If $\sup \{|f_n(x) - f(x)|: a \leqslant x \leqslant b\} < \varepsilon/(b - a)$ for $n > n_0$, then $|F_n(x) - F(x)| < \varepsilon$ for $n < n_0$.

32 The pointwise limit function is given by $f(x) = 0$. $f_n'(0) = 1$, $f'(0) = 0$. Yes.

33 $f(x) = 0$. $|f_n(x) - f(x)| \leqslant 1/n$, so convergence is uniform. $f_n'(x) = -\sin nx$ which is not pointwise convergent for any $x \neq 0$. $\int_0^x f_n = (\sin nx)/n^2$. Limit of integral = integral of limit by qn 30.

34 (i) qn 20, (ii) qn 10.42, (iii) qn 24, (iv) qn 10.54, (v) qn 3.44(iii), (vi) Fundamental Theorem of Calculus, qn 10.57, (vii) qns 24 and 31.

35 (i) qn 5.81.

(ii) $|e_{n+m}(x) - e_n(x)| = |x^{n+1}/(n + 1)! + \ldots + x^{n+m}/(n + m)!|$

$$\leqslant |x^{n+1}/(n + 1)!| + \ldots + |x^{n+m}/(n + m)!|$$

$$\leqslant a^{n+1}/(n + 1)! + \ldots + a^{n+m}/(n + m)!$$

$$\leqslant e_{n+m}(a) - e_n(a).$$

(iii) qn 3.64.

(iv) $\left(e(a) - e_n(a)\right) \to 0$ as $n \to \infty$.

(vi) let $f_n = e_n$, then $f_n' = e_{n-1}$. $(f_n') \to e$ uniformly and $(f_n) \to e$ uniformly. From qn 34(vi), $e = e'$.

36 $\sum u_n$ is absolutely convergent by (i), (ii) and the first comparison test, qn 5.23. So by qn 5.63 the series is convergent for all $x \in A$ and f is well defined. The inequality with m terms is obtained by repeated application of the triangle inequality (qn 2.65), the second inequality is only a rewrite of the first, and the third inequality comes from qn 3.64. As $N \to \infty$ the right-hand side of the last inequality tends to 0; see qn 5.15.

37 (i) Take $M_n = 1/n^2$, (ii) $M_n = (\frac{1}{2})^n$, (iii) $M_n = 1/n^2$, (iv) $M_n = a/n^2$,
   (v) $M_n = 1/n^2 + a^2/n^3$, (vi) $M_n = (n+1)a^n$ using Cauchy's nth root
   test, qn 5.32, or d'Alembert 5.40, (vii) $M_n = 1/2n^2$ using differentiation
   and qn 2.51.

38 $f(x) = 1/(1-x)$. As $x \to 1^-$, $|f_n(x) - f(x)| \to \infty$.
   If $-a \leqslant x \leqslant a < 1$, $|x^n| \leqslant a^n = M_n$.

39 (i) When $x = n^2$, there is a term in the series, after the nth which is
   greater than $\frac{1}{2}$. So although the series is convergent for each value of
   x by comparison with $\sum 1/n^2$, the greatest difference between the
   nth partial sum (i.e. f_n) and the limit function is not null.
   (ii) Likewise when $x = 1/\sqrt[2n]{(2n)}$, there is a term in the series, after the
   nth which is greater than 1. So although the series is convergent for
   each value of x in $(-1, 1)$, the greatest difference between the nth
   partial sum and the limit function is greater than 1.

40 The limit function is continuous by qn 20.

41 A partial sum of a power series is a polynomial which is continuous. The
   uniform convergence of the series on $[0, x]$ gives the limit of the integral
   equal to the integral of the limit from qn 24.
   The radius of convergence of the two series is the same by qn 5.102.

42 Partial sums are polynomials, so derivatives of partial sums are also
   polynomials. Both series have the same circle of convergence by qn 5.102
   so, within that circle, both converge uniformly. This is the context in
   which qn 34(vi) may be applied.

43 $(1-x)^{-1} = 1 + x + x^2 + \ldots + x^n + \ldots$

$$= 1 + \binom{-1}{1}(-x) + \binom{-1}{2}(-x)^2 + \ldots + \binom{-1}{n}(-x)^n + \ldots$$

$$= \sum_{n=0}^{\infty} \binom{-1}{n}(-x)^n$$

The circle of convergence is $|x| = 1$.

Suppose $(1-x)^{-m} = \sum_{n=0}^{\infty} \binom{-m}{n}(-x)^n$.

Then $m(1-x)^{-m-1} = \sum_{n=0}^{\infty} -(n+1)\binom{-m}{n+1}(-x)^n$

and hence $(1-x)^{-m-1} = \sum_{n=0}^{\infty} -\frac{n+1}{m}\binom{-m}{n+1}(-x)^n$

thus $(1 - x)^{-m-1} = \sum_{n=0}^{\infty} \binom{-m-1}{n}(-x)^n$.

This establishes the result by induction.

44 Radius of convergence for both series $|x| = 1$ by qn 5.102.

45 By qn 5.93, the series is convergent for $|x| < 1$, so f is a pointwise limit function. Within the circle of convergence the convergence is uniform so, using qn 42,

$$f'(x) = \sum_{n=0}^{\infty} (n + 1)\binom{a}{n + 1}x^n.$$

Now $n\dfrac{a(a - 1) \ldots (a - n + 1)}{n!} + (n + 1)\dfrac{a(a - 1) \ldots (a - n)}{(n + 1)!}$

$$= \dfrac{a(a - 1) \ldots (a - n + 1)}{(n - 1)!}\left(1 + \dfrac{a - n}{n}\right).$$

So $(x + 1) f'(x) = af(x)$.

$g'(x) = -a(1 + x)^{-a-1}f(x) + (1 + x)^{-a}f'(x)$

$\quad = (1 + x)^{-a-1}\big(-af(x) + (1 + x) f'(x)\big)$

$\quad = 0.$

By the Mean Value Theorem g is a constant function.

Since $f(0) = 1$, $g(0) = 1$, so $f(x) = (1 + x)^a$.

Appendix 1
Geometry and intuition

Sometimes authors and lecturers on analysis insist that students must not use geometrical intuition in developing the fundamental concepts of analysis or in constructing proofs in analysis. Such a prohibition appears to be consonant with Felix Klein's description in 1895 of the developments due to Weierstrass, Cantor and Dedekind, namely the *Arithmetisation of Analysis*.

However, such advice is impossible to implement and is in any case untrue to the origins of the subject. We can hardly conceive of a Dedekind cut, for example, without imagining a 'real line', and such imagining was certainly part of Dedekind's own thought. We have eyes, and we have imaginations with which to visualise, and such visualisation is central to much of the development of analysis. Every development of the real number system is a way of formalising our intuitions of the points on an endless straight line. We cannot conceive how the theory of real functions could have developed had there been no graphs drawn.

However, geometric intuition is not always reliable, and knowing when it should be trusted and when it should not is part of the mathematical maturity which should develop during an analysis course.

There are contexts in which geometrical intuition is misleading.

1 When comparing infinities: because there is a one-to-one correspondence between the points of the segment $[0, 1]$ and the points on the segment $[0, 2]$, there appear to be the 'same' number of points on both segments. The conflict with intuition here is simply to do with infinity, not to do with rationals and irrationals, because the same paradox arises if we restrict our attention to rational points.

2 When comparing denseness with completeness: because there is an infinity of rationals between any two points on the line there are rationals as close as we like to any point. That most of the cluster points of $\mathbb{Q}$ are not in $\mathbb{Q}$ again seems paradoxical. Even the terminating decimals are dense on the line and will give us measurements as accurate as we may wish, yet they do not even include all the rational numbers.

3 Continuity: when using a pencil to make a 'continuous' line the Intermediate Value Theorem appears to be unfalsifiable. This is closely related to the denseness–completeness issue in 2.

4 What a continuous function may be like is not obvious. One of the most unexpected functions, from the segment [0, 1] to the unit square $\{(x, y)|\ 0 \leqslant x \leqslant 1,\ 0 \leqslant y \leqslant 1\}$ constructed by mapping the number $0.a_1a_2a_3a_4a_5a_6 \ldots$ to the point $(0.a_1a_3a_5 \ldots,\ 0.a_2a_4a_6 \ldots)$.
is continuous. Its image, called the Peano curve, passes through all the points of the unit square just once (provided terminating decimals are always represented by recurring nines).

5 The connection between continuity and differentiability seems straightforward enough (with non-differentiability at the occasional sharp point on the curve) until one considers a function which is everywhere continuous and nowhere differentiable, such as the one in qn 12.46 or David Tall's *Blancmange Function*.

The examples 2–5 above all show the inadequacy of considering a mark made by a pencil, without lifting it from the paper, as the illustrative model of the graph of a continuous function. It is tantalising to contrast these illustrations of suspect visualisation with an example where the intuitive, pencil and paper, point of view is sound enough, as at the beginning of chapter 7. Try finding the kind of set A which can be the range of a continuous function $\mathbb{R} \to A \subseteq \mathbb{R}$. Here even quite rough work with pencil and paper leads to a precise and accurate formulation.

I would certainly concur with the judgement of J. E. Littlewood who wrote in 1953 (*Bollobás*, 1986, p. 54):

> My pupils *will* not use pictures, even unofficially and when there is no question of expense. This practice is increasing. I have lately discovered that it has existed for 30 years or more, and also why. A heavy warning used to be given

that pictures are not rigorous; this has never had its bluff called and has permanently frightened its victims into playing for safety. Some pictures, of course, are not rigorous, but I should say most are (and I use them wherever possible myself). An obviously legitimate case is to use a graph to define an awkward function (*e.g.* behaving differently in successive stretches).

Appendix 2

Multiple-choice questions for student discussion

1 $a^2 + b^2 + c^2 \geq bc + ca + ab$ is true:
   A. always;   B. sometimes;   C. never

2 If $(a_n) \to a$, must $\left((\sum a_n)/n\right) \to a$? A. Yes B. No
   If $\left((\sum a_n)/n\right) \to a$, must $(a_n) \to a$? A. Yes B. No

3 The sequence (a_n) is bounded. Which of the following propositions are sufficient to guarantee its convergence?
   (i) $(a_n) \to 0$;
   (ii) (a_n) is positive and bounded below;
   (iii) (a_n) is bounded above and increasing;
   (iv) for all n, $a_n^2 < a_n$;
   (v) none of the above.

4 Between two rationals there is an irrational. Between two irrationals there is a rational. So the rationals and the irrationals alternate on the line.
   A. This is a valid argument.
   B. The result is true but the argument is invalid.
   C. The result is false.

5 True or false?
   Whatever the values of a or b, the set $[a, b] \cap \mathbb{Q}$ contains its own supremum and infimum.
   Whatever the values of a or b, the set $(a, b) \cap \mathbb{Q}$ never contains its own supremum or infimum.

6 A real function f is continuous on $(0, 1)$ and takes positive values. Which **one** of the following statements must be true?

    (i) f is bounded on $(0, 1)$;

    (ii) $f(x)$ tends to a non-negative limit as $x \to 0^+$;

    (iii) f attains a minimum though not necessarily a maximum, value;

    (iv) the Riemann integral $\int_0^1 f$ exists;

    (v) none of the four above.

7   A real function f is continuous on the open interval $(0, 1)$.

    (i) If f is uniformly continuous, must f be bounded?

    (ii) If f is bounded, must f be uniformly continuous?

    (iii) May f be neither bounded, nor uniformly continuous?

8   A differentiable function $f: \mathbb{R} \to \mathbb{R}$ is strictly monotonic increasing. Must its derivative be positive?

Problems for corporate or individual investigation

9   In many books Archimedean order is claimed by saying that, if two positive numbers are given, then some positive integer multiple of the first exceeds the second. Is this claim equivalent to the Axiom of Archimedean order which we have used in chapter 3?

10   On the set of formal power series $\sum_{n=-m}^{\infty} a_n x^n$ can you construct

$+$, $\times$ and $<$ so that the set is an ordered field?

Is the field then Archimedean ordered?

Does every Cauchy sequence of rational numbers converge?

11   Find the cluster points of the bounded sets

(i) $\{\sin n \mid n \in \mathbb{N}\}$, (ii) $\{n\sqrt{2} - [n\sqrt{2}] \mid n \in \mathbb{N}\}$. Is there a subsequence of $(\sin n)$ which tends to 0? If so, can you find one?

12   Does $(\sqrt[n]{|\sin n|})$ have a limit?

13   Can you give meaning to the following:

    (i) $1 + \cfrac{1}{2 + \cfrac{1}{3 + \cfrac{1}{4 + \ldots}}}$,

    (ii) $\sqrt{(1 + \sqrt{(2 + \sqrt{(3 + \ldots)}))}}$,

    (iii) $2\sqrt{2}$, $4\sqrt{(2 - \sqrt{2})}$, $8\sqrt{(2 - \sqrt{(2 + \sqrt{2})})}$, $16\sqrt{(2 - \sqrt{(2 + \sqrt{(2 + \sqrt{2})})})}$, $\ldots$

14 If for a sequence of positive terms, $\left(\sqrt[n]{a_n}\right) \to k$, does it follow that $(a_{n+1}/a_n) \to k$ What about the converse?

15 (*Cantor*, 1874) A real number is said to be *algebraic* if it is the solution of a polynomial equation with integer coefficients:
$$a_0 + a_1 x + a_2 x^2 + \ldots + a_n x^n = 0$$
Calling $|a_0| + |a_1| + |a_2| \ldots + |a_n| + n$, the 'weight' of the polynomial show that the set of algebraic numbers is countable. A real number which is not algebraic is said to be *transcendental*.

16 Does a real sequence (a_n) necessarily converge if, given $\varepsilon > 0$, there exists an integer N (depending on p and ε), such that $|a_{n+p} - a_n| < \varepsilon$ when $n > N$, for each integer p?

17 Is there a non-empty set of real numbers which is both closed (contains all its limit points) and open (contains a neighbourhood of each of its points)?

18 Under what circumstances does
$$\lim_{x \to \infty} \left(\sqrt{(ax^2 + bx + c)} - \sqrt{(Ax^2 + Bx + C)}\right)$$
exist?

19 (*Peano*, 1884) Investigate the values of the function $g \colon \mathbb{R} \to \mathbb{R}$ defined by $g(x) = \lim f(\sin n!\pi x)$ as $n \to \infty$, where f is the function in qn 6.10,
(a) when x is rational,
(b) when x is irrational.

20 If a, b, c and d are irrational numbers with $a < b$ and $c < d$, can you construct a bijection with domain $[a, b] \cap \mathbb{Q}$ and range $[c, d] \cap \mathbb{Q}$? A continuous bijection? A differentiable bijection?

21 Let $a_1 < a_2 < a_3 < \ldots < a_n$.
A function $f \colon [a_1, a_n] \to \mathbb{R}$ is defined by
$f(x) = |x - a_1| + |x - a_2| + \ldots + |x - a_n|$. Is f continuous? Is f bounded? Where does it attain its bounds?
A function $g \colon [a_1, a_n] \to \mathbb{R}$ is defined by
$g(x) = (x - a_1)^2 + (x - a_2)^2 + \ldots + (x - a_n)^2$. Is g continuous? Is g bounded? Where does it attain its bounds?

22 *The waterfall function*
If the function $f \colon [a, b] \to \mathbb{R}$ is continuous at every point of its domain, must the function $g \colon [a, b] \to \mathbb{R}$ defined by
$g(t) = \sup \{f(x) | a \leqslant x \leqslant t\}$ be continuous at every point?.

23 A function f is defined on $[a, b]$. $x_i = a + i(b - a)/n$. Under what conditions will the sequence with nth term $\left(\sum_{i=1}^{i=n} f(x_i)\right)/n$ be convergent as $n \to \infty$?

24 Under what circumstances will a sequence (a_n) defined by the relation $a_{n+1} = a_n - f(a_n)/f'(a_n)$ converge?

25 (*Cauchy*, 1821) What properties can be established for a function $f: \mathbb{R} \to \mathbb{R}$ such that $f(x + y) = f(x) \cdot f(y)$? What further properties can you claim if f is continuous with $f(1) = 2$?

Bibliography

Armitage, J. V. and Griffiths, H. B., 1969, *Companion to Advanced Mathematics*, vol. 1, Cambridge University Press.
Part II of this book contains a concise but illuminating overview of first-year undergraduate analysis from a second-year/metric space point of view.

Artmann, B. 1988, *The Concept of Number*, Ellis Horwood.
An advanced and thorough treatment.

Ausubel, D. P., 1968, *Educational Psychology: a Cognitive View*, Holt, Rienhart and Winston.

Baylis, J. and Haggarty, R., 1988, *Alice in Numberland*, Macmillan Education.
A prosy, jokey, but none the less serious introduction to the real numbers.

Beckenbach, E. F. and Bellman, R., 1961, *An Introduction to Inequalities*, Mathematical Association of America.
Inequalities for the sixth-former.

Boas, R. P., 1960, *A Primer of Real Functions*, Mathematical Association of America.
Excellent further reading.

Bollobás, B. (ed.), 1986, *Littlewood's Miscellany*, Cambridge University Press.

Bolzano, B., 1950, *Paradoxes of the Infinite*, Routledge and Kegan Paul.
First published posthumously in 1851; a remarkable testament to one of the finest analytical thinkers of the period 1800–50.

Boyer, C. B., 1959, *The History of the Calculus*, Dover reprint of *The Concepts of the Calculus*, 1939.

Bryant, V., 1990, *Yet Another Introduction to Analysis*, Cambridge University Press.
A good collection of expository ideas.

Burkill, J. C., 1960, *An Introduction to Mathematical Analysis*, Cambridge University Press.
A concise accessible treatment which establishes the irrationality of π in the last exercise of chapter 7.

Burkill, J. C. and Burkill, H., 1970, *A Second Course in Mathematical Analysis*, Cambridge University Press.
Contains a very good introduction to uniform convergence.

Burn, R. P., 1982, *A Pathway to Number Theory*, Cambridge University Press.
Gives the Fundamental Theorem of Arithmetic in chapter 1.

Burn, R. P., 1985, *Groups: A Path to Geometry*, Cambridge University Press.
Gives an account of equivalence relations in chapter 6.

Burn, R. P., 1990, Filling holes in the real line, *Math. Gaz.*, **74**, 228–32.

Cohen, D., 1988, *Calculus By and For Young People*. Published by the author at 809 Stratford Drive, Champaign, Ill. 61821, USA.
A numerical approach to sequences, series and graphs, said to be for those aged 7 and upwards. Just right for the sixth form!

Cohen, L. W. and Ehrlich, G., 1963, *The structure of the Real Number System*, van Nostrand Reinhold.
A detailed, thorough and modern account of the number system from the Peano postulates to the various forms of completeness.

Courant, R. and John, F., 1965, *Introduction to Calculus and Analysis*, vol. 1, New York: Wiley. Reprinted 1989, Berlin: Springer.
Excellent text and diagrams. Style of discussion consonant with a good sixth form text. Gives integration before differentiation. Strong on applications.

Dedekind, R., 1963, *Essays on the theory of numbers*, Dover.
First published 1872 and 1887, these are two of the historically seminal documents in the arithmetisation of analysis.

Dieudonné, J., 1960, *Foundations of Modern Analysis*, Academic Press.
For 'further reading' only.

Dugac, P., 1978, Sur les fondements de l'Analyse de Cauchy à Baire, doctoral thesis, Université Pierre et Marie Curie, Paris.
A masterly survey of the foundations of analysis during the nineteenth century, particularly informative on the contribution of Weierstrass.

Edwards, C. H., 1979, *The Historical Development of the Calculus*, Springer.
Strong on the seventeenth and eighteenth centuries. Written to generate interest in mathematics.

Ernest, P., 1982, Mathematical Induction: a recurring theme. *Math. Gaz.*, **66**, 120–5.
This article establishes the equivalence of finite induction and the well-ordering of the natural numbers.

Ferrar, W. L., 1938, *Convergence*, Oxford University Press.

Gardiner, A., 1982, *Infinite Processes*, Springer.
An attempt to bridge the gap between sixth form and undergraduate thinking on three fronts: numbers, geometry and functions.

Gelbaum, B. R. and Olmsted, J. M. H., 1964, *Counterexamples in Analysis*, Holden-Day.
A rich collection of correctives to the untutored intuition.

Grabiner, J. V., 1981, *The Origins of Cauchy's Rigorous Calculus*, MIT.
The definitive introduction to Cauchy's analysis and calculus. Strong on the late eighteenth and early nineteenth century.

Grattan-Guinness, I. (ed.), 1980, *From the Calculus to Set Theory, 1630–1910*, Duckworth.
Strong on the nineteenth century, but you must know the mathematics before you start reading this history book.

Hardy, G. H., 1st edition 1908, *Pure Mathematics*, Cambridge University Press.
Until about 1960, this (2nd edition, 1914, onwards) was the leading English language text on analysis. The text is hard going but the exercises offer a rich and

broad diet.

Hardy, G. H., Littlewood, J. E., and Pólya, G., 1952, *Inequalities*, Cambridge University Press.
A graduate text.

Hardy, G. H. and Wright, E. M., 1960, *An Introduction to the Theory of Numbers*, Oxford University Press.
The standard reference on number theory. Contains a full discussion of decimals in chapter 9.

Hauchart, C. and Rouche, N., 1987, *Apprivoiser l'Infini*, Ciaco.
A thesis on the beginnings of analysis in the secondary school.

Heath, T. L. (trans.), 1956, *The thirteen books of Euclid's Elements*, Dover.

Heijenoort, J. van, 1967, *From Frege to Gödel*, Harvard University Press.
Historical source material.

Hemmings, R. and Tahta, D., 1984, *Images of Infinity*, Leapfrogs – Tarquin.
A rich stimulus to the intuition. Ideal for a sixth-former.

Hight, D. W., 1977, *A Concept of Limits*, Dover.
Unique for its geometrical illustrations of limits.

Kazarinoff, N. D., 1961, *Analytic Inequalities*, Holt, Rinehart and Winston.
Suitable for a sixth-former.

Klambauer, G., 1979, *Problems and Propositions in Analysis*, Marcel Dekker.
For the lecturer's bookshelf: substantial problems with solutions.

Klein, F., 1924, *Elementary Mathematics from an Advanced Standpoint*. Part I. *Arithmetic, Algebra and Analysis*, 3rd edition trans. E. R. Hedrick and C. A. Noble, Dover reprint.
Contains an interesting discussion of the elementary functions and proofs of the transcendence of e and π.

Knopp, K., 1928, *Theory and application of Infinite Series*, Blackie.
Thorough, humane and with a wealth of historical reference.

Körner, T. W., 1991, Differentiable functions on the rationals. *Bull. Lond. Math. Soc.*, **23**, 557–62.

Korovkin, P. P., 1961, *Inequalities*, Pergamon Press (contained in *Popular Lectures in Mathematics*, vols 1–6, trans. H. Moss).
Very good for sixth-formers.

Landau, E., 1960, *Foundations of Analysis*, Chelsea, New York.
First published in German 1930; this was the first published account of the development of the number system from the Peano Postulates to Dedekind sections for undergraduates.

Leavitt, T. C. J., 1967, *Limits and Continuity*, McGraw-Hill.
Useful diagrams for beginners, a partly programmed text.

Levi, H., 1961, *Elements of Algebra*, Chelsea, New York.
The number system developed from notions of cardinality.

Lieber, L. P., 1953, *Infinity*, Holt, Rinehart and Winston.
An artistic and poetic education for the mathematical intuition.

Moise, E. E., 1982, *Introductory Problem Courses in Analysis and Topology*, Springer.

Niven, I., 1961, *Numbers: rational and irrational*, Mathematical Association of America.
A superb introduction for the sixth-former.

Northrop, E. P., 1960, *Riddles in Mathematics: A Book of Paradoxes*, Pelican. (1st UK edition, 1945, English Universities Press).
Excellent high-school material. Fascinating on series.

O'Brien, K. E., 1966, *Sequences*, Houghton-Mifflin.
For beginners.

Osgood, W. F., 1907, *Lehrbuch der Funktionentheorie*, Teubner.

Pólya, G., 1954, *Induction and Analogy in Mathematics (Mathematics and Plausible Reasoning*, vol. 1), Princeton University Press.
Rich with insight into how mathematics works. Gives historical illumination.

Pólya, G. and Szegö, G., 1976, *Problems and Theorems in Analysis*, vol. 1, Springer.
Hard problems with solutions: a very rich diet.

Quadling, D. A., 1955, *Mathematical Analysis*, Oxford University Press.

Reade, J. B., 1986, *An Introduction to Mathematical Analysis*, Oxford University Press.
As readable as a conventional text can be.

Rosenbaum, L. J., 1966, *Induction in Mathematics*, Houghton-Mifflin.
Good sixth-form material.

Rudin, W., 1953, *Principles of Mathematical Analysis*, McGraw-Hill.
Concise and purposeful. A good second course.

Scott, D. B. and Tims, S. R., 1966, *Mathematical Analysis*, Cambridge University Press.

Smith, W. K., 1964, *Limits and Continuity*, Macmillan.
A concrete development of the neighbourhood definition of limit.

Sominskii, I. S., 1961, *The Method of Mathematical Induction*, Pergamon Press (contained in *Popular Lectures in Mathematics*, vols 1–6 trans. H. Moss).
Excellent sixth-form material for the student willing to do some work!

Spivak, M., 1967, *Calculus*, Benjamin.
The modern lecturer's favourite text: beautifully written, copiously illustrated, and with an extensive collection of exercises.

Stewart, I. and Tall, D., 1977, *The Foundations of Mathematics*, Oxford University Press.
A serious introduction to university mathematics.

Swann, H. and Johnson, J., 1977, *Prof. E. McSquared's Original, Fantastic and Highly Edifying Calculus Primer*, William Kaufmann.
A comic-strip approach to functions and limits (by neighbourhoods).

Tall, D., 1982, The blancmange function: Continuous everywhere but differentiable nowhere. *Math. Gaz.*, **66**, 11–22.
Gives the mathematical discussion behind the function obtainable on a BBC computer using David Tall's software entitled *Graphic Calculus I*, published by Glentop.

Tall, D., 1992, The transition to advanced mathematical thinking: functions, limits, infinity and proof. Article in Grouws, D. A. (ed.), *Handbook of Research on Mathematics Teaching and Learning*, Macmillan.
A useful survey of recent research.

Thurston, H. A., 1967, *The Number-system*, Dover.
A useful combination of logic and rationale in the development of the number system.

Toeplitz, O., 1963, *The Calculus: A Genetic Approach*, University of Chicago Press.
  History: the key to a humane approach to analysis.
Wheeler, D., 1974, *R is for Real*, Open University Press.
  A problem-motivated introduction to the real numbers.
Yarnelle, J. E., 1964, *An Introduction to Transfinite Mathematics*, D. C. Heath.
  The most readable account of countability that I know.
Zippin, L., 1962, *Uses of Infinity*, Mathematical Association of America.

Index

2.53 means that the reference is to be found in qn 53 of chapter 2.
2.53$^-$ means that the reference is to be found just before that question.
2.53$^+$ means that the reference is to be found just after that question.
2.53–8 refers to questions 53 to 58 of chapter 2.
2.53, 64 refers to questions 53 and 64 of chapter 2.
2H means that the reference is to be found in the historical note to chapter 2.